Lignin Utilization Strategies: From Processing to Applications

ACS SYMPOSIUM SERIES **1377**

Lignin Utilization Strategies: From Processing to Applications

Chang Geun Yoo, Editor
State University of New York College of Environmental Science and Forestry
Syracuse, New York, United States

Arthur Ragauskas, Editor
Center of Bioenergy Innovation, Oak Ridge National Laboratory
Oak Ridge, Tennessee, United States
University of Tennessee
Knoxville, Tennessee, United States

Sponsored by the
ACS Division of Energy and Fuels

American Chemical Society, Washington, DC

Library of Congress Cataloging-in-Publication Data

Names: Yoo, Chang Geun, editor. | Ragauskas, Arthur J. (Arthur Jonas), 1957- editor.
Title: Lignin utilization strategies : from processing to applications / Chang Geun Yoo, Editor, State University of New York College of Environmental Science and Forestry, Syracuse, New York, United States, Arthur Ragauskass, Editor, Center of Bioenergy Innovation, Oak Ridge National Laboratory, Oak Ridge, Tennessee, United States, University of Tennessee, Knoxville, Tennessee, United States ; sponsored by the ACS Division of Energy and Fuels.
Description: Washington, DC : American Chemical Society, [2021] | Series: ACS symposium series ; 1377 | Includes bibliographical references and index.
Identifiers: LCCN 2021002973 (print) | LCCN 2021002974 (ebook) | ISBN 9780841298460 (hardcover) | ISBN 9780841298453 (ebook other)
Subjects: LCSH: Lignin--Biotechnology.
Classification: LCC TP248.65.L54 L5345 2021 (print) | LCC TP248.65.L54 (ebook) | DDC 676/.5--dc23
LC record available at https://lccn.loc.gov/2021002973
LC ebook record available at https://lccn.loc.gov/2021002974

The paper used in this publication meets the minimum requirements of American National Standard for Information Sciences—Permanence of Paper for Printed Library Materials, ANSI Z39.48n1984.

Foreword

The ACS Symposium Series is an established program that publishes high-quality volumes of thematic manuscripts. For over 40 years, the ACS Symposium Series has been delivering essential research from world leading scientists, including 36 Chemistry Nobel Laureates, to audiences spanning disciplines and applications.

Books are developed from successful symposia sponsored by the ACS or other organizations. Topics span the entirety of chemistry, including applications, basic research, and interdisciplinary reviews.

Before agreeing to publish a book, prospective editors submit a proposal, including a table of contents. The proposal is reviewed for originality, coverage, and interest to the audience. Some manuscripts may be excluded to better focus the book; others may be added to aid comprehensiveness. All chapters are peer reviewed prior to final acceptance or rejection.

As a rule, only original research papers and original review papers are included in the volumes. Verbatim reproductions of previous published papers are not accepted.

ACS Books

Contents

Indexes

Chapter 1

Opportunities and Challenges of Lignin Utilization

Chang Geun Yoo[1,*] and Arthur J. Ragauskas[2,3]

[1]Department of Chemical Engineering,
State University of New York College of Environmental Science and Forestry,
Syracuse, New York 13210, United States

[2]Center for Bioenergy Innovation, Biosciences Division,
Oak Ridge National Laboratory, Oak Ridge, Tennessee 37831, United States

[3]Department of Chemical and Biomolecular Engineering,
Department of Forestry, Wildlife, and Fisheries, University of Tennessee,
Knoxville, Tennessee 37996, United States

*Email: cyoo05@esf.edu.

Lignin is the most abundant non-carbohydrate component of plant cell walls and the largest aromatic source on earth. It is gaining more attention due to its potentials for many industrial applications. Compared to the applications of carbohydrate components like cellulose and hemicellulose, the utilization of lignin is challenging due to its natural rigidity, heterogeneity, and complexity. On the other hand, the successful utilization of lignin can broaden the spectrum of renewable resources in producing chemicals, fuels, and materials. In addition, value-added lignin products can enhance the economic competitivity of many biofuels and biochemicals mainly produced from cellulose and hemicellulose. In the last few decades, there have been notable improvements in fundamental understandings of the plant cell wall, lignin chemistry, catalysis, and other engineering principles, as well as the development of diverse conversion and characterization technologies. Prior to the detailed discussions of recent lignin utilization strategies through decomposition approaches such as thermochemical, biological, photoelectrochemical and electrochemical depolymerization and material approaches including carbon materials, adsorbents, thermosets, thermoplastics, hydrogels and nanoparticles, this chapter briefly discusses the types of lignin and its potentials and limitations based on its characteristics and applied processes.

Introduction

Lignin has been reported as an important component of lignocellulosic biomass in different roles. In nature, lignin provides the rigidity to the plant cell wall by binding with other major components like cellulose and hemicellulose and forming lignin-carbohydrate complexes (*1*). It also facilitates water transportation in the cell wall, defenses against pathogen attack, and response to stresses from the environments (*2*). In traditional biomass utilization processes in pulp and bioethanol industries, lignin is a major impurity and represents one of the most important recalcitrance factors in the bioconversion of biomass to biofuels, so preprocessing such as pulping and pretreatment is necessary for its removal (*3*, *4*).

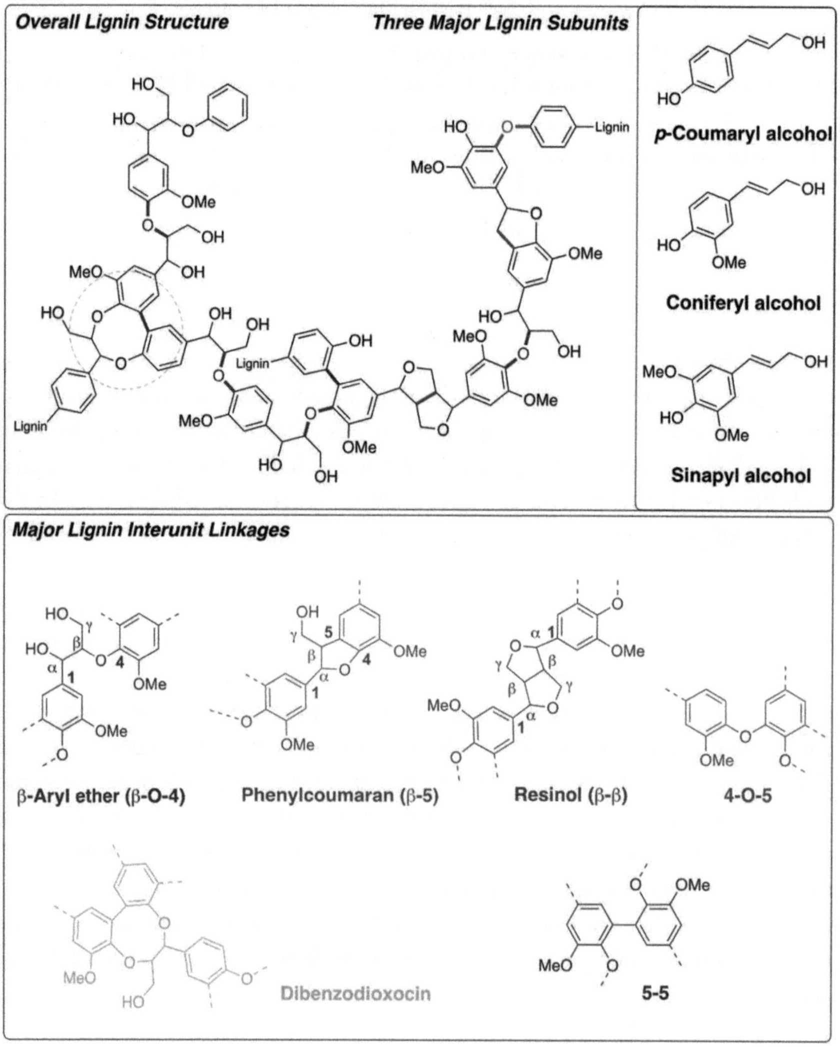

Figure 1. A representative structure of lignin and its major subunits and interunit linkage motifs.

Lignin contributes 15 – 40% of the plant cell wall and is mainly composed of three aromatics (syringyl (S), guaiacyl (G), and *p*-hydroxyphenyl (H) units) through various interunit linkages such as β-O-4, β-β, β-1, β-5, 4-O-5, α-O-4, 5-5, and others (Figure 1). The composition of lignin is different depending on its species. For instance, softwood lignin is mainly composed of G lignin, while hardwood lignin mostly consists of S and G lignins. Herbaceous lignin has a relatively large quantity of H unit compared to wood lignin. Hydroxycinnamates such as ferulic acids and *p*-coumaric acids in herbaceous plants and *p*-hydroxybenzoic acids in other plants are also important components. They can cross-link between lignin and polysaccharides or exist as acylated lignin units (*5*). Many characteristics of plant cell walls are directly or indirectly influenced by the composition and abundance of linkages (*6–8*). A better understanding of the aforementioned lignin properties does not only provide the characteristics of lignin but also elucidate the effects of applied catalysts and/or processes through the observation of its structural changes.

Lignin has great potentials for many industrial applications because of its aromatic structure, relatively high energy content, and sustainability. It is the largest natural source of aromatics, which can potentially replace petroleum-based chemicals. It has higher energy content (~25.5 kJ/g) than carbohydrates have (17.0 – 17.5 kJ/g for cellulose and 17.2 – 17.3 kJ/g for starch) (*9*). Industrially generated lignin represents a large portion of a waste stream/byproduct in the biomass processes. In 2015, about 100 million tons of lignin was produced, and the annual lignin production is expected to increase by 225 million tons per year in 2030 because of the increase of biofuel production (*10, 11*).

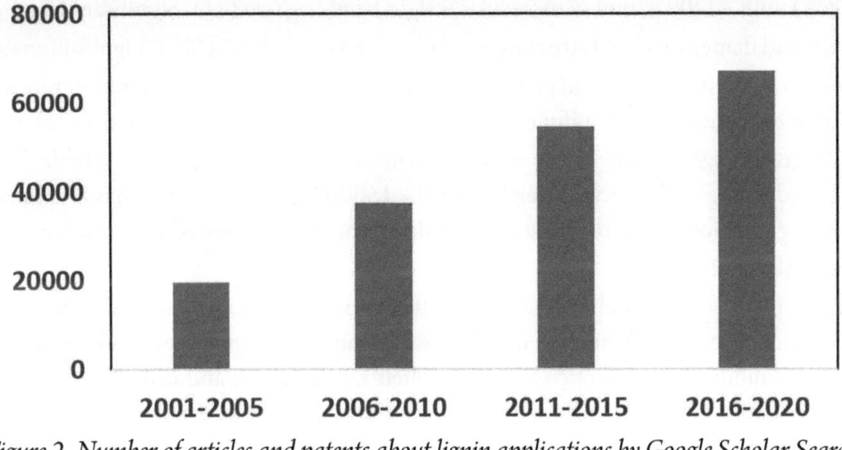

Figure 2. Number of articles and patents about lignin applications by Google Scholar Search.

Lignin has been mostly used as low-value waste products in the traditional pulp processes. Only less than 2% of lignin was used as a dispersant, surfactant, and wood adhesive, while the rests were consumed as a direct combustion fuel to supply internal energy in the biomass processes (*11*). For this reason, effective lignin utilization strategies can improve the sustainability of biorefinery processes and make current biomass-based products such as biofuels and chemical-pulps economically more competitive by producing a new revenue stream and reducing downstream costs. Recent advancements in lignin characterization and conversion technologies allow a deeper understanding of the fundamentals in lignin biosynthesis, characteristics (physicochemical, thermal, morphological properties), and its biological conversion and thermochemical conversion mechanisms as well as improvements in its conversion efficiency and spectrum of its applications. Many recent articles have advanced our understanding of the biosynthesis of lignin, genetic modification of lignin, fractionation/separation processes, inhibitory effects on biomass conversion, depolymerization approaches, and material applications (*12–17*). These studies often highlight that lignin valorization would be a key for accomplishing a successful integrated biorefinery that is competitive in the marketplace. As Figure 2 shows, after 2000, the interests in lignin applications increased according to the Google Scholar search. Between 2001 and 2005, about 19,700 articles and patents were reported about lignin applications, and this number keeps increasing: 37,500 in 2006-2010, 54,600 in 2011-2015, and 67,000 in 2016-2020. However, despite people's recent growing interest in lignin, only a few applications are currently available in industries.

Types of Lignin and Their Physicochemical Properties

As biomass conversion technologies developed, diverse types of lignin have been produced. Table 1 summarizes various lignins produced through different fractionation/isolation methods and their characteristics. In traditional pulp and paper industries, kraft lignin, soda lignin, and lignosulfonate have been produced. Kraft lignin is one of the common types of lignin recovered from the black liquor in the pulping process (*18*). This type of lignin is depolymerized and transformed into complex structures through several reactions including β-aryl ether linkage cleavage, demethylation, and vinyl ethers and stilbene formation by γ-elimination of the primary alcohol, and condensation of aromatic unit's C5 position (*19*). The molecular weight of kraft lignin ranged from 2000 to 7000 g/mol as a result of the kraft pulping (*19*, *20*). Soda lignin is produced from the sulfur-free pulping process using sodium hydroxide. Mousavioun and Doherty got high purity soda lignin (~96%) with ~2200 g/mol of molecular weight from bagasse (*21*). Significant cleavage of aryl ether linkage and demethoxylated structure were observed in this lignin (*22*). Lignosulfonates are the other common lignin in the pulp and paper processes by sulfite with calcium, ammonium, sodium or magnesium. It has relatively high sulfur (3.5 – 8.0 %) and ash (4.0 – 8.0 %) contents (*22*). Interaction of sulfonate groups with lignin at Cα position made this lignin more water-soluble. This lignin showed a broader range of molecular weight (1000 – 150000 g/mol). The lignin is fractionated from biomass through the formation of quinone methide intermediate followed by sulfitolytic cleavage of β-aryl ether linkage (*23*).

With the increased research interest in biofuels and biomaterials, several pretreatment and fractionation processes have generated new types of lignin using hot-water, acid, alkaline, organic solvents, ionic liquids, deep eutectic solvents, molten salt hydrates, and cellulose-derived solvents (e.g., γ-valerolactone (GVL), Cyrene). Dilute acid pretreatment has been applied in many biomass conversions due to the low solvent cost and availability to diverse biomass feedstocks. However, low lignin recovery yield and significant condensation of lignin structures often make the utilization of this lignin unrealistic (*24*). Redeposit of the dissolved lignin and pseudo lignin formation under acidic conditions (*25*) also negatively affect the recovery and utilization of the lignin. Hot-water extracted lignin can be recovered 10 – 40% depending on the reaction conditions and biomass species. It preserved a relatively large quantity of β-O-4 linkages and had 5000 – 10000 g/mol molecular weight values (*26*). Ammonia pretreatments are an effective method for herbaceous plants like corn stover. It retains most of the cellulose and a large quantity of hemicellulose while removes a significant amount of lignin under mild conditions (*27*). The recovered lignin showed ~4000 g/mol of molecular weight with large quantity of aryl ether linkage (*28*). Interestingly, the ammonia extracted lignin showed the transformation of hydroxycinnamates (i.e., ferulate, p-coumarate) to the amides (i.e., ferulolyl amides, p-coumaramides).

Organosolv lignin has been investigated in many lignin utilization studies due to its high recovery yield and purity (*29*, *30*). Ethanol organosolv lignin is a popular organosolv lignin. It had some condensed structures including a stilbene unit with significant cleavage of aryl ether linkages. The size of the recovered lignin was 1000 – 4000 g/mol depending on the pretreatment conditions and species (*30, 31*). Co-solvent enhanced lignocellulosic fractionation (CELF) pretreatment using tetrahydrofuran (THF) and water is one type of organosolv pretreatment strategies resulting in high lignin purity and yield (*32*). The isolated lignin, termed as CELF lignin, exhibited similar properties to that of ethanol organosolv lignin including the degree of linkage cleavage, degree of condensation, and molecular weights (*31, 33, 34*). GVL has been introduced as a promising biomass-derived solvent. About 70 – 80% of lignin was removed under mild temperature (*35*). The recovered lignin (GVL lignin) has a relatively large quantity of β-O-4 linkage and less condensed structure compared to typical organosolv and CELF lignins (*31*). Also, the molecular weight of GVL lignin was higher than the lignins from ethanol organosolv and CELF pretreatments with the same hardwood. Cyrene is recently reported as a cellulose-derived green pretreatment solvent (*36*). The fractionated lignin from Cyrene pretreatment preserved a large quantity of aryl ether linkage with the limited condensation. More than 60% of lignin was fractionated with 4000 – 6300 g/mol of molecular weight. The authors discussed the possible suppression of acidolysis cleavage of ether linkage and condensation due to the formation of diol structures in Cyrene solution.

Ionic liquid (IL) has been applied in many biomass conversions as an effective medium for the dissolution of biomass and its components like lignin (*37*). About 30 – 50% of lignin was fractionated from switchgrass by an aqueous ionic liquid, and the fractionated lignin had 3200 – 3400 g/mol of molecular weight (*38*). Based on its abundance of linkages and aromatic structures, the IL lignin preserved its original structures well compared to other solvent pretreatments. Another study reported that the preservation of linkages depended on the reaction severity; however, these lignins still kept a relatively high quantity of the intact structure (*39*). Deep eutectic solvents are introduced as an alternative of conventional ILs with similar properties but have easier synthesis with relatively inexpensive components. Lignin fractionation yield varied by the combination of hydrogen bond donor (HBD) and hydrogen bond acceptor (HBA) (*40*). Also, the existence of co-catalyst affected the lignin separation. Up to 95% of lignin purity was obtained with ~60% of yield from Douglas fir, while the fractionation yield was up to 78% from poplar (*41*). However, to obtain the high purity of lignin, further purification by repeated washing with ethanol/water mixture was necessary. The recovered lignin showed some amount of Hibbert's ketone without subsequent condensed structures. Wang et al. also fractionated about 70% lignin using lignin-derived deep eutectic solvent, and the separated lignin has ~1400 g/mol for molecular weight with a narrow polydispersity index (<1.6) (*42*). Most β-O-4 linkages were cleaved, and a certain degree of the condensed structure was observed depending on the reaction time.

In addition, a few lignin isolation methods were examined to recover the lignin close to its native form. Milled wood lignin (MWL), cellulolytic enzyme lignin (CEL), and enzymatic mild acidolysis lignin (EMAL) have been introduced as a representative for native lignin with minimal modification of the original structure (*43, 44*). These lignins are normally used as a control to monitor the transformation of lignin during the applied conversion processes.

Table 1. A Summary of Diverse Lignins and Their Characteristics

Types of lignin	*Characteristics*	*Ref.*
Kraft lignin	High lignin recovery High purity (>90%) 2000-7000 g/mol Mw Significant cleavage of β-O-4 linkage Condensation at C5 in the aromatic ring Formation of vinyl ethers and stilbene	(*19, 20*)
Soda lignin	High lignin recovery High purity (~93%) 1000-3000g/mol Mw Demethoxylated structure Significant cleavage of β-O-4 linkage	(*21, 22*)
Lignosulfonate	High lignin recovery 1000-150000 g/mol Mw Sulfonate groups bound at Cα	(*22, 23, 45*)
Hot-water extracted lignin	Relatively low lignin recovery (10 - 40%) 5000-10000 g/mol Mw Relatively mild cleavage of β-O-4 linkage	(*26, 46*)
Ammonia lignin	50-70% lignin recovery Relatively well preserved β-O-4 linkage Ferulates and coumarates are transformed to ferulolyl amides and coumaramides ~4000 g/mol Mw	(*28*)
Ethanol organosolv lignin	High purity and recovery of lignin Severe cleavage of β-O-4 linkage Relatively large condensed structure 1000 - 4000 g/mol Mw	(*29–31*)
CELF lignin	High purity and recovery of lignin Severe cleavage of β-O-4 linkage Relatively large condensed structure 1500 - 3000 g/mol Mw	(*31, 33, 34*)
GVL lignin	70-80% of lignin recovery Relatively well preserved β-O-4 linkage Relatively less condensed structure ~5500 g/mol Mw	(*31, 35*)
Cyrene lignin	>60% lignin recovery 4000 - 6300 g/mol Mw Moderate preservation of β-O-4 linkage Limited condensation reactions	(*36*)
Ionic liquid lignin	High purity and recovery of lignin Relatively well preserved β-O-4 linkage Relatively less condensed structure 3200 - 3400 g/mol Mw	(*38*)

Table 1. (Continued). A Summary of Diverse Lignins and Their Characteristics

Types of lignin	*Characteristics*	*Ref.*
DES lignin	Lignin recovery vary depending on HBA and HBD Some DES remained in the recovered lignin Degree of condensation vary by conditions Low Mw (500 - 2600 g/mol) due to significant cleavage of β-O-4 linkage	(*41, 42*)
MWL	Close to native lignin Lignin recovery is relatively low (up to 50%)	
CEL	Close to native lignin Compared to MWL, more lignin can be recovered. Depending on the concentration of dioxane, carbohydrate contents are different.	(*43, 44*)
EMAL	Close to native lignin Mild acidolysis cleaves lignin-carbohydrate bonds and results in high yield and purity.	

Factors in Lignin Utilization

Most physical, chemical, and thermal properties of lignin are important parameters affecting its utilization. These properties are closely correlated to the structural properties of lignin; therefore, a deep understanding of lignin structures provides a guidance to anticipate and explain their characteristics. Among many factors, (i) the chemical reactivity of lignin is an essential information. Most lignin fractionation and their further applications involve chemical processes. For instance, α-O-4 linkages are the most reactive structure in phenolic unit but stable in non-phenolic (*47, 48*). The β-O-4 linkages are relatively stable compared to α-O-4. The position and number of functional groups in the aromatic ring and side-chain affect the chemical reactivity of β-O-4 linkages. Gierer and Ljunggren reported three major factors affecting the kinetics of a β-O-4 bond cleavage: the degree of ionization of the group, nucleophilicity of the anion generated by ionization, and a formation rate of a product or intermediate (*47, 49*).

(ii) The solubility of lignin in the process media is another key factor affecting its applications. The content of hydroxyl groups in lignin controls the hydrophilicity of lignin to some extent. For example, the increase of aliphatic hydroxyl groups typically improves the solubility of lignin in aqueous solution. The cleavage of interunit linkage and demethylation produce free phenolic hydroxyl groups and increased the dissolution of lignin (*47*). For these reasons, depending on the applied solvent in lignin processing, abundance and types of functional groups are important factors. Hildebrand parameter and Hansen parameter were used to interpret the lignin solubility in different solvents (*50, 51*).

(iii) Thermal properties like glass transition temperature and melting temperature are also important factors in the application of lignin. These properties influence both product properties, in particular, the thermal stability of lignin-based materials and processing conditions like extrusion temperature. The aromatic nature of lignin gives moderate thermal stability, while the thermal properties can be changed through modification of its structure. For instance, oxidative processing of lignin enhanced the thermal stability by forming carbonyl groups and eliminating volatile compounds (*52, 53*).

(iv) The compatibility of lignin with other materials also plays an important role in its material applications. It was reported that better interfacial compatibility between lignin and polylactic acid (PLA) by the hydrogen bonding interaction resulted in higher tensile strength and storage modulus of the composite (*54*). Chemical structure, particle shape, surface, and size of lignin were investigated to understand the compatibility of lignin with other materials (*54*, *55*).

Perspectives and Other Challenges in Lignin Utilization

Valorization of lignin is an essential task for the successful development of biomass utilization approach. It can contribute to the sustainability of biorefinery processes by generating new revenue and reducing the cost of the waste stream. Application strategies of lignin have been studied in broad scope from its decomposition to value-added fuels and chemicals to the formation of advanced materials such as nanoparticle, hydrogel, and carbon fiber. This book introduces recent lignin valorization strategies through thermochemical, biological, and photoelectrochemical depolymerization pathways and by material applications such as carbon material, adsorbents, thermosets, thermoplastics, hydrogel, and nanoparticles.

Besides the aforementioned lignin valorization approaches, downstream process such as separation and purification of lignin and its products is also important. Many lignins are considered as a co-product/by-product from cellulose-centered biorefinery processes; therefore, the fractionated lignin stream contains process solvent and chemicals as well as other decomposed components like inorganic components and extractives. Several membrane separation and drying methods are used in lab-scale studies, but their practical application in industrial is still in question. Also, residual carbohydrates bound to lignin need further purification to prevent the devaluation of lignin. Moreover, the conversion of lignin, in particular, in its decomposition processes generates multiple products and unknown components; thus, separation of the targeted products is essential. Unfortunately, compared to the utilization strategies of lignin, its downstream approaches are not investigated much. Therefore, further studies for pre- and post-processing of lignin and its products will be essential topics in future lignin valorization strategies.

Acknowledgments

Oak Ridge National Laboratory is managed by UT Battelle, LLC, under contract DE-AC05-00OR22725 with the U.S. Department of Energy (DOE). This work was supported and performed as part of the Center for Bioenergy Innovation (CBI). The CBI is a U.S Department of Energy Bioenergy Research Center supported by the Office of Biological and Environmental Research in the DOE Office of Science.

The views and opinions of the authors expressed herein do not necessarily state or reflect those of the United States Government or any agency thereof. Neither the United States Government nor any agency thereof, nor any of their employees, makes any warranty, expressed or implied, or assumes any legal liability or responsibility for the accuracy, completeness, or usefulness of any information, apparatus, product, or process disclosed, or represents that its use would not infringe privately owned rights.

References

1. Yoo, C. G.; Meng, X.; Pu, Y.; Ragauskas, A. J. The critical role of lignin in lignocellulosic biomass conversion and recent pretreatment strategies: A comprehensive review. *Bioresource Technology* **2020**, *301*, 122784.
2. Liu, Q.; Luo, L.; Zheng, L. Lignins: biosynthesis and biological functions in plants. *International journal of molecular sciences* **2018**, *19* (2), 335.
3. Santos, R. B.; Capanema, E. A.; Balakshin, M. Y.; Chang, H.-M.; Jameel, H. Effect of hardwoods characteristics on kraft pulping process: emphasis on lignin structure. *BioResources* **2011**, *6* (4), 3623–3637.
4. Yoo, C. G.; Li, M.; Meng, X.; Pu, Y.; Ragauskas, A. J. Effects of organosolv and ammonia pretreatments on lignin properties and its inhibition for enzymatic hydrolysis. *Green Chemistry* **2017**, *19* (8), 2006–2016.
5. Ralph, J. Hydroxycinnamates in lignification. *Phytochemistry Reviews* **2010**, *9* (1), 65–83.
6. Yoo, C. G.; Dumitrache, A.; Muchero, W.; Natzke, J.; Akinosho, H.; Li, M.; Sykes, R. W.; Brown, S. D.; Davison, B.; Tuskan, G. A. Significance of lignin S/G ratio in biomass recalcitrance of *Populus trichocarpa* variants for bioethanol production. *ACS Sustainable Chemistry & Engineering* **2018**, *6* (2), 2162–2168.
7. Yoo, C. G.; Yang, Y.; Pu, Y.; Meng, X.; Muchero, W.; Yee, K. L.; Thompson, O. A.; Rodriguez, M.; Bali, G.; Engle, N. L.; Lindquist, E.; Singan, V.; Schmutz, J.; DiFazio, S. P.; Tschaplinski, T. J.; Tuskan, G. A.; Chen, J.-G.; Davison, B.; Ragauskas, A. J. Insights of biomass recalcitrance in natural *Populus trichocarpa* variants for biomass conversion. *Green Chemistry* **2017**, *19* (22), 5467–5478.
8. Li, M.; Yoo, C. G.; Pu, Y.; Biswal, A. K.; Tolbert, A. K.; Mohnen, D.; Ragauskas, A. J. Downregulation of pectin biosynthesis gene GAUT4 leads to reduced ferulate and lignin-carbohydrate cross-linking in switchgrass. *Communications biology* **2019**, *2* (1), 1–11.
9. Kienzle, E.; Schrag, I.; Butterwick, R.; Opitz, B. Calculation of gross energy in pet foods: new data on heat combustion and fibre analysis in a selection of foods for dogs and cats. *Journal of Animal Physiology and Animal Nutrition* **2001**, *85* (5-6), 148–157.
10. Hockstad, L.; Hanel, L. Inventory of U.S. greenhouse gas emissions and sinks. *Environmental System Science Data Infrastructure for a Virtual Ecosystem* **2018**DOI:10.15485/1464240.
11. Bajwa, D.; Pourhashem, G.; Ullah, A.; Bajwa, S. A concise review of current lignin production, applications, products and their environmental impact. *Industrial Crops and Products* **2019**, *139*, 111526.
12. Vanholme, R.; Demedts, B.; Morreel, K.; Ralph, J.; Boerjan, W. Lignin biosynthesis and structure. *Plant Physiology* **2010**, *153* (3), 895–905.
13. Bryant, N. D.; Pu, Y.; Tschaplinski, T. J.; Tuskan, G. A.; Muchero, W.; Kalluri, U. C.; Yoo, C. G.; Ragauskas, A. J. Transgenic poplar designed for biofuels. *Trends in Plant Science* **2020**, *25* (9), 881–896.
14. Toledano, A.; Serrano, L.; Garcia, A.; Mondragon, I.; Labidi, J. Comparative study of lignin fractionation by ultrafiltration and selective precipitation. *Chemical Engineering Journal* **2010**, *157* (1), 93–99.

15. Rinaldi, R.; Jastrzebski, R.; Clough, M. T.; Ralph, J.; Kennema, M.; Bruijnincx, P. C.; Weckhuysen, B. M. Paving the way for lignin valorisation: recent advances in bioengineering, biorefining and catalysis. *Angewandte Chemie International Edition* **2016**, *55* (29), 8164–8215.
16. Rahikainen, J. L.; Martin-Sampedro, R.; Heikkinen, H.; Rovio, S.; Marjamaa, K.; Tamminen, T.; Rojas, O. J.; Kruus, K. Inhibitory effect of lignin during cellulose bioconversion: the effect of lignin chemistry on non-productive enzyme adsorption. *Bioresource technology* **2013**, *133*, 270–278.
17. Stewart, D. Lignin as a base material for materials applications: Chemistry, application and economics. *Industrial crops and products* **2008**, *27* (2), 202–207.
18. Froass, P. M.; Ragauskas, A. J.; Jiang, J.-e. Nuclear magnetic resonance studies. 4. Analysis of residual lignin after kraft pulping. *Industrial & engineering chemistry research* **1998**, *37* (8), 3388–3394.
19. Hu, Z.; Du, X.; Liu, J.; Chang, H.-m.; Jameel, H. Structural characterization of pine kraft lignin: BioChoice lignin vs Indulin AT. *Journal of Wood Chemistry and Technology* **2016**, *36* (6), 432–446.
20. Nagy, M.; Kosa, M.; Theliander, H.; Ragauskas, A. J. Characterization of CO_2 precipitated kraft lignin to promote its utilization. *Green Chemistry* **2010**, *12* (1), 31–34.
21. Mousavioun, P.; Doherty, W. O. Chemical and thermal properties of fractionated bagasse soda lignin. *Industrial Crops and Products* **2010**, *31* (1), 52–58.
22. Patil, V.; Adhikari, S.; Cross, P.; Jahromi, H. Progress in the solvent depolymerization of lignin. *Renewable and Sustainable Energy Reviews* **2020**, *133*, 110359.
23. Matsushita, Y. Conversion of technical lignins to functional materials with retained polymeric properties. *Journal of Wood Science* **2015**, *61* (3), 230–250.
24. Cao, S.; Pu, Y.; Studer, M.; Wyman, C.; Ragauskas, A. J. Chemical transformations of *Populus trichocarpa* during dilute acid pretreatment. *RSC Advances* **2012**, *2* (29), 10925–10936.
25. Shinde, S. D.; Meng, X.; Kumar, R.; Ragauskas, A. J. Recent advances in understanding the pseudo-lignin formation in a lignocellulosic biorefinery. *Green Chemistry* **2018**, *20* (10), 2192–2205.
26. Wang, K.-T.; Jing, C.; Wood, C.; Nagardeolekar, A.; Kohan, N.; Dongre, P.; Amidon, T. E.; Bujanovic, B. M. Toward complete utilization of miscanthus in a hot-water extraction-based biorefinery. *Energies* **2018**, *11* (1), 39.
27. Yoo, C. G.; Nghiem, N. P.; Hicks, K. B.; Kim, T. H. Maximum production of fermentable sugars from barley straw using optimized soaking in aqueous ammonia (SAA) pretreatment. *Applied Biochemistry and Biotechnology* **2013**, *169* (8), 2430–2441.
28. Yoo, C. G.; Kim, H.; Lu, F.; Azarpira, A.; Pan, X.; Oh, K. K.; Kim, J. S.; Ralph, J.; Kim, T. H. Understanding the physicochemical characteristics and the improved enzymatic saccharification of corn stover pretreated with aqueous and gaseous ammonia. *BioEnergy Research* **2016**, *9* (1), 67–76.
29. Pan, X.; Xie, D.; Yu, R. W.; Lam, D.; Saddler, J. N. Pretreatment of lodgepole pine killed by mountain pine beetle using the ethanol organosolv process: fractionation and process optimization. *Industrial & Engineering Chemistry Research* **2007**, *46* (8), 2609–2617.

30. Pan, X.; Xie, D.; Yu, R. W.; Saddler, J. N. The bioconversion of mountain pine beetle-killed lodgepole pine to fuel ethanol using the organosolv process. *Biotechnology and Bioengineering* **2008**, *101* (1), 39–48.
31. Meng, X.; Bhagia, S.; Wang, Y.; Zhou, Y.; Pu, Y.; Dunlap, J. R.; Shuai, L.; Ragauskas, A. J.; Yoo, C. G. Effects of the advanced organosolv pretreatment strategies on structural properties of woody biomass. *Industrial Crops and Products* **2020**, *146*, 112144.
32. Cai, C. M.; Zhang, T.; Kumar, R.; Wyman, C. E. THF co-solvent enhances hydrocarbon fuel precursor yields from lignocellulosic biomass. *Green Chemistry* **2013**, *15* (11), 3140–3145.
33. Meng, X.; Parikh, A.; Seemala, B.; Kumar, R.; Pu, Y.; Christopher, P.; Wyman, C. E.; Cai, C. M.; Ragauskas, A. J. Chemical transformations of poplar lignin during cosolvent enhanced lignocellulosic fractionation process. *ACS Sustainable Chemistry & Engineering* **2018**, *6* (7), 8711–8718.
34. Meng, X.; Parikh, A.; Seemala, B.; Kumar, R.; Pu, Y.; Wyman, C. E.; Cai, C. M.; Ragauskas, A. J. Characterization of fractional cuts of co-solvent enhanced lignocellulosic fractionation lignin isolated by sequential precipitation. *Bioresource technology* **2019**, *272*, 202–208.
35. Shuai, L.; Questell-Santiago, Y. M.; Luterbacher, J. S. A mild biomass pretreatment using γ-valerolactone for concentrated sugar production. *Green Chemistry* **2016**, *18* (4), 937–943.
36. Meng, X.; Pu, Y.; Li, M.; Ragauskas, A. J. A biomass pretreatment using cellulose-derived solvent Cyrene. *Green Chemistry* **2020**, *22* (9), 2862–2872.
37. Pu, Y.; Jiang, N.; Ragauskas, A. J. Ionic liquid as a green solvent for lignin. *Journal of Wood Chemistry and Technology* **2007**, *27* (1), 23–33.
38. Liu, E.; Li, M.; Das, L.; Pu, Y.; Frazier, T.; Zhao, B.; Crocker, M.; Ragauskas, A. J.; Shi, J. Understanding lignin fractionation and characterization from engineered switchgrass treated by an aqueous ionic liquid. *ACS Sustainable Chemistry & Engineering* **2018**, *6* (5), 6612–6623.
39. Sathitsuksanoh, N.; Holtman, K. M.; Yelle, D. J.; Morgan, T.; Stavila, V.; Pelton, J.; Blanch, H.; Simmons, B. A.; George, A. Lignin fate and characterization during ionic liquid biomass pretreatment for renewable chemicals and fuels production. *Green Chemistry* **2014**, *16* (3), 1236–1247.
40. Wang, Y.; Kim, K. H.; Jeong, K.; Kim, N.-K.; Yoo, C. G. Sustainable biorefinery processes using renewable deep eutectic solvents. *Current Opinion in Green and Sustainable Chemistry* **2020**, 100396.
41. Alvarez-Vasco, C.; Ma, R.; Quintero, M.; Guo, M.; Geleynse, S.; Ramasamy, K. K.; Wolcott, M.; Zhang, X. Unique low-molecular-weight lignin with high purity extracted from wood by deep eutectic solvents (DES): a source of lignin for valorization. *Green chemistry* **2016**, *18* (19), 5133–5141.
42. Wang, Y.; Meng, X.; Jeong, K.; Li, S.; Leem, G.; Kim, K. H.; Pu, Y.; Ragauskas, A. J.; Yoo, C. G. Investigation of a lignin-based deep eutectic solvent using *p*-hydroxybenzoic acid for efficient woody biomass conversion. *ACS Sustainable Chemistry & Engineering* **2020**, *8* (33), 12542–12553.
43. Guerra, A.; Filpponen, I.; Lucia, L. A.; Argyropoulos, D. S. Comparative evaluation of three lignin isolation protocols for various wood species. *Journal of Agricultural and Food Chemistry* **2006**, *54* (26), 9696–9705.

44. Guerra, A.; Filpponen, I.; Lucia, L. A.; Saquing, C.; Baumberger, S.; Argyropoulos, D. S. Toward a better understanding of the lignin isolation process from wood. *Journal of Agricultural and Food Chemistry* **2006**, *54* (16), 5939–5947.
45. Aro, T.; Fatehi, P. Production and application of lignosulfonates and sulfonated lignin. *ChemSusChem* **2017**, *10* (9), 1861–1877.
46. Nagardeolekar, A.; Ovadias, M.; Wang, K. T.; Bujanovic, B. Willow lignin recovered from hot-water extraction for the production of hydrogels and thermoplastic blends. *ChemSusChem* **2020**, *13* (17), 4702–4721.
47. Evstigneyev, E. I.; Shevchenko, S. M. Structure, chemical reactivity and solubility of lignin: a fresh look. *Wood Science and Technology* **2019**, *53* (1), 7–47.
48. Ljunggren, S. The significance of aryl ether cleavage in kraft delignification of softwood. *Svensk Papperstidn* **1980**, *83* (13), 363–369.
49. Gierer, J.; Ljunggren, S. Comparative studies of the participation of different neighbouring groups in the alkaline cleavage of [beta]-aryl ether bonds in lignins. *Svensk Papperstidning (Sweden)* **1983**.
50. Lê, H. Q.; Zaitseva, A.; Pokki, J. P.; Ståhl, M.; Alopaeus, V.; Sixta, H. Solubility of organosolv lignin in γ-valerolactone/water binary mixtures. *ChemSusChem* **2016**, *9* (20), 2939.
51. Zhang, Q.; Tan, X.; Wang, W.; Yu, Q.; Wang, Q.; Miao, C.; Guo, Y.; Zhuang, X.; Yuan, Z. Screening solvents based on Hansen solubility parameter theory to depolymerize lignocellulosic biomass efficiently under low temperature. *ACS Sustainable Chemistry & Engineering* **2019**, *7* (9), 8678–8686.
52. Sen, S.; Patil, S.; Argyropoulos, D. S. Thermal properties of lignin in copolymers, blends, and composites: a review. *Green Chemistry* **2015**, *17* (11), 4862–4887.
53. Brodin, I.; Sjöholm, E.; Gellerstedt, G. The behavior of kraft lignin during thermal treatment. *Journal of Analytical and Applied Pyrolysis* **2010**, *87* (1), 70–77.
54. Guo, J.; Chen, X.; Wang, J.; He, Y.; Xie, H.; Zheng, Q. The influence of compatibility on the structure and properties of PLA/lignin biocomposites by chemical modification. *Polymers* **2020**, *12* (1), 56.
55. Mimini, V.; Sykacek, E.; Syed Hashim, S. N. A.; Holzweber, J.; Hettegger, H.; Fackler, K.; Potthast, A.; Mundigler, N.; Rosenau, T. Compatibility of kraft lignin, organosolv lignin and lignosulfonate with PLA in 3D printing. *Journal of Wood Chemistry and Technology* **2019**, *39* (1), 14–30.

Chapter 2

Tailoring Lignin Structure to Maximize the Value from Lignin

Jae Hoon Lee,[1] Joon Weon Choi,[2] and Kwang Ho Kim[3,*]

[1]Department of Agriculture, Forestry and Bioresources, Seoul National University, Seoul 08826, Republic of Korea

[2]Graduate School of International Agricultural Technology, Seoul National University, Pyeongchang 25354, Republic of Korea

[3]Clean Energy Research Center, Korea Institute of Science and Technology, Seoul 02792, Republic of Korea

[*]Email: kwanghokim@kist.re.kr.

Lignin valorization is one of the most important topics across the biorefinery research community. The overdependence on petroleum-derived products has raised increasing concerns about the energy crisis and environmental pollution, which has warranted significant research on producing renewable chemicals and materials from lignin, replacing petroleum-derived carbons. Despite the potential promise of lignin derivatives, there are still major technical obstacles that need addressing to make lignin valorization more feasible and sustainable. In addition to developing processes, recent studies have focused more on tailoring lignin structures to make lignin a process-friendly feedstock. This review introduces recent studies on chemoselective alteration of lignin structures in an effort to maximize the value from lignin.

Introduction

Lignin is the second most abundant biopolymer next to cellulose, and nature's dominant aromatic polymer. Most lignin is currently produced from pulp and paper industries, and a large amount of technical lignin is expected from emerging biorefineries. Although lignin is an abundant and natural aromatic polymer, the heterogeneous, reactive, and inherent intractability of lignin limit its potential for various industrial applications. Until now, lignin is considered as one of the most challenging materials to work with. However, lignin valorization is still a key research area in the scientific community because the success of future biorefineries highly depends on lignin valorization (*1*, *2*).

Lignin valorization typically includes two different approaches; deconstruction and reconstruction, as shown in Figure 1. In the deconstruction approach, macromolecular lignin is thermochemically or biologically depolymerized into low-molecular-weight phenols to produce fuels and aromatic chemicals. However, it is often discussed that lignin depolymerization is challenging because lignin contains various C-O and C-C linkages with the broad distribution of bond strengths (*3*). Besides, the primary products from the thermal decomposition of lignin readily undergo irreversible recondensation reactions, which lead to the formation of recalcitrant species (*4–7*).

Reconstruction approaches utilize lignin as a raw material or an additive for the production of polymeric materials. For example, technical lignins obtained from lignocellulosic biomass can be used to produce plant-derived plastic and composites. Reducing the omnipresence of disposable plastics made from petroleum and the high demand for bio-derived materials have warranted intensive research on lignin valorization to produce green and renewable materials. The superficial similarities between lignin and petroleum-derived carbon fiber have encouraged efforts to utilize lignin for carbon fiber composite materials. In addition to the carbon fiber, other promising lignin application areas include electrochemicals (*8–10*) and pharmaceuticals (*11–13*). Despite the potential promise of lignin-derived materials, the nonuniformity and several reactive linkages of lignin structure make it process-unfriendly, resulting in poor material properties (*14*). Recent studies have focused on tailoring lignin structures to make lignin more suitable feedstocks for the production of polymeric materials. This chapter presents an overview of the recent efforts toward tailoring lignin structures for successful lignin valorization. Two different, but not competing, approaches on lignin utilization will be highlighted.

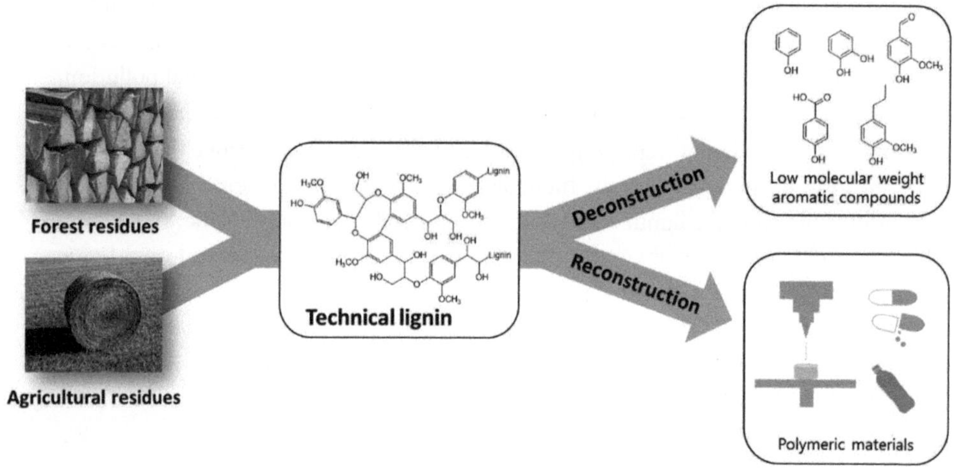

Figure 1. Two approaches in the lignin valorization.

Tailoring Lignin Structure for Depolymerization Approach

There have been significant efforts to produce low molecular weight aromatic compounds from technical lignins. In a typical thermochemical conversion process, heat and catalyst are employed to efficiently break down lignin to phenolic compounds. Besides, using a proper solvent(s) is important in solvent-mediated depolymerization (i.e., solvolysis). One major problem of lignin depolymerization is the recondensation of the lignin fragments, forming aromatic C-C bonds and undesirable condensed products, which significantly reduces the yield of low molecular weight phenolics. This unwanted and irreversible reaction significantly occurs between the primary products of lignin. To maximize the lignin conversion and yield of phenolic monomers, repolymerization has to be minimized. From the process development perspectives, the secondary condensation reactions can be quenched by employing hydrogen-donor solvents (e.g., formic acid, isopropyl alcohol, and tetralin), molecular hydrogen, and scavenging agents. More recently, tailoring lignin structure has been focused on preventing the repolymerization reaction during the lignin depolymerization. In the following text, recent studies on the modification of lignin structures for enhancing lignin conversion to phenols are presented.

Protection of Reactive Sites in Lignin Structure

Altering specific sites in lignin structure can make lignin more amenable to thermochemical depolymerization, minimizing the undesirable condensation reactions. To this end, the selective blocking strategy of reactive functional groups in lignin was introduced. The addition of formaldehyde prevents lignin condensation during extraction by forming 1,3-dioxane structures with C_α-OH and C_γ-OH in the lignin sidechains (*15*). The protected lignin with a dioxane unit produced near theoretical yields of 47% lignin monomers after hydrogenolysis. Formaldehyde-incorporated lignin, forming a stable six-membered acetal structure, can suppress the formation of benzylic cations responsible for the condensation reactions and blocks reactive aromatic positions with hydroxyl methyl groups. Recently, a density functional theory (DFT) calculation study provided mechanistic insights into the reactions of formaldehyde-blocked lignin (*16*). Lignin preferably reacts with formaldehyde with a low energy barrier to form a 1,3-dioxane structure, which protects the reactive sites of lignin, including benzylic hydroxyl groups and aromatic rings. The formaldehyde-protected lignin eventually suppresses lignin condensation reactions, resulting in an enhanced yield of low-molecular-weight products.

The strategy of protecting benzylic hydroxyl groups was further extended with other protection groups. In addition to formaldehyde, Lan et al. reported that acetaldehyde and propionaldehyde could effectively stabilize the C_α-OH and C_γ-OH during the lignin extraction. The subsequent hydrogenolysis of protected lignin in the presence of Pd/C catalyst produced lignin-derived monomeric products at near-theoretical yields (*17*). The protected lignin was subjected to oxidation using 2,3-dichloro-5,6-dicyano-1,4-benzoquinone (DDQ) as oxidant and catalyst to selectively oxidize the C_α-OH to C_α=O. The oxidized lignin was depolymerized with a formic acid and sodium formate, resulting in a 36 mol% of aromatic monomers with high selectivity of syringyl propane dione and guaiacyl propane dione.

Selective oxidation of the C_α alcohol into a ketone in lignin was previously found to enhance the lignin conversion into low molecular weight phenolics. Aspen-derived cellulolytic enzyme lignin (CEL) was oxidized using 4-acetamido-2,2,6,6-tetramethylpiperidine-*N*-oxyl (TEMPO) before the depolymerization. Formic acid-induced depolymerization of the modified lignin under relatively mild conditions (110 °C) yielded 61.2 wt% of low-molecular-weight aromatics (*18*). A density functional theory study revealed that the conversion proceeds sequentially via formylation and elimination at C_γ, followed by hydrolysis of aryl-ether linkages (*19*). Wang et al. used DDQ to oxidize C_α alcohol of birch lignin (*20*). Hydrogenolysis of oxidized lignin with carbon-modified nickel catalyst promoted selective cleavage of aryl-ether linkages, resulting in a 22 wt% yield of aromatics.

The conversion of the active C_α-OH groups to C_α=O was also subjected to oxidative deconstruction (*21*). TEMPO-assisted oxidized lignin was depolymerized by polyoxometalate oxidation at 150 °C with 10 bar O_2. The extractable lignin-derived oil yield was 74.5%, with 32.8% of total monomer yields. Lignin oxidation to block the highly reactive C_α-OH can minimize undesirable condensation reactions and weaken the β-O-4 ether bond, promoting the overall lignin degradation with a high yield of monomers.

Besides the oxidation of benzylic hydroxyl groups, stabilizing agents have been used to stabilize reactive sites in lignin. The addition of 2-naphthol as a carbocation scavenger to biomass fractionation can minimize lignin condensation reactions by attaching to the C_α position (*22*), providing opportunities to utilize intact lignin as a feedstock for the production of phenolic compounds (*23*). In addition, phenolic acids, including 4-hydroxybenzoic acid, vanillic acid, and syringic acid, are also electron-rich, lignin-derived, and reactive toward electrophilic carbocations, inhibiting secondary condensation reactions (*24*).

Zhang et al. reported a stepwise method to strategically defunctionalize the benzylic sidechain of the lignin (*25*). The ethanosolv process resulted in α-ethoxylated β-O-4 motif, followed by aerobic oxidation and subsequent decarbonylation of the primary hydroxyl group in the C_γ position. The resulting partially defunctionalized β-O-4 lignin motif showed enhanced C-O bond cleavage.

In addition to protecting sidechains in lignin, modification of functional groups in the aromatic ring was also studied. Roberts et al. investigated the effect of the addition of boric acid during hydrothermal lignin depolymerization (*26*). They observed inhibition of concurrent polymerization reactions because boric acid, as a capping agent, significantly suppressed these reactions by capping phenolic hydroxyl groups with the formation of strongly coordinated boric esters. It was found that the phenolic hydroxyl groups play an important role in the addition and polymerization reactions, which eventually lead to the formation of unwanted polymerization of primary phenolic products.

Kim et al. also investigated the effect of selective masking of the phenolic hydroxyl group on the depolymerization of lignin (*27*). The phenolic hydroxyl groups were blocked by simple methylation. Thermal degradation of the structurally altered lignin, where the reaction was coupled with the *in-situ* EPR monitoring system, showed a decrease in the formation of reactive quinone methide intermediates responsible for repolymerization reactions. Figure 2 summarizes recent efforts to increase the lignin depolymerization efficiency by structural alteration of lignin.

It is widely discussed that methoxylation is helpful in lignin degradation under anaerobic conditions. Because the methoxyl group is the main point of attack for many anaerobic microbes, having more methoxyl groups in lignin structure by pretreatment would benefit the biological depolymerization of lignin (*28*).

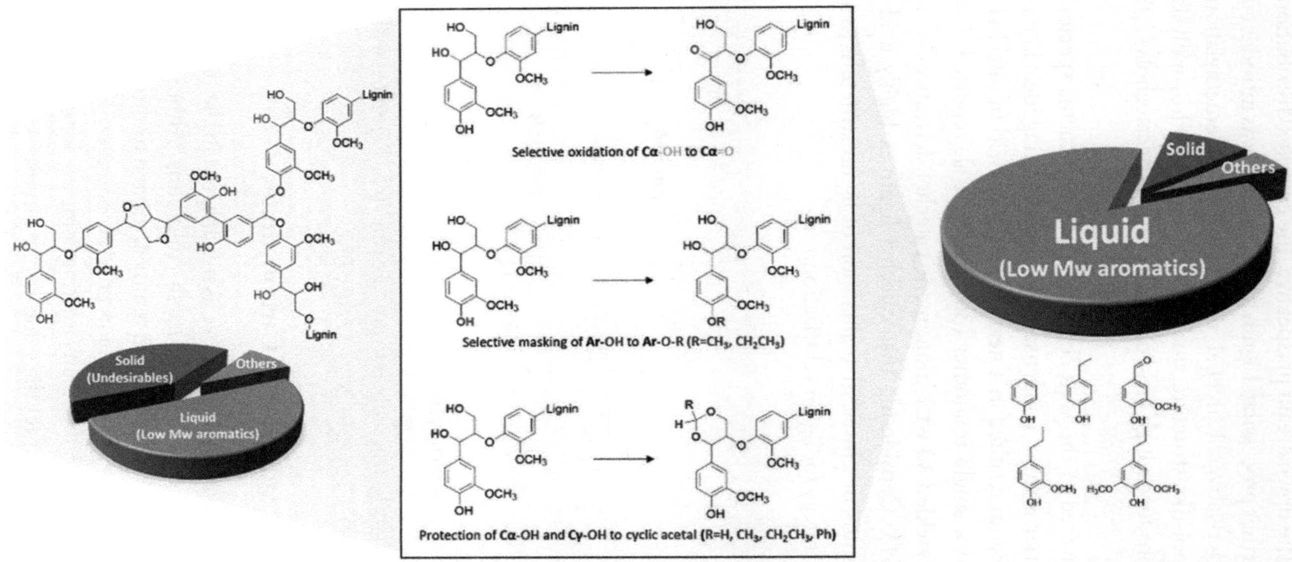

Figure 2. Tailoring lignin structure to enhance the yield of low-molecular weight phenols. Reprinted with permission from reference (1). Copyright 2018 Frontiers.

Lignin Modification to Increase the Selectivity of Final Products

Recently, many efforts to tailor the lignin structure have been made to increase the selectivity of the final products. Considering the heterogeneous structure of lignin, the depolymerized product is a mixture of hydroxyphenyl- (H-), guaiacyl- (G-), and syringyl- (S-) units with various sidechains, depending on the reaction conditions. From the downstream perspective, the complex distribution of the products is technically challenging to further processing. Lignin demethylation was introduced to selectively produce catechol and catechol derivatives. Kim et al. reported that the demethylation of softwood-derived lignin produces a catechol-like structure, and the subsequent hydrogenolysis reaction using a hydrogen donor solvent can produce a high yield of catechol and alkylcatechols (*29*) (Figure 3).

More recently, catechyl lignin was discovered in the seed coats of vanilla and cactus species. Catechyl lignin is a benzodioxane homopolymer without condensed units, representing ideal lignin for valorization. Hydrogenolysis of catechyl lignin resulted in a near-quantitative yield of catechyl-type monomers with a selectivity of 90% to a single monomer (*30*). Wang et al. reported that catalytic depolymerization of catechyl lignin yielded 30 wt% monomeric catechol derivatives with 86% selectivity toward catechylpropanol (*31*). Considering that catechyl lignin is highly acid-resistant with structural uniformity (*30*), chemo- or bioselective alteration of lignin structure to a catechyl-like structure would be a promising strategy for lignin valorization.

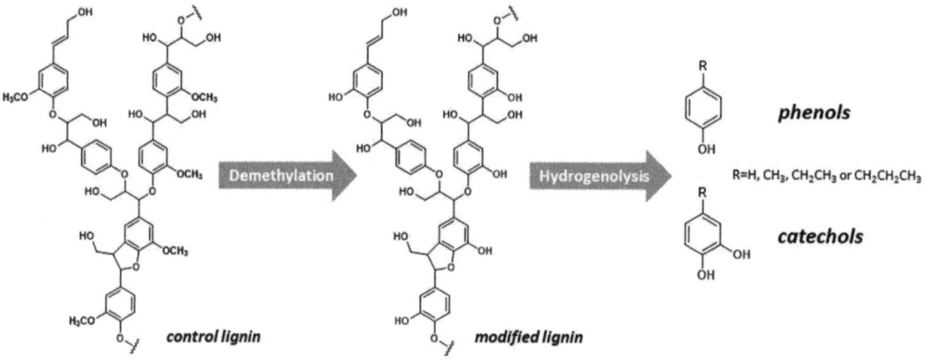

Figure 3. Selective demethylation of lignin followed by hydrogenolysis to produce catechols. Reprinted with permission from reference (29). Copyright 2021 ELSEVIER.

Lignin Modification for the Production of Carbohydrates

Traditionally, biomass pretreatment technologies tried to remove as much lignin in the biomass cell wall typically using strong and toxic acid or base catalyst at harsh conditions. Although the conventional approach has been successfully demonstrated in lignin removal, it often results in poor lignin quality, devaluing the overall process efficiency. Modern biorefineries, however, have been developing a milder treatment process to modify lignin rather than targeting complete lignin removal (*32*), which enables better utilization of lignin and maximizes the value from biomass.

Alkali-oxygen treatment was found to effectively modify lignin structure, producing a high yield of fermentable sugars (*33, 34*). Biomass chips treated with sodium carbonate and oxygen before mechanical refining resulted in modified lignin with a greater amount of carboxylic acid groups retaining most of the carbohydrates in the solid fraction. The modified substrates showed increased swelling and a decrease in the tendency of the lignin to bind enzymes, resulting in a high yield of fermentable sugars.

Lai et al. developed an integrated pretreatment process combining an in-situ lignin modification with 2-naphthol and poly (ethylene glycol) diglycidyl ether (PEGDE) (*35*). Lignin modification with 2-naphthol and PEGDE showed suppressed lignin repolymerization and lowered the non-productive binding on lignins of enzymes, respectively, synergistically facilitating enzymatic hydrolysis of the substrates. Similarly, 2-naphthol-7-sulfonate was used as a carbocation scavenger in the biomass pretreatment (*36*). The addition of scavenger to the acid pretreatment resulted in acid group-incorporated lignin with enlarged biomass porosity, and the hydrolysis yield increased by more than 48% after dilute acid pretreatment.

Tailoring Lignin Structure for the Production of Lignin-Derived Materials

Chemical modification of lignin to develop new materials is another promising way of lignin value-addition. Lignin has plenty of reactive functional groups, such as hydroxyl, carbonyl, benzyl alcohol, and carboxyl groups. However, the structural complexity and heterogeneity of lignin, attributed to the various functional groups, decrease its processability toward making lignin-derived materials. Therefore, proper structural transformation to introduce new functionalities on lignin has been performed. One way is the incorporation of functional groups that have not existed in raw lignin material. Another is to modify the hydroxyl group, the most frequent and reactive functional group in lignin polymer, for further development. In addition, controlled functionalization by polymer chemistry improves the performance of lignin-based materials. In the next sections, several technical approaches on the functionalization of the lignin structure for material production are presented. Also, the material properties of lignin-derived products are compared with those of commercial products where applicable.

Incorporation of New Functional Groups on Lignin Aromatic Region

Several types of chemical reactions have been used to synthesize new functional groups onto lignin; hydroxyalkylation, nitration, amination, etc. (Table 1) Each modification method can contribute to the change of physicochemical properties of lignin.

Hydroxyalkylation is a reaction that can replace the carboxyl group and hydroxyl group of lignin using formaldehyde and strong bases, such as NaOH or Na_2CO_3. This reaction includes the condensation of an aldehyde with activated lignin aromatic regions for further functionalization or improving solubility and mechanical properties of the lignin-derived resin. Several studies were conducted to use hydroxyalkylated lignin-derived resin as a high-performance and greener binder replacement in particleboard or plywood. Yang et al. (*37*) synthesized hydroxymethyl lignin from four kinds of technical lignin (corncob lignin, two different poplar wood lignins, and wheat straw lignin) obtained from black liquor or biorefinery mills. In this study, an aromatic region of lignin and formaldehyde were reacted in the presence of NaOH (pH 9–10) at 80 °C to form a methylol group on the *para* or *ortho* positions of a phenolic hydroxyl group. These functionalized lignins were then copolymerized with phenol to produce lignin-phenol-formaldehyde resin adhesive, which showed similar structure and curing behavior with the commercial one, in the preparation of plywood. A similar experiment of lignin modification was described by Wang and Chen. (*38*). Alkaline lignin solution extracted from corn stalk (pH 10.5–12.0) was methylolated at 60–80 °C. The authors noticed that high temperatures at over 90 °C could accelerate the self-condensation process of methylolated lignin. The methylol groups in the aromatic C_5 position of lignin were subsequently condensed with phenolic resin as plywood adhesives under hot pressing conditions (0.8 MPa at 160 °C). Another study of lignin hydroxyalkylation was conducted by Jiao et al. (*39*). In their study, corncob lignin was methylolated at 80 °C and pH 11.5 to enhance the reactivity of lignin toward making a hydroxymethylurea polymer. This three-dimensional lignin polymer network captures hydroxymethylurea and slowly releases nitrogenous compounds (NH_4^+ and NO_3^-). As a result, the authors suggested that the hydroxymethylated lignin could act as a diffusional barrier of controlled-release nitrogen fertilizer.

Table 1. Examples of Incorporation of New Functionalities to Lignin Aromatic Structure

<table>
<tr><th>Reactions</th><th>Scheme</th><th>References</th></tr>
<tr><td>Hydroxyalkylation</td><td>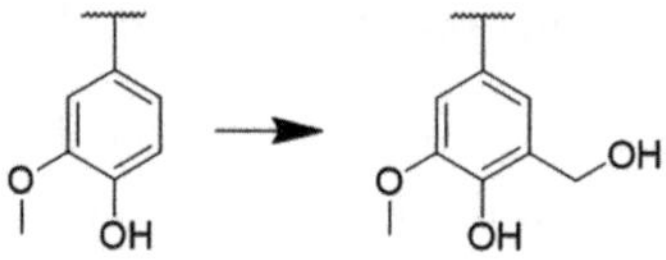
</td><td>(37-39)</td></tr>
<tr><td>Nitration</td><td>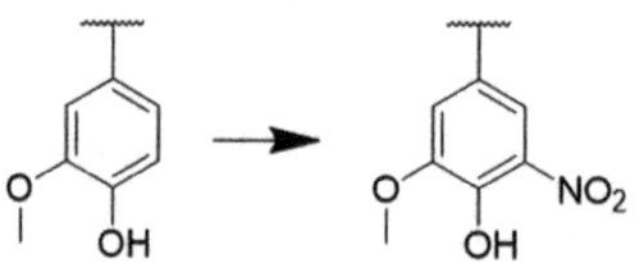
</td><td>(40)</td></tr>
<tr><td>Mannich reaction</td><td>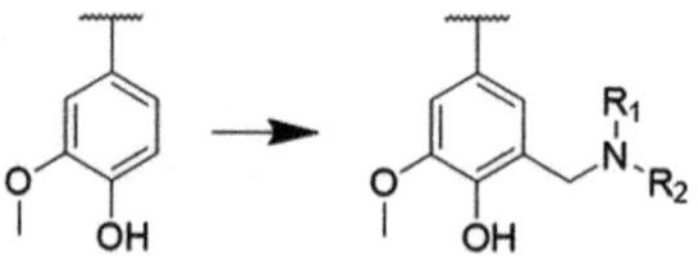

For dimethylamine, $R_1 = R_2 = CH_3$
For diethylamine, $R_1 = R_2 = CH_2CH_3$
For triethylenetetramine,
$R_1 = H$, $R_2 = CH_2CH_2(NHCH_2CH_2)_2NH_2$</td><td>(41, 43, 44)</td></tr>
</table>

Nitration and Mannich reaction are used to add nitrogen-containing functional groups in an aromatic region of lignin structure. Lignin nitration is used not only for further substitution reactions or add directing group that enables the incoming electrophile to react at specific positions on the aromatic ring, but also to produce nitrated aromatics, which is an important intermediate in chemical industries. Graglia et al. (*40*) reported the preparation of nitrogen-doped carbon using functionalized lignin via aromatic nitration followed by carbonization in molten eutectic $KCl/ZnCl_2$. It was reported that the nitrogen doping enhances the electroactivity of carbon. Residual lignin obtained from lactic acid production was reacted with inorganic acid catalysts in acetic anhydride and quenched at 0 °C. The obtained acetylated nitrogen-functionalized lignin (NL) with 6.7% nitrogen content was employed for the synthesis of nitrogen-doped carbon via carbonization in molten eutectic $KCl/ZnCl_2$. Compared to further deacetylated NL and aminated lignin, NL was reported to show slightly better electrocatalytic activity for the oxygen reduction reaction when carbonized.

The Mannich reaction is a condensation reaction of ammonia, primary or secondary amine in the presence of formaldehyde. This reaction introduces amine groups to the unsubstituted *ortho* position in the lignin and is used for the synthesis of building blocks to prepare nitrogen-containing compounds or nitrogen-doped materials.

Du et al. (*41*) performed the amination of the lignin model compound (apocynin) and LignoBoost softwood kraft lignin (*42*) by the Mannich reaction. In this research, the lignin was dissolved in 1,4-dioxane, then 10 molar equivalents of dimethylamine and formaldehyde were subsequently introduced in the solution. The Mannich reaction between lignin, amine, and formaldehyde was carried out under sealing conditions at 60 °C for 4 h. Du et al. identified a new $ArCH_2N$ signal in ^{13}C NMR and 2D-HSQC NMR from the resulting aminated lignin and determined that 28 amine groups were successfully introduced per 100 aromatic rings. Furthermore, 42 amine groups were introduced per 100 aromatic rings by the Mannich reaction after phenolation. The aminated lignin products were hydrophilic and showed higher zeta potential (max. 31.6 mV) and charge density (max. 1.6×10^{-7} equiv./mL) that could form surfactant chemicals or polycationic materials.

Work by Li et al. (*43*) applied the Mannich reaction to synthesize the aminated lignin-polylactic acid (PLA) blend. Softwood kraft lignin was reacted with diethylamine and formaldehyde at 60 °C for 4 h under sealing and stirring. ^{31}P NMR spectra showed the complete disappearance of the guaiacyl hydroxyl group peak and rise of C_5 substituted (dialkylaminomethyl) guaiacyl hydroxyl group peak, which indicates the complete introduction of dialkylaminomethyl group at the C_5 position of the lignin aromatic ring. They reported that blend coating of PLA-aminated lignin on the urea pellet fertilizer for food production provides additional bound nitrogen (high molecular amine groups) to the fertilizer to supply a slow-release effect. On the other hand, Mendis et al. (*44*) performed the Mannich reaction of lignin to increase dispersion in epoxy through lignin–amine interaction. Kraft lignin was introduced to triethylenetetramine and quickly homogenized for 5 min. Then formaldehyde solution (38 wt%) was introduced at a rate of 0.1 g/min to the lignin-triethylenetetramine solution. Increased hydroxyl and tertiary amine functional groups in the Mannich-functionalized lignin enhanced the dispersion ability of the lignin in an epoxy matrix but discouraged thermomechanical properties.

Modification of Lignin Functional Groups

Lignin functionalization has been performed to increase the chemical reactivity and processability of lignin. Among the modification strategies, alteration of the hydroxyl group is considered to be the simplest to increase the potential of lignin applications in the polymer industry, which includes allylation, esterification, etherification and silylation. Table 2 presents several examples of structural modification of lignin.

In the allylation, the hydroxyl groups of lignin are converted into allyl ether groups which can be further functionalized or directly applied to various chemical reactions, including epoxidation, thiol-end click chemistry and bromination.

Llevot et al. (*45*) reported an allylation of bio-based phenols (guaiacol, eugenol, vanillyl alcohol, and syringol) and organosolv lignin in the Tsuji–Trost reaction with a Pd nanoparticle catalyst. The study revealed that selective allylation of the phenolic hydroxyl groups was employed by 1 mol% of Pd catalyst in an H_2O/dimethylacetamide solvent mixture (44% allylated) and an H_2O/ethyl acetate solvent mixture (83% allylated). Dumont et al. (*46*) also described the Pd-catalyzed selective allylation of kraft lignin for the production of malleable materials. Dumont et al. described that the telomerized lignin showed a decreased glass-transition temperature of 50 °C, and can be easily shaped. Furthermore, ^{1}H NMR analysis determined the presence of unsaturated sidechains in the telomerized lignin, which could be further functionalized to synthesize fillers, macromonomers, and compatibilizing agents.

Traditionally, lignin esterification with acid derivatives has been widely used for structural analysis of lignin since it functionalizes the lignin hydroxyl group, decreases intermolecular hydrogen bonding, and thus increases lignin solubility in nonpolar solvents. Also, lignin esters with larger-chain derivatives showed increased thermal mobility compared to unmodified lignin, which has lead to research for several applications in carbon fibers, thermoset plastics, and coating polymers.

Koivu et al. (*47*) carried out a detailed study of lignin esterification with different acids (C2–C16) to determine the effect of chain length and substitution degree on the thermal properties of the lignin ester. In this study, softwood kraft lignin was dissolved with the mixture of tetrahydrofuran (THF), dimethylformamide (DMF), and pyridine. Acyl chlorides with four different fatty acid chains (C2, C8, C12, and C16) were introduced in the lignin solution and further reacted. All the synthesized esters showed a decrease in the glass-transition temperature (ΔTg max. -100 °C) and melting temperature with a high loading of fatty acids, which suggests the potential of the lignin ester as additives for polymer blends by melt extrusion.

Lignin has been viewed as a promising feedstock of renewable carbon fiber. However, there has been critical issues of decreasing tensile properties (instability or surface defects) for lignin-based carbon fibers. Zhang & Ogale (*48*), to overcome the problem, synthesized partially acetylated lignin and used to prepare dry-spinning carbon fiber. The acetylated lignin with low acetic anhydride concentration (0.66 mL/g lignin) was dissolved in acetone and successfully dry-spun into lignin as-spun fibers (i.d. 30 μm). Finally carbonized fibers obtained in this study (i.d. 6-8 μm) displayed higher tensile modulus (52 ± 2 GPa), larger surface area, and higher reactivity due to its less graphitic surfaces.

In the study by García-Mateos et al. (*49*), carbonyl and hydroxyl groups of Alcell lignin were esterified with phosphoric acid to generate phosphate and/or polyphosphate esters during electrospinning lignin fiber preparation in ethanol. Product fibers containing phosphorus-functionalities have shown an increased glass transition temperature and a much shorter stabilization time (105 min) than the pure lignin fibers (more than 90 h) without fiber fusion issue. The fiber also has submicron size (i.d. ≤ 1 μm), higher surface area, and uniformly distributed O and P surface functionalities that has potential for use as heterogeneous catalyst, adsorbent, and electrode. In contrast, Ding et al. (*50*) described that the butylation of organosolv lignin in the presence of 1-methylimidazole catalyst at 60 °C for 3 h decreased the glass transition temperature of lignin-polyacrylonitrile blend when compared to pure lignin-polyacrylonitrile blend. Also, a carbon fiber mat from a butylated lignin blend at 1,000 °C showed higher tensile strength and Young's modulus than that from the pure lignin-polyacrylonitrile blend.

Li et al. (*43*) additionally performed both acetylation and palmitoylation of the hydroxyl group of lignin to coat urea pellet fertilizer. The esterification of lignin was conducted using acetic anhydride or palmitic anhydride together with pyridine as a both solvent and catalyst at room temperature for 24 h. The esterification increased the hydrophobicity of the esterified lignin, especially with long-chain palmitic acid. Compared to the aminated lignin blend, PLA-palmitoylated lignin blend coating displayed a longer nitrogen release delay due to largely increased hydrophobicity.

Table 2. Examples of Structural Modifications of Lignin Functional Groups

Reactions	Scheme	References
Allylation	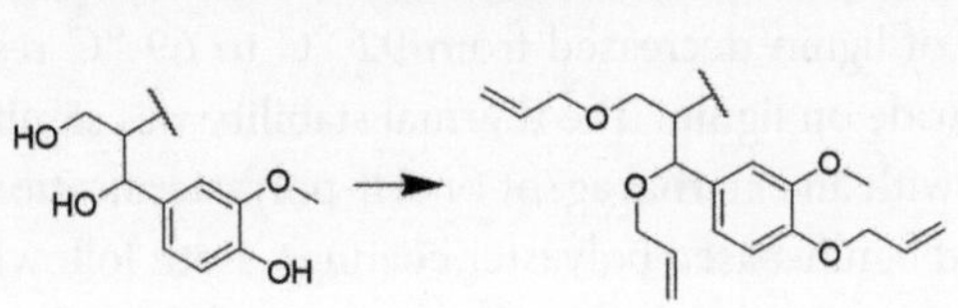	(45, 46)
Esterification	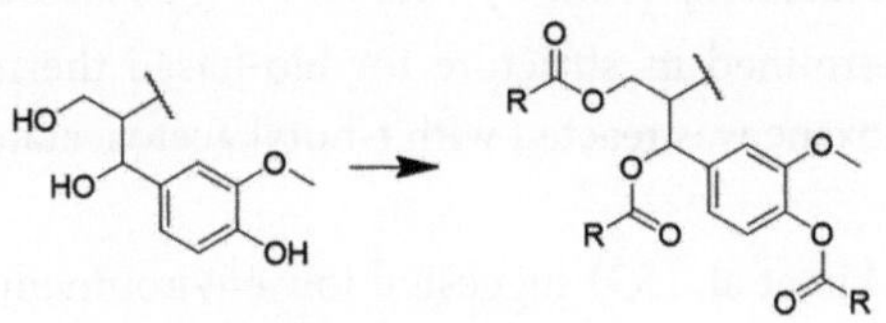 For carboxylic acids, R = $CH_3(CH_2)_n$ For *t*-butyl acetoacetate, R = CH_3COCH_2	(43, 47-53)
Etherification	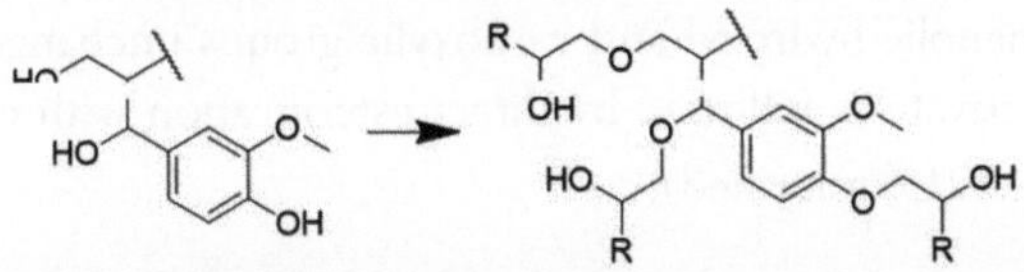 For styrene oxide, R = C_6H_6	(54)
Silylation	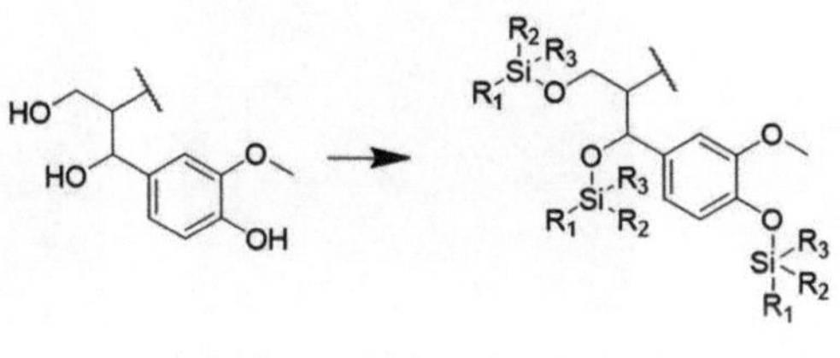 For divinyldisilazane, R_1 = CH_3CH, $R_2 = R_3 = CH_3$ For *t*-butyldimethylsilylchloride, $R_1 = (CH_3)_3C$, $R_2 = R_3 = CH_3$	(55, 56)
Tosylation	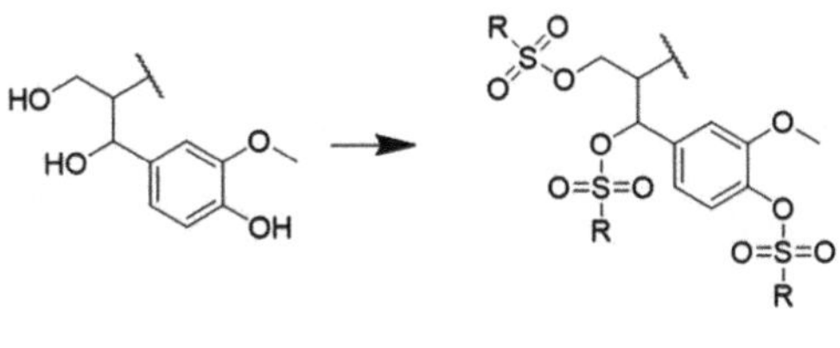 For 4-toluenesulfonyl chloride, R = $C_6H_6C_3$	(65, 66, 76)
Demethylation	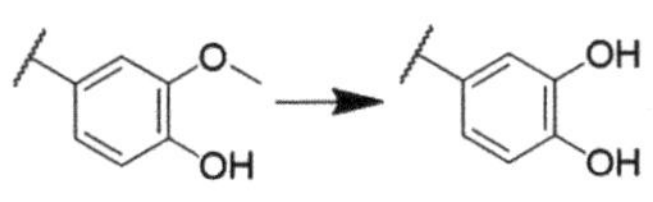	(57)

Scarica et al. (*51*) performed aliphatic hydroxyl group-specific esterification of Indulin AT kraft lignin using succinic anhydride. Lignin and succinic anhydride with different weight ratios were dissolved in THF in the presence of 1-methylimidazole catalyst. The authors reported that glass-transition temperature of lignin decreased from 92 °C to 69 °C resulting from the plasticization effect of succinic anhydride on lignin. The thermal stability was significantly increased by following reactions (crosslinking with an external agent or self-polyesterification of the succinylated lignin) to obtain fully cross-linked lignin-based polyester coatings. Both following reactions were successfully performed and the high water repellence, hardness, and adhesion strength of the lignin-based thermoset coating were evaluated. Work by Krall et al. (*52*) also esterified Indulin AT kraft lignin without catalyst and determined its structure for bio-based thermoset coating application. Kraft lignin dissolved in 1,4-dioxane was reacted with *t*-butyl acetoacetate at 130 °C for 3 h to be directly acetoacetylated.

On the other hand, Liu et al. (*53*) suggested low environmental impact esterification of lignin using a weak organic acid as the solvent and reagent rather than a strong acid. Softwood kraft lignin and hardwood organosolv lignin were mixed with propionic acid and magnetically stirred at 90–120 °C for 24–96 h. Interestingly, this reaction only functionalized the aliphatic hydroxyl group in the lignin, keeping the phenolic hydroxyl and carboxylic groups unchanged. The authors synthesized hydroxyethyl lignin derivatives followed by direct esterification with organic acid to obtain highly substituted lignin (max. 91% esterified).

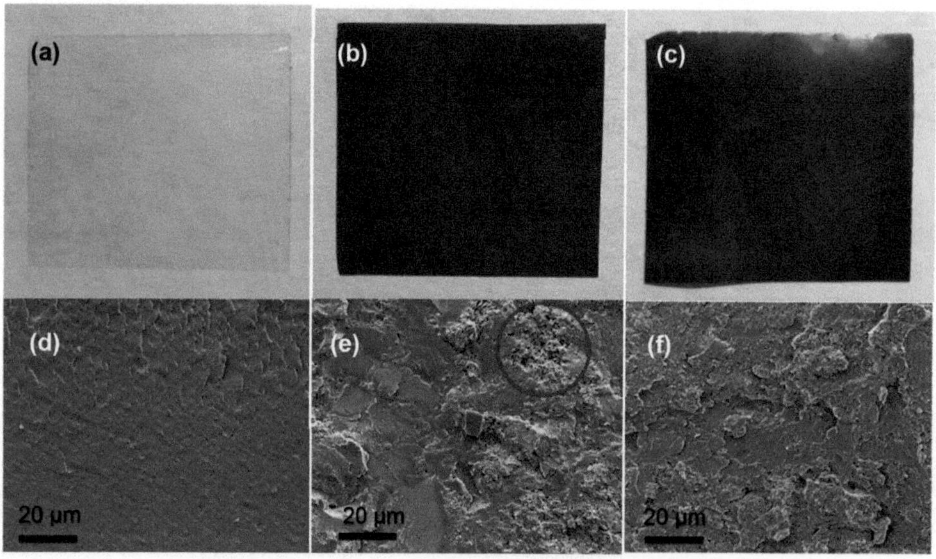

Figure 4. Optical images of (a) PAN, (b) lignin-g-PAN, and (c) vinyl silylated lignin-g-PAN films. Cross-section SEM images (1000×) of (d) PAN, (e) lignin-g-PAN, and (f) vinyl silylated lignin-g-PAN films. Reprinted with permission from reference (55). Copyright 2018 American Chemical Society.

Functionalization with an alkylene oxide applied to lignin is the most well-known etherification process, giving lignin not only higher solubility in organic solvents but also additional active sites. Duval et al. (*54*) applied epoxide ring-opening etherification with the lignin hydroxyl group to give the benzene ring site to lignin for further modifications. Phenolic hydroxyl groups of kraft lignin were selectively reacted with ring-opened styrene oxide in an alkaline solution (NaOH) at 50 °C. The new benzene rings in this benzylated lignin were further reacted with diazotized *N,N*-dimethyl-*p*-

phenylenediamine to contain diazobenzene groups, granting pH- and light-responsive properties to it.

Silylation is also one of the candidate modifications to improve the mechanical properties of lignin as polymer alternatives. Li et al. (55) performed an experiment of lignin modification with 1,1,3,3-tetramethyl-1,3-divinyldisilazane, which increases film flexibility and can partially replace polyacrylonitrile (PAN) in applications. Given its properties as light weight and stiffness, PAN is the most commonly used precursor for the production of carbon film. Pine kraft lignin was mixed with 1,1,3,3-tetramethyl-1,3-divinyldisilazane using a twin-screw extruder at 120 °C for 15 min (31 % sylilated). This vinyl silylated lignin was then copolymerized with acrylonitrile to form the PAN-grafted vinyl silylated lignin with improved thermal stability. Moreover, the films based on lignin-*graft* (g)-PAN showed much higher tensile strength (18 MPa) and strain (55% by a dynamic mechanical analyzer) than that of unmodified lignin-based copolymer (10 MPa and 10%, respectively) (Figure 4).

Buono et al. (56) produced silylated lignin and its thermal, physicochemical, and blending properties were compared to acetylated lignin. Soda lignin (Protobind 1000^{TM}) was silylated with *t*-butyldimethylsilylchloride using imidazole as a catalyst and base at room temperature. The authors noted that the whole lignin phenolic hydroxyl group was functionalized quickly; however, around 15% of the aliphatic hydroxyl group could not be reacted possibly due to the steric hindrance. The silylated lignin was well melt-blended with low-density polyethylene (LDPE) at 10% loading and showed better thermal stability and compatibility with LDPE matrix than the neat or acetylated lignins.

So far, approaches to modify the hydroxyl group in the lignin has been increasingly studied. Sawamura et al. (57) investigated the effect of demethylation of the lignin methoxyl group to convert it to hydroxyl group with three different chemicals (1-dodecanethiol, hydroiodic acid, and iodocyclohexane). Synthetic lignin, guaiacyl-type dehydrogenation polymer (GDHP), was used in this study. Among the chemicals used, iodocyclohexane found to be the most effective for increasing the number of phenolic hydroxyl groups.

Strategies for Lignin Copolymerization

The incorporation of lignin into new polymeric materials could be achieved by derivatization using polymer chemistry. There are diverse processing techniques to form chain or three-dimensional networks by combination of monomers/polymers; free radical addition polymerization, ionic polymerization, vinyl polymerization with complex coordination catalysts, step-reaction polymerization, ring-opening polymerization, cycloaddition polymerization, atom transfer polymerization, block and graft copolymer formation, etc. Various types of copolymerization processes have been applied to modify lignin into a useful bio-based polymer. Among them, graft copolymerization that uses the lignin as the polymer core unit (backbone) and covalently bonded monomers (branch), is the most common process to control the architecture and use of the resulting copolymer.

Panesar et al. (58) synthesized lignin-g-vinyl acetate copolymer with potassium persulfate as an initiator and ammonium iron(II) sulfate hexahydrate as a catalyst at 60–80 °C. The authors noted that the highest grafting efficiency (60%) was observed for 2:3 ratio of lignin and vinyl acetate at 70 °C. The copolymer displayed decreased glass transition temperature from 118 °C to 103 °C and increased hydrophobicity. In another study of lignin copolymerization by Hu et al. (59), grafting of polyacrylic acid with alkali lignin was performed by adding an ammonium persulfate/DMF solution as initiators and *N,N*-methylenebisacrylamide/DMF solution as a cross-linker at room temperature. The highly increased content of carboxylic acid group in lignin after copolymerization could contribute to improving the biosorption capacity of heavy metals, such as Cu(II) or Cd(II).

Several researches modified lignin before copolymerization to change the lignin as a proper macroinitiator, which enhances the reactivity or dispersion in the monomer mixture. Abdollahi et al. (*60*) carried out ring-opening polymerization of ε-caprolactone with three different lignin (kraft lignin, silica-lignin hybrid, and hydroxymethylated lignin) to prepare lignin-derived macroinitiators for thermoplastic synthesis. Lignin-g-polycaprolactone (PCL) with different polymer chain length was synthesized in the presence of a $ZnCl_2$ catalyst at 150 °C for 24 h with stirring, which showed relatively low grafting efficiency (52–55%). The authors suggested that water generation induced by $ZnCl_2$ contributed to reduced efficiency; therefore, $ZnCl_2$ can be replaced with another catalyst. Thermal properties of the lignin-g-PCL thermoplastics depended on the grafted chain length rather than the types of lignin. In another way, grafting of poly-L-lactide (PLLA) on fractionated lignin was conducted by Park et al. (*61*). Soda lignin was fractionated by sequential extraction with five different solvents (ethyl acetate, 2-butanone, methanol, acetone, and 1,4-dioxane). Obtained lignin fractions were reacted with L-lactide and triazabicyclodecene (ring-opening catalyst) to produce lignin-g-PLLA copolymer at 130 °C for 4 h under stirring. Finally, the lignin-PLA composite was synthesized by mixing each lignin-g-PLLA and PLA. The authors mentioned that high molecular weight lignin fraction, showing lower hydroxyl group content, achieved a slightly lower rate of conversion L-lactide to PLLA chains (min. 88.3%) than low molecular weight one (max. 91.2%). But the PLA composite with lower conversion rate presented better elastic modulus and UV blocking ability.

Cannatelli and Ragauskas (*62, 63*) synthesized kraft lignin-based copolymer by laccase-catalyzed grafting with low molecular weight quinone compounds to form the hyperbranched polymer. Methylhydroquinone and tris(2-mercaptoethyl)amine were introduced to the lignin/1,4-dioxane solution, followed by laccase-assisted thiol-Michael addition at 50 °C for 20 h. The resulting product, which has potential as a lignin-based biocomposite thermoplastic, showed moderate glass transition temperature (50–60 °C) and good thermostability.

Ring-opening chain-growth polymerization is utilized to obtain a grafted thermoset polymer of lignin with small molecules. Hydroxyl or other functional groups in lignin are etherified to produce polymers in processes such as the curing of epoxy resin. Li et al. (*64*) conducted the synthesis of epoxy resin by the formation of an ether bond between hydroxyl groups in lignin and glycerol diglycidyl ether (GDE)-derived crosslinker epoxide without the use of modifiers or catalysts. A mixture of lignin and GDE was cured at 130–150 °C. Lignin-GDE copolymer showed comparable characteristics (water-tolerance, fast curing, and adhesion performance) as conventional formaldehyde-based resins. Mahata et al. (*65*), formed hydrogel for the use of anti-infective ointment. A hydrophilic polyoxazoline chain was grafted through ring-opening polymerization between tosylated lignin macroinitiator (*66*) and 2-ethyl oxazoline in dimethyl sulfoxide at 100 °C for 10 h under nitrogen atmosphere. This copolymer has undergone further copolymerization with 3-amino-1H-1,2,4 triazole to achieve antimicrobial and antibiofilm properties.

Free-radical polymerization (FRP) is a kind of polymerization method of successive addition induced by free-radical monomers. In work by Gupta et al. (*67*), FRP of kraft lignin and acrylamide was conducted. First, hydroxyl groups in kraft lignin was functionalized by ring-opening polymerization of epoxide ring in glycidyl methacrylate (GM). Then lignin-GM macromolecule was reacted with acrylamide dissolved in DMF in the presence of azobisisobutyronitrile (AIBN) at 70 °C for 1 h. The authors compared the plasticizing ability of lignin-polyacrylamide copolymer with a commercial superplasticizer (polycarboxylate ether) in cement suspensions. Lignin-polyacrylamide by FPR showed comparable viscosity at low superplasticizer concentration and compressive strength but poor slump flow spread value.

One of the major drawbacks of FRP is uncontrolled molecular weight distribution due to the rapid irreversible termination process. On the other hand, controlled radical polymerization through a degenerative chain transfer process can be performed by reversible addition-fragmentation chain transfer (RAFT) of a wide range of monomers and RAFT agents which mediate the polymerization. In this reaction, lignin is usually transformed into a RAFT agent. Gupta et al. (*67*) also performed RAFT to synthesize polyacrylamide-grafted kraft lignin. Lignin-based RAFT macroinitiator, obtained by reaction of lignin and acyl chloride xanthate at room temperature for 12 h (*68*), was then grafted with acrylamide in the presence of AIBN initiator. Almost comparable slump spread to commercial polycarboxylate ether was reported in synthesis of lignin-polyacrylamide by RAFT.

Xu et al. (*69*) converted organosolv lignin into the RAFT agent by reacting to lignin hydroxyl groups with 4-cyano-4-(phenylcarbonothioylthio)pentanoic acid at 25 °C for 48 h in the presence of *N,N*-dicyclohexylcarbodiimide, 4-dimethylaminopyridine, and dichloromethane. This lignin-derived RAFT agent and soybean oil-derived methacrylate monomers (SBMAs) were grafted as a RAFT reaction at 70 °C for 48 h under nitrogen gas. The authors cured a mixture of lignin-g-poly(SBMAs), maleic anhydride, and tri-*n*-propylamine at 120 °C for 12 h to form lignin epoxy resin. The presence of hydrogen bonding between secondary amides resulted in grafted copolymers with different tensile behaviors, lower stress at the break with higher strain at break.

Atom transfer radical polymerization (ATRP) is another popular method that enables control of polymer structure and composition while grafting various types of monomer. Among the important components of ATRP, lignin usually acts as an initiator. In the study by Liu et al. (*70*), kraft lignin-derived ATPR macroinitiator was produced first by esterification of the lignin hydroxyl group with 2-bromoisobutyryl bromide. Next, ATRP of 2-dimethylaminoethyl methacrylate (DMAEMA) with the lignin macroinitiator was carried out in the presence of copper(I) bromide (catalyst) and 1,1,4,7,10,10-hexamethyltriethylenetetramine as a ligand. Lignin-g-poly(DMAEMA) copolymer had a hyperbranched structure with a hydrophobic backbone (lignin) and multiple cationic hydrophilic arms (poly(DMAEMA)) which can bind and condense plasmid DNA to form a nano-sized gene delivery system . Kim et al. (*71*) also synthesized a graft polymer by ATRP of styrene with a bromoisobutyrylated kraft lignin macroinitiator in the presence of copper(I) bromide and *N,N,N′,N″,N″*-pentamethylenediethylenetriamine as a ligand. Thermal decomposition properties of lignin-g-polystyrene copolymer was improved with increase of number average molecular weight, while the radical polymerization process were less controlled.

Polyurethanes are the most commonly used polymers to make foams, synthesized by the reaction of a polyol (an organic compound containing multiple hydroxyl groups) with a polyisocyanate. Since lignin is a natural polyol, the urethanization of lignin to produce bio-based polyurethane has been steadily explored. Zhang et al. (*72*) reported the preparation of rigid polyurethane (RPU) foams from kraft lignin via surface functionalization followed by urethane bonding. A mixture of kraft lignin and diisocyanate was heated at 80 °C for 1 h under 800 RPM stirring to form urethane moiety by interaction of lignin hydroxyl group and highly reactive isocyanate group. This functionalized lignin was reacted with a commercial polyester polyol, R-23-015, in the presence of HFC-365mfc (blowing agent) and an amine catalyst to crosslink the RPU system. The authors revealed that lignin-based RPU foams with 30% of lignin substitution of the commercial polyol showed comparable thermal and mechanical properties to conventional RPU foams. Specific compressive strength and modulus were increased by increasing lignin ratio from 0 to 20% but decreased by further increasing the ratio. In a similar study from Gómez-Fernández et al. (73), hydroxyl groups in kraft lignin were reacted with secondary isocyanate groups of isophorone diisocyanate and dibutyltin dilaurate (catalyst) at 60 °C under 600 RPM stirring and nitrogen atmosphere. Then RPU was subsequently synthesized by blending the lignin-diisocyanate copolymer, castor oil derived polyether polyol, water (blowing agent), and additives at room temperature under stirring.

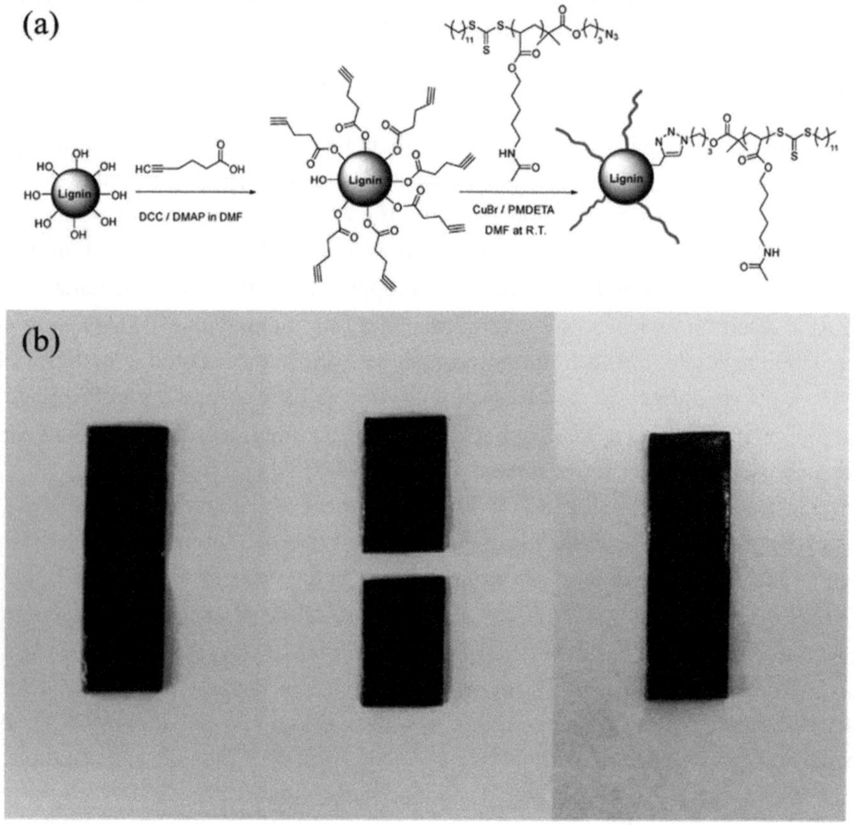

Figure 5. (a) Alkynylation and copper-catalyzed click reaction of poly(5-acetylaminopentyl acrylate) and natural lignin. (b) Self-healing test of 20 wt % lignin containing lignin-g-PAA under ambient conditions (23 °C and 55% humidity). Reprinted with permission from reference (75). Copyright 2018 American Chemical Society.

Work by de Haro et al. (74) developed lignin-based polyurethane coatings by urethanization of vanillic acid-derived diisocyanate (VA-NCO) and hydroxyl groups in lignin. Three different lignins (Indulin AT kraft lignin, acetone organosolv lignin, and soda lignin) were used in this study. VA-NCO and lignin were dissolved at room temperature in THF at 20 wt% concentration under stirring. The authors estimated a maximum lignin content in the obtained PU coatings in the 92–96 wt % range in a desirable condition.

Liu and Chung (75) utilized a copper-catalyzed azide-alkyne cycloaddition or "click" reaction, forming a triazole ring in very high yields with little byproduct, of alkyne-functionalized softwood lignin and poly(5-acetylaminopentyl acrylate) (PAA). Alkyne functionalization of lignin was performed using 5-hexynoic acid in the presence of *N,N′*-dicyclohexylcarbodiimide and DMF for 48 h followed by further purification processes. Next, alkyne groups of functionalized lignin and azide terminal groups of PAA were reacted by copper(I) bromide-catalyzed alkyne-azide cycloaddition at room temperature (Figure 5). The resulting copolymer showed superior mechanical properties (maximum tensile strength, energy-to-break, and Young's modulus) with longer PAA chains. Interestingly, acetylamino groups on the polymers led directly to autonomic healing properties in bulk lignin-g-PAA, showing 93% of maximum strength recovery after the self-healing process when severely damaged.

Work by Panovic et al. (76) selectively modified the β-O-4 γ-position of organosolv lignin with a high level of α-alkoxylated β-O-4 units (walnut shell butanol alkoxasolv lignin) via tosylation, azidation, and copper-catalyzed azide-alkyne triazole formation. First, hydroxyl groups in organosolv lignin was tosylated with 3 equivalents of toluenesulfonyl chloride at room temperature for 6 h. Next, azidation of the tosylated group was achieved with 2 equivalents of NaN_3 at 50 °C for 24 h, followed by copper-catalyzed click chemistry with several alkynes, $CuSO_4{\cdot}5H_2O$, and ascorbic acid dissolved in DMF/water at room temperature for 16 h . They described that the click reactions can only occur from the γ-position of β-O-4 linkage in this process because phenolic end groups are tosylated and the α-hydroxyl groups are replaced with butyl ethers. The authors concluded that this grafting method has potential to meet suitable and novel applications since it is a highly regioselective reaction.

Diels–Alder reaction is a versatile reaction linking an azido and a thermally reversible alkynyl group. Duval et al. (77) carried out a reversible crosslinking of lignin via Diels–Alder reaction. Both furan-functionalized moieties and maleimide-functionalized moieties were synthesized based on softwood kraft lignin. To prepare furan moieties, kraft lignin dissolved in NaOH solution was reacted with furfuryl glycidyl ether at 50 °C overnight. Maleimide moieties were synthesized by the reaction between the lignin/DMF solution and 6-maleimidohexanoic acyl chloride in the presence of K_2CO_3 at 50 °C for 1 h. Diels-Alder reaction of each functionalized lignins has occurred in DMSO at 70 °C until the resulting gelatin was formed. By heating up to 120 °C, the gel turned back into the liquid state due to a retro-Diels-Alder reaction (Figure 6). This new reversible crosslinking ability of lignin-based copolymer can be applied to manufacture material with recyclability or self-healing ability.

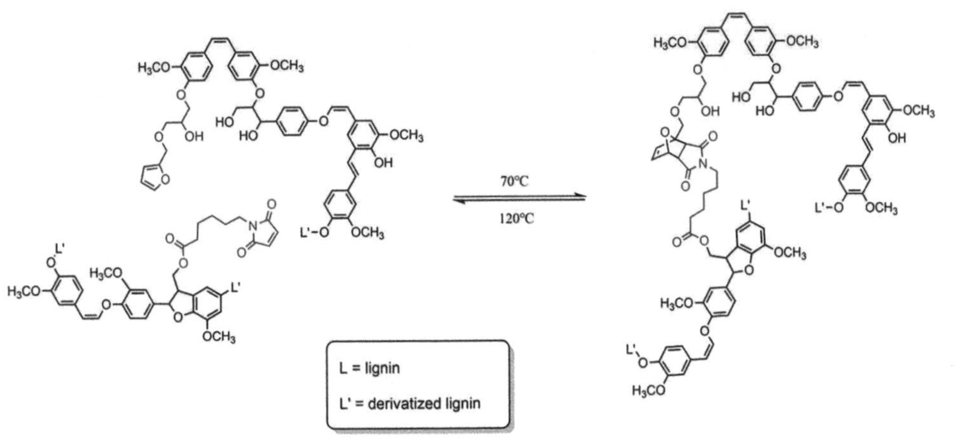

Figure 6. Crosslinking of the furfurylated lignin with furfuryl glycidyl ether (blue) and esterified lignin with 6-maleimidohexanoic acyl chloride (red) via the Diels–Alder reaction. Reprinted with permission from reference (77). Copyright 2015 Royal Society of Chemistry.

Future Perspectives

In the past years, lignin valorization has become a hot topic of developing sustainable biorefineries. The network map of keywords in the recent 500 scientific publications on "lignin valorization" was created using the full counting method of VOSViewer (*78, 79*), and the resulting keyword network map is shown in Figure 7. As shown in the figure, the major keywords are "valorization," "biomass," "lignin," "conversion" "depolymerization," "lignocellulosic biomass," "pretreatment," "kraft lignin," and "chemicals." This network map shows what recent studies focused on under the area of lignin valorization. However, we recognize that, despite efforts to utilize lignin in many applications, significant improvement in conversion technologies is needed to meet the rigorous requirements of the market of the products. It is often discussed that inherent physical and chemical properties of lignin negatively impact properties of final materials, limiting the amount of lignin in the products. Also, several technical challenges, including separation and recovery of the final products and catalyst recycling, need to be tackled to make lignin valorization more sustainable. Considering the long history of petroleum refining, lignocellulosic biorefineries are still in their early-stage, yet are the best non-petroleum alternatives for sustainable development. Lignin valorization, as a key research area in this field, requires continued, but more strategic efforts to maximize the value from lignin. It is also important to note that product cost and quality were the primary criteria of the past; future products need to meet environmental and social sustainability for achieving a sustainable carbon economy.

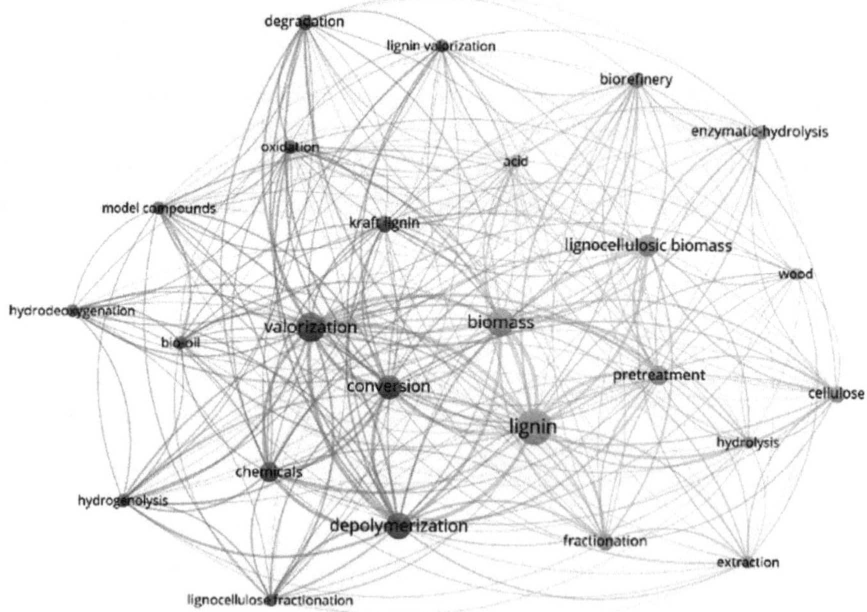

Figure 7. The keyword network of lignin valorization.

Acknowledgments

This work is supported by the Korea Institute of Science and Technology – the University of British Columbia (KIST-UBC) Biorefinery On-site Laboratory project and the Basic Science Research Program (NRF-2019R1A2C2086328) of the National Research Foundation funded by the Ministry of Science and ICT.

References

1. Kim, K. H.; Kim, C. S. Recent efforts to prevent undesirable reactions from fractionation to depolymerization of lignin: Toward maximizing the value from lignin. *Front Energy Res* **2018**, *6*, 92.
2. Schutyser, W.; Renders, a. T.; Van den Bosch, S.; Koelewijn, S.-F.; Beckham, G.; Sels, B. F. Chemicals from lignin: an interplay of lignocellulose fractionation, depolymerisation, and upgrading. *Chem Soc Rev* **2018**, *47* (3), 852–908.
3. Ragauskas, A. J.; Beckham, G. T.; Biddy, M. J.; Chandra, R.; Chen, F.; Davis, M. F.; Davison, B. H.; Dixon, R. A.; Gilna, P.; Keller, M. Lignin valorization: improving lignin processing in the biorefinery. *Science* **2014**, *344* (6185), 1246843.
4. Toledano, A.; Serrano, L.; Labidi, J. Improving base catalyzed lignin depolymerization by avoiding lignin repolymerization. *Fuel* **2014**, *116*, 617–624.
5. Huang, X.; Korányi, T. I.; Boot, M. D.; Hensen, E. J. Ethanol as capping agent and formaldehyde scavenger for efficient depolymerization of lignin to aromatics. *Green Chem* **2015**, *17* (11), 4941–4950.
6. Trajano, H. L.; Engle, N. L.; Foston, M.; Ragauskas, A. J.; Tschaplinski, T. J.; Wyman, C. E. The fate of lignin during hydrothermal pretreatment. *Biotechnol Biofuels* **2013**, *6* (1), 110.
7. Huang, X.; Korányi, T. I.; Boot, M. D.; Hensen, E. J. Catalytic depolymerization of lignin in supercritical ethanol. *ChemSusChem* **2014**, 7 (8), 2276–2288.
8. Khan, A.; Colmenares, J. C.; Gläser, R., Lignin-Based Composite Materials for Photocatalysis and Photovoltaics. In *Lignin Chemistry*, Springer: 2020; pp 1-31.
9. Chen, T.; Zhang, Q.; Pan, J.; Xu, J.; Liu, Y.; Al-Shroofy, M.; Cheng, Y.-T. Low-temperature treated lignin as both binder and conductive additive for silicon nanoparticle composite electrodes in lithium-ion batteries. *ACS applied materials & interfaces* **2016**, *8* (47), 32341–32348.
10. Ajjan, F.; Casado, N.; Rębiś, T.; Elfwing, A.; Solin, N.; Mecerreyes, D.; Inganäs, O. High performance PEDOT/lignin biopolymer composites for electrochemical supercapacitors. *Journal of Materials Chemistry A* **2016**, *4* (5), 1838–1847.
11. Zhang, Y.; Jiang, M.; Zhang, Y.; Cao, Q.; Wang, X.; Han, Y.; Sun, G.; Li, Y.; Zhou, J. Novel lignin–chitosan–PVA composite hydrogel for wound dressing. *Materials Science and Engineering: C* **2019**, *104*, 110002.
12. Farhat, W.; Venditti, R.; Mignard, N.; Taha, M.; Becquart, F.; Ayoub, A. Polysaccharides and lignin based hydrogels with potential pharmaceutical use as a drug delivery system produced by a reactive extrusion process. *Int J Biol Macromol* **2017**, *104*, 564–575.
13. Domínguez-Robles, J.; Stewart, S. A.; Rendl, A.; González, Z.; Donnelly, R. F.; Larrañeta, E. Lignin and cellulose blends as pharmaceutical excipient for tablet manufacturing via direct compression. *Biomolecules* **2019**, *9* (9), 423.
14. Saha, D.; Payzant, E. A.; Kumbhar, A. S.; Naskar, A. K. Sustainable mesoporous carbons as storage and controlled-delivery media for functional molecules. *ACS applied materials & interfaces* **2013**, *5* (12), 5868–5874.
15. Shuai, L.; Amiri, M. T.; Questell-Santiago, Y. M.; Heroguel, F.; Li, Y. D.; Kim, H.; Meilan, R.; Chapple, C.; Ralph, J.; Luterbacher, J. S. Formaldehyde stabilization facilitates lignin monomer production during biomass depolymerization. *Science* **2016**, *354* (6310), 329–333.

16. Mu, X.; Han, Z.; Liu, C.; Zhang, D. Mechanistic Insights into formaldehyde-blocked lignin condensation: a DFT study. *The Journal of Physical Chemistry C* **2019**, *123* (14), 8640–8648.
17. Lan, W.; Amiri, M. T.; Hunston, C. M.; Luterbacher, J. S. Protection group effects during α, γ-diol lignin stabilization promote high-selectivity monomer production. *Angewandte Chemie* **2018**, *130* (5), 1370–1374.
18. Rahimi, A.; Ulbrich, A.; Coon, J. J.; Stahl, S. S. Formic-acid-induced depolymerization of oxidized lignin to aromatics. *Nature* **2014**, *515* (7526), 249–252.
19. Qu, S.; Dang, Y.; Song, C.; Guo, J.; Wang, Z.-X. Depolymerization of oxidized lignin catalyzed by formic acid exploits an unconventional elimination mechanism involving 3c–4e bonding: a DFT mechanistic study. *Acs Catal* **2015**, *5* (11), 6386–6396.
20. Wang, M.; Zhang, X.; Li, H.; Lu, J.; Liu, M.; Wang, F. Carbon modification of nickel catalyst for depolymerization of oxidized lignin to aromatics. *Acs Catal* **2018**, *8* (2), 1614–1620.
21. Yang, W.; Du, X.; Liu, W.; Tricker, A. W.; Dai, H.; Deng, Y. Highly Efficient Lignin Depolymerization via Effective Inhibition of Condensation during Polyoxometalate-Mediated Oxidation. *Energy & Fuels* **2019**, *33* (7), 6483–6490.
22. Li, J.; Gellerstedt, G. Improved lignin properties and reactivity by modifications in the autohydrolysis process of aspen wood. *Industrial crops and products* **2008**, *27* (2), 175–181.
23. Li, J.; Henriksson, G.; Gellerstedt, G. Lignin depolymerization/repolymerization and its critical role for delignification of aspen wood by steam explosion. *Bioresource Technol* **2007**, *98* (16), 3061–3068.
24. Zhai, R.; Hu, J.; Saddler, J. N. Minimizing cellulase inhibition of whole slurry biomass hydrolysis through the addition of carbocation scavengers during acid-catalyzed pretreatment. *Bioresource Technol* **2018**, *258*, 12–17.
25. Zhang, Z.; Lahive, C. W.; Zijlstra, D. S.; Wang, Z.; Deuss, P. J. Sequential catalytic modification of the lignin α-ethoxylated β-O-4 motif to facilitate C–O bond cleavage by ruthenium-Xantphos catalyzed hydrogen transfer. *ACS Sustainable Chemistry & Engineering* **2019**, *7* (14), 12105–12116.
26. Roberts, V. M.; Stein, V.; Reiner, T.; Lemonidou, A.; Li, X.; Lercher, J. A. Towards quantitative catalytic lignin depolymerization. *Chemistry–A European Journal* **2011**, *17* (21), 5939–5948.
27. Kim, K. H.; Dutta, T.; Walter, E. D.; Isern, N. G.; Cort, J. R.; Simmons, B. A.; Singh, S. Chemoselective Methylation of Phenolic Hydroxyl Group Prevents Quinone Methide Formation and Repolymerization During Lignin Depolymerization. *Acs Sustain Chem Eng* **2017**, *5* (5), 3913–3919.
28. Khan, M. U.; Ahring, B. K. Lignin degradation under anaerobic digestion: Influence of lignin modifications-A review. *Biomass and Bioenergy* **2019**, *128*, 105325.
29. Kim, K. H.; Jeong, K.; Zhuang, J.; Jeong, H. J.; Kim, C. S.; Koo, B.; Yoo, C. G. Tandem conversion of lignin to catechols via demethylation and catalytic hydrogenolysis. *Industrial Crops and Products* **2021**, *159*, 113095.
30. Li, Y.; Shuai, L.; Kim, H.; Motagamwala, A. H.; Mobley, J. K.; Yue, F.; Tobimatsu, Y.; Havkin-Frenkel, D.; Chen, F.; Dixon, R. A. An "ideal lignin" facilitates full biomass utilization. *Science advances* **2018**, *4* (9), eaau2968.

31. Wang, S.; Su, S.; Xiao, L.-P.; Wang, B.; Sun, R.-C.; Song, G. Catechyl Lignin Extracted from Castor Seed Coats Using Deep Eutectic Solvents: Characterization and Depolymerization. *ACS Sustainable Chemistry & Engineering* **2020**, *8* (18), 7031–7038.
32. Chandra, R. P.; Chu, Q.; Hu, J.; Zhong, N.; Lin, M.; Lee, J.-S.; Saddler, J. The influence of lignin on steam pretreatment and mechanical pulping of poplar to achieve high sugar recovery and ease of enzymatic hydrolysis. *Bioresource Technol* **2016**, *199*, 135–141.
33. Wu, J.; Chandra, R.; Saddler, J. Alkali–oxygen treatment prior to the mechanical pulping of hardwood enhances enzymatic hydrolysis and carbohydrate recovery through selective lignin modification. *Sustain Energ Fuels* **2019**, *3* (1), 227–236.
34. Chu, Q.; Chandra, R. P.; Kim, C.-S.; Saddler, J. N. Alkali-oxygen impregnation prior to steam pretreating poplar wood chips enhances selective lignin modification and removal while maximizing carbohydrate recovery, cellulose accessibility, and enzymatic hydrolysis. *ACS Sustainable Chemistry & Engineering* **2017**, *5* (5), 4011–4017.
35. Lai, C.; Jia, Y.; Zhou, C.; Yang, C.; Shen, B.; Zhang, D.; Yong, Q. Facilitating enzymatic digestibility of larch by in-situ lignin modification during combined acid and alkali pretreatment. *Bioresource Technol* **2020**, 123517.
36. Chu, Q.; Song, K.; Wang, J.; Hu, J.; Chen, X. Improving enzymatic saccharification of hardwood through lignin modification by carbocation scavengers and the underlying mechanisms. *Bioresource Technol* **2019**, *294*, 122216.
37. Yang, S.; Zhang, Y.; Yuan, T.-Q.; Sun, R.-C. Lignin-phenol-formaldehyde resin adhesives prepared with biorefinery technical lignins. *Journal of Applied Polymer Science* **2015**, *132* (36)
38. Wang, G.; Chen, H. Carbohydrate elimination of alkaline-extracted lignin liquor by steam explosion and its methylolation for substitution of phenolic adhesive. *Industrial Crops and Products* **2014**, *53*, 93–101.
39. Jiao, G.-J.; Xu, Q.; Cao, S.-L.; Peng, P.; She, D. Controlled-release fertilizer with lignin used to trap urea/hydroxymethylurea/urea-formaldehyde polymers. *BioResources* **2018**, *13* (1), 1711–1728.
40. Graglia, M.; Pampel, J.; Hantke, T.; Fellinger, T.-P.; Esposito, D. Nitro lignin-derived nitrogen-doped carbon as an efficient and sustainable electrocatalyst for oxygen reduction. *ACS Nano* **2016**, *10* (4), 4364–4371.
41. Du, X.; Li, J.; Lindström, M. E. Modification of industrial softwood kraft lignin using Mannich reaction with and without phenolation pretreatment. *Industrial Crops and Products* **2014**, *52*, 729–735.
42. Axegård, P.; STFI-Packforsk, A. The kraft pulp mill as a biorefinery. *Division Fiber, Pulp Energy and Chemicals. STFI-Packforsk PO Box* **2007**, *5604*, 1–6.
43. Li, J.; Wang, M.; She, D.; Zhao, Y. Structural functionalization of industrial softwood kraft lignin for simple dip-coating of urea as highly efficient nitrogen fertilizer. *Industrial Crops and Products* **2017**, *109*, 255–265.
44. Mendis, G. P.; Hua, I.; Youngblood, J. P.; Howarter, J. A. Enhanced dispersion of lignin in epoxy composites through hydration and mannich functionalization. *Journal of Applied Polymer Science* **2015**, *132* (1)

45. Llevot, A.; Monney, B.; Sehlinger, A.; Behrens, S.; Meier, M. Highly efficient Tsuji–Trost allylation in water catalyzed by Pd-nanoparticles. *Chemical Communications* **2017**, *53* (37), 5175–5178.

46. Dumont, C.; Gauvin, R. M.; Belva, F.; Sauthier, M. Palladium-Catalyzed Functionalization of Kraft Lignin: Ether Linkages through the Telomerization Reaction. *ChemSusChem* **2018**, *11* (10), 1649–1655.

47. Koivu, K. A.; Sadeghifar, H.; Nousiainen, P. A.; Argyropoulos, D. S.; Sipilä, J. Effect of fatty acid esterification on the thermal properties of softwood kraft lignin. *ACS Sustainable Chemistry & Engineering* **2016**, *4* (10), 5238–5247.

48. Zhang, M.; Ogale, A. A. Carbon fibers from dry-spinning of acetylated softwood kraft lignin. *Carbon* **2014**, *69*, 626–629.

49. García-Mateos, F.; Berenguer, R.; Valero-Romero, M.; Rodríguez-Mirasol, J.; Cordero, T. Phosphorus functionalization for the rapid preparation of highly nanoporous submicron-diameter carbon fibers by electrospinning of lignin solutions. *Journal of Materials Chemistry A* **2018**, *6* (3), 1219–1233.

50. Ding, R.; Wu, H.; Thunga, M.; Bowler, N.; Kessler, M. R. Processing and characterization of low-cost electrospun carbon fibers from organosolv lignin/polyacrylonitrile blends. *Carbon* **2016**, *100*, 126–136.

51. Scarica, C.; Suriano, R.; Levi, M.; Turri, S.; Griffini, G. Lignin functionalized with succinic anhydride as building block for biobased thermosetting polyester coatings. *ACS Sustainable Chemistry & Engineering* **2018**, *6* (3), 3392–3401.

52. Krall, E. M.; Serum, E. M.; Sibi, M. P.; Webster, D. C. Catalyst-free lignin valorization by acetoacetylation. Structural elucidation by comparison with model compounds. *Green Chem* **2018**, *20* (13), 2959–2966.

53. Liu, L.-Y.; Hua, Q.; Renneckar, S. A simple route to synthesize esterified lignin derivatives. *Green Chem* **2019**, *21* (13), 3682–3692.

54. Duval, A.; Lange, H.; Lawoko, M.; Crestini, C. Modification of Kraft lignin to expose diazobenzene groups: Toward pH-and light-responsive biobased polymers. *Biomacromolecules* **2015**, *16* (9), 2979–2989.

55. Li, S.; Xie, W.; Wilt, M.; Willoughby, J. A.; Rojas, O. J. Thermally stable and tough coatings and films using vinyl silylated lignin. *ACS Sustainable Chemistry & Engineering* **2018**, *6* (2), 1988–1998.

56. Buono, P.; Duval, A.; Verge, P.; Averous, L.; Habibi, Y. New insights on the chemical modification of lignin: acetylation versus silylation. *ACS Sustainable Chemistry & Engineering* **2016**, *4* (10), 5212–5222.

57. Sawamura, K.; Tobimatsu, Y.; Kamitakahara, H.; Takano, T. Lignin functionalization through chemical demethylation: preparation and tannin-like properties of demethylated guaiacyl-type synthetic lignins. *ACS Sustainable Chemistry & Engineering* **2017**, *5* (6), 5424–5431.

58. Panesar, S. S.; Jacob, S.; Misra, M.; Mohanty, A. K. Functionalization of lignin: Fundamental studies on aqueous graft copolymerization with vinyl acetate. *Industrial Crops and Products* **2013**, *46*, 191–196.

59. Hu, L.-Q.; Dai, L.; Liu, R.; Si, C.-L. Lignin-graft-poly(acrylic acid) for enhancement of heavy metal ion biosorption. *Journal of Materials Science* **2017**, *52* (24), 13689–13699.

60. Abdollahi, M.; Habashi, R. B.; Mohsenpour, M. Poly (ε-caprolactone) chains grafted from lignin, hydroxymethylated lignin and silica/lignin hybrid macroinitiators: synthesis and characterization of lignin-based thermoplastic copolymers. *Industrial Crops and Products* **2019**, *130*, 547–557.
61. Park, S. Y.; Kim, J. Y.; Youn, H. J.; Choi, J. W. Utilization of lignin fractions in UV resistant lignin-PLA biocomposites via lignin-lactide grafting. *Int J Biol Macromol* **2019**, *138*, 1029–1034.
62. Cannatelli, M. D.; Ragauskas, A. J. Conversion of lignin into value-added materials and chemicals via laccase-assisted copolymerization. *Applied Microbiology and Biotechnology* **2016**, *100* (20), 8685–8691.
63. Cannatelli, M. D.; Ragauskas, A. J. Laccase-mediated synthesis of lignin-core hyperbranched copolymers. *Applied Microbiology and Biotechnology* **2017**, *101* (16), 6343–6353.
64. Li, R. J.; Gutierrez, J.; Chung, Y.-L.; Frank, C. W.; Billington, S. L.; Sattely, E. S. A lignin-epoxy resin derived from biomass as an alternative to formaldehyde-based wood adhesives. *Green Chem* **2018**, *20* (7), 1459–1466.
65. Mahata, D.; Jana, M.; Jana, A.; Mukherjee, A.; Mondal, N.; Saha, T.; Sen, S.; Nando, G. B.; Mukhopadhyay, C. K.; Chakraborty, R. Lignin-graft-polyoxazoline conjugated triazole a novel anti-infective ointment to control persistent inflammation. *Scientific Reports* **2017**, *7*, 46412.
66. Diop, A.; Awada, H.; Zerrouki, R.; Daneault, C.; Montplaisir, D. Tosylation and characterization of lignin in water. *Industrial & Engineering Chemistry Research* **2014**, *53* (43), 16771–16776.
67. Gupta, C.; Sverdlove, M. J.; Washburn, N. R. Molecular architecture requirements for polymer-grafted lignin superplasticizers. *Soft Matter* **2015**, *11* (13), 2691–2699.
68. Gupta, C.; Washburn, N. R. Polymer-grafted lignin surfactants prepared via reversible addition–fragmentation chain-transfer polymerization. *Langmuir* **2014**, *30* (31), 9303–9312.
69. Xu, Y.; Yuan, L.; Wang, Z.; Wilbon, P. A.; Wang, C.; Chu, F.; Tang, C. Lignin and soy oil-derived polymeric biocomposites by "grafting from" RAFT polymerization. *Green Chem* **2016**, *18* (18), 4974–4981.
70. Liu, X.; Yin, H.; Zhang, Z.; Diao, B.; Li, J. Functionalization of lignin through ATRP grafting of poly (2-dimethylaminoethyl methacrylate) for gene delivery. *Colloids and Surfaces B: Biointerfaces* **2015**, *125*, 230–237.
71. Kim, Y. S.; Youe, W.-J.; Kim, S. J.; Lee, O.-K.; Lee, S.-S. Preparation of a thermoplastic lignin-based biomaterial through atom transfer radical polymerization. *Journal of Wood Chemistry and Technology* **2015**, *35* (4), 251–259.
72. Zhang, X.; Jeremic, D.; Kim, Y.; Street, J.; Shmulsky, R. Effects of surface functionalization of lignin on synthesis and properties of rigid bio-based polyurethanes foams. *Polymers* **2018**, *10* (7), 706.
73. Gómez-Fernández, S.; Ugarte, L.; Calvo-Correas, T.; Peña-Rodríguez, C.; Corcuera, M. A.; Eceiza, A. Properties of flexible polyurethane foams containing isocyanate functionalized kraft lignin. *Industrial Crops and Products* **2017**, *100*, 51–64.
74. de Haro, J. C.; Allegretti, C.; Smit, A. T.; Turri, S.; D'Arrigo, P.; Griffini, G. Biobased polyurethane coatings with high biomass content: tailored properties by lignin selection. *ACS Sustainable Chemistry & Engineering* **2019**, *7* (13), 11700–11711.

75. Liu, H.; Chung, H. Self-healing properties of lignin-containing nanocomposite: Synthesis of lignin-graft-poly (5-acetylaminopentyl acrylate) via RAFT and click chemistry. *Macromolecules* **2016**, *49* (19), 7246–7256.

76. Panovic, I.; Montgomery, J. R.; Lancefield, C. S.; Puri, D.; Lebl, T.; Westwood, N. J. Grafting of Technical Lignins through Regioselective Triazole Formation on β-O-4 Linkages. *ACS Sustainable Chemistry & Engineering* **2017**, *5* (11), 10640–10648.

77. Duval, A.; Lange, H.; Lawoko, M.; Crestini, C. Reversible crosslinking of lignin via the furan–maleimide Diels–Alder reaction. *Green Chem* **2015**, *17* (11), 4991–5000.

78. Perianes-Rodriguez, A.; Waltman, L.; Van Eck, N. J. Constructing bibliometric networks: A comparison between full and fractional counting. *Journal of Informetrics* **2016**, *10* (4), 1178–1195.

79. Ubando, A. T.; Felix, C. B.; Chen, W.-H. Biorefineries in circular bioeconomy: A comprehensive review. *Bioresource Technol* **2020**, *299*, 122585.

Chapter 3

Reductive Catalytic Fractionation: From Waste Wood to Functional Phenolic Oligomers for Attractive, Value-Added Applications

Gil Van den Bossche,[a] Thijs Vangeel,[a] Korneel Van Aelst,[a] Wouter Arts,[a] Laura Trullemans,[a] Kranti Navare,[b,c] Sander Van den Bosch,[a,*] Karel Van Acker,[b,d] and Bert F. Sels[a,*]

[a]Centre for Sustainable Catalysis and Engineering (CSCE), KULeuven, Celestijnenlaan 200 F, 3001 Heverlee, Belgium
[b]Department of Materials Engineering, KULeuven, Kasteelpark Arenberg 44, 3001 Heverlee, Belgium
[c]Sustainable Materials, VITO, Boeretang 200, 2400 Mol, Belgium
[d]Research Centre for Economics and Corporate Sustainability, KULeuven, Warmoesberg 26, 1000 Brussels, Belgium

***Email: bert.sels@kuleuven.be. (B.F.S.)**

***Email: sander.vandenbosch@kuleuven.be. (S.V.d.B.)**

Reductive catalytic fractionation (RCF) is proven to be a promising biorefinery strategy for the valorization of lignocellulosic biomass. A large part of past and present RCF research has been directed at the conversion of freshly harvested biomass instead of recycled lignocellulose fractions. Moreover, the focus of RCF research often lies in obtaining high phenolic monomer yields, while the product mixture also consists of dimers and oligomers. However, this paradigm is shifting. Following the cascading principle, the use of sustainable and inexpensive post-consumer wood (PCW) forms an alluring alternative to freshly harvested wood. In parallel, the applicability of the dimers and oligomers in the phenolic product mixture is receiving increased attention, which was previously overshadowed by the monomer fraction. This chapter discusses the potential of post-consumer wood (PCW) as a feedstock for RCF biorefining, aiming to produce functional, 'intermediate molecular weight' phenolics. The first part discusses the framework of cascading wood use in the EU, with an emphasis on the collection of PCW and its sorting based on (in)organic contaminants. The second part reviews biorefining strategies that specifically target PCW. The third part highlights RCF as potential biorefinery strategy, while critically discussing catalyst deactivation as a major challenge. The last part outlines the possible applications of the targeted phenolic oligomers, providing value to this intermediate molecular weight product fraction.

Introduction: Cascading Principles in the Circular Bioeconomy, Unlocking the Potential of Post-Consumer Wood

Depletion of fossil resources and their environmental impact, drive the necessity to explore sustainable alternatives for energy, fuels and chemicals. Lignocellulose biomass, a cheap and abundant biopolymer, for ages used for construction and heat generation, is regarded as an interesting alternative for the manufacturing of fuels and chemicals, providing a CO_2-neutral alternative for the current petrochemical routes. Several lignocellulose biorefinery strategies have been developed over the last decades, each having their advantages and limitations. Nevertheless, overexploitation of fresh harvested wood can lead to resource scarcity and both global and local environmental problems related to deforestation. Sustainable forest management might tackle this issue as the freshly harvested wood could answer the lignocellulose demand for the production of materials, chemicals and fuels. However, the principles of sustainable forest management are not fully implemented and environmental problems related to deforestation still exist. Besides, growing forests have the capacity of capturing CO_2 from the atmosphere and could serve as a carbon sink, hereby reducing environmental problems related to global warming (*1*, *2*). When freshly harvested wood is used in biorefineries, the majority of the fixed carbon is eventually lost to the system as CO_2 *via* burning of fuels or solid waste incineration. Closing the material loops by moving towards circular economy models for wood materials, can greatly reduce the loss of carbon, improve resource (wood) availability and lower the need of raw materials, *i.e.* freshly harvested wood, in the system (*3–5*). An important principle is 'cascading use', which is considered the efficient use of resources, through reuse and recycle of residues, prior to incineration or disposal. In this perspective, wood materials are preferentially reused, prior to being recycled into lower grade materials. For example, solid wood products (primary products) can be recycled to particleboards (secondary products) and subsequently to fiberboards (final products) (Figure 1). This way, wood-based products can be regarded as an additional carbon sink, becoming increasingly efficient with each additional cycle or cascade that increases their lifetime and reduces the need for fresh carbon in the system. Ultimately, these low-grade wood products (*e.g.* fiberboards), are no longer suited for recycling and incinerated for energy recovery.

Key to the cascading of wood materials is an international framework for quantification, characterization and classification of 'wood flows' and identifying the barriers impeding collection and recycling of wood wastes (*5*). Such barriers can be of technical nature *e.g.* the large variation in the content of organic (*6–8*) and inorganic (*9–12*) contaminants and the necessary cleaning pretreatment. Compared to fresh woods, wood wastes contain increased amounts of heavy (Pb, Zn, Cd, Hg) and toxic (Cu, As, Cr) metals, attributed to surface pretreatment methods and industrial preservative pretreatments, respectively. However, cleaning and sorting methods have proven able to effectively decrease the metal contamination in the upcycled fraction. The large variation of (in)organic contaminants, and the necessity for cleaning prior to reuse and recycle form a technical barrier which has to be overcome. Other substantial barriers are of economical nature, *e.g.* the dependence of upstream products or of governance nature, *e.g.* integrated approaches in the classification of wood wastes and waste directives (*vide infra*) (*5*).

Within the EU, 'post-consumer wood' (PCW) is regarded as an overarching term for waste wood streams like 'packaging wood', 'construction and demolition wood' (C&D), 'commercial and industrial wood' (C&I) and 'municipal solid waste' (MSW), the latter typically originating from households. 52.3 Mm^3 of PCW is generated annually by all EU member states, of which 15.9 Mm^3 (30.4%) is directly disposed of, 19.6 Mm^3 (37.5%) is incinerated for energy recovery and 16.8 Mm^3

(32,1%) is recycled to materials (*5*). The percentage of PCW being disposed differs substantially between EU countries. For example, in northern European countries like Sweden, Denmark and Finland only 5% of PCW is disposed of, while in Poland and Portugal this exceeds 85% (*13*). Overall, there is still plenty of room for optimizing the cascading use of wood materials. One strategy consists of designing and manufacturing robust wood products with enhanced reuse and recycle potential. Further optimizing the cascading of wood materials can be achieved by increasing the reuse and recycling rates for every product category, thus minimizing the wood flows to landfills or incineration facilities. Design and manufacture of robust products with increased reuse and recyclability can contribute to this strategy as well as better collection and cleaning processes for PCW. Another possibility is to increase the number of cascade stages prior to incineration. In this perspective, biorefining of PCW towards materials, fuels and chemicals, can be considered as an additional cascade for wood materials, prior to incineration (Figure 1) (*14*).

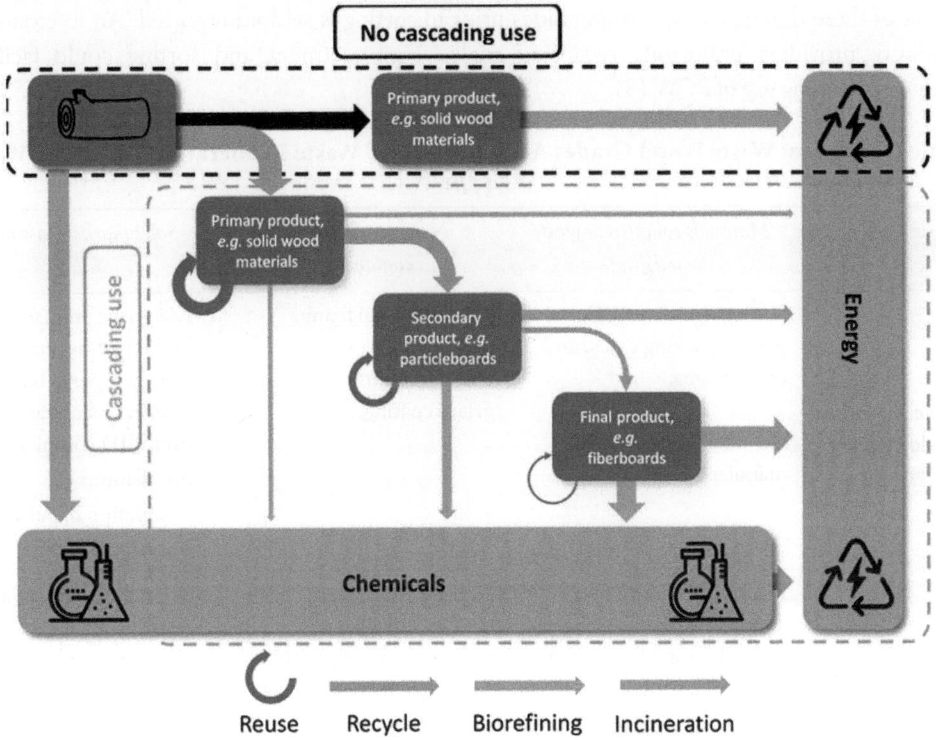

Figure 1. Schematic overview of cascading use of wood materials: post-consumer wood (PCW) is reused, or recycled prior to incineration. Biorefining, i.e. conversion of lignocellulose material to chemicals, poses an alternative to current petroleum-based refineries for the production of chemicals. Adapted with permission from reference (15). Copyright 2013 Elsevier.

Classification of Post-Consumer Wood

Based on the EU waste incineration directive (WID), post-consumer wood can be sorted in 4 different fractions or 'grades': grade A, B, C and D (*16–18*). As summarized in Table 1, grade A waste wood mostly consists of clean wood from packaging and off-cuts from joinery or manufacture, containing only a minor amount of metals, from nails and other fixings, and/or organic contaminants such as paint, glue, coatings etc. Waste grade A is generally recycled to materials (*e.g.* panel boards, animal bedding) or can be combusted outside WID compliant installations, usually as pellets and briquettes. Waste grade B is characterized by higher amounts of C&D wood, plus domestic wood waste and furniture. As a result, it contains increased amounts of non-wood material, such as paints, coatings, glues, glass, grit etc. (Table 1). The majority of B grade wood material is recycled towards materials, mainly wood-panel (*e.g.* particleboards). Further increased amounts of coatings and preservatives are present in C grade wood, comprising fencing material and flat-pack furniture, largely made from particleboards: MDF, plywood, OSB (oriented strand board). Consequently, C grade wood contains very large amounts of non-wood materials and is mainly used for energy recovery. In addition to the high non-wood content, the toxicity of the chemicals used in the coatings and preservatives impedes the reuse and recycling of C grade wood. Wood subdued to preservative pretreatment with creosote, copper, chrome and arsenic is excluded from C grade waste wood and is graded as D wood. D grade wood comprises fencing wood, railway sleepers, transmission poles and cooling towers, and is disposed at special facilities. Currently, sorting rules are merely based on the origin, content and visual selection of the different PCW fractions at sorting sites. A thorough analysis of these different waste wood grades prior to sorting is seldom reported. An international framework providing uniformity regarding analysis, pretreatment and sorting could facilitate increased cascading use of PCW (5).

Table 1. Different Waste Wood Grades According to EU Waste Incineration Directive (WID) (*16*)

Waste wood grade	*Materials collected in waste wood grade*	*Non-wood materials/ contaminants*	*Sales market/cascading routes*
Grade A: 'Clean' recycled wood	Packaging waste, scrap pallets, packing cases, and cable drums. Off-cuts from joinery/ manufacturing.	Nails and metal fixing. Minor amounts of paint, and surface coatings.	Materials: animal bedding, horticultural mulches, panel board. Energy: Incineration in non WID compliant installations or production of pellets/ briquettes.
Grade B: Industrial feedstock grade	May contain up to 60% Grade A material as above, with addition of construction and demolition materials and domestic solid wood furniture.	Nails and metal fixings. Paints, coatings and glues, plastics and binders, glass, grit. Limited amounts of treated or coated materials.	Recycling to materials: panel board, including chipboard and medium density fiberboard (MDF).
Grade C: Fuel grade	All of the above, with addition of fencing products, flat-pack furniture made from board products and DIY materials. High content of panel products: chipboard, MDF, plywood, OSB and fiberboard.	Nails and metal fixings. Paints coatings and glues, paper, plastics and rubber, glass, grit. Coated and treated timber, with exception of creosote, copper, chrome and arsenic ('CCA wood').	Energy recovery in WID compliant installations.
Grade D: Hazardous waste	Fencing, transmission poles, railway sleepers, cooling towers.	Wood subdued to preservative treatment containing creosote, copper, chrome and arsenic.	Disposal at special facilities.

Post-Consumer Wood Streams as Feedstock for Biorefining

The term 'biorefining' covers a wide range of valorization routes for biomass to materials, fuels and chemicals. Analogous to petrorefineries, which separate and convert crude fossil oil into different useable fractions, biorefineries fractionate (lignocellulose) biomass into different product streams. Lignocellulose is a solid, highly oxygenated material, composed out of lignin, a heterogeneous aromatic biopolymer, and two sugar biopolymers, cellulose and hemicellulose (*19*). This diverse but rigid composition provides structural strength and resistance to (bio)chemical conversion, referred to as biomass recalcitrance (*20, 21*).

Overcoming this recalcitrance is a key challenge for lignocellulose valorization beyond its caloric value, but today no universal process exists fully tackling this challenge (*22*). Instead, different biorefinery strategies exist, often subdivided by technology (thermal, chemical, microbial) (*23*) or by the targeted fraction (cellulose, hemicellulose or lignin) (*24*). Most biorefinery strategies focus on the valorization of the carbohydrate fraction, especially cellulose but also hemicellulose, while lignin is regarded as a mere obstacle (*24*). Various fractionation processes have therefore been developed, rendering a pure and lignin free carbohydrate pulp, readily available for the production of materials (*e.g.* paper & cardboard), fuels (*e.g.* bioethanol) & chemicals (*e.g.* furanics) (*19*). Next to lignocellulose recalcitrance, the presence of non-wood materials and (in)organic contaminants (*vide supra*), poses an additional challenge in post-consumer wood biorefining. The heavy- or toxic metals and chlorinated compounds present in PCW will possibly impact the thermal, chemical or biological reactions taking place during the fractionation processes and eventually migrate to one (or multiple) of the product stream(s). Therefore PCW, compared to virgin wood, is not as widely investigated in biorefining processes. The following section sheds light upon the strategies that exist today.

A well-known example of PCW biorefining is pyrolysis, *i.e.* the thermal conversion of biomass in absence of oxygen yielding gaseous (bio-gas), liquid (bio-oil) and solid (char) product streams (*25*). Through history, pyrolysis was merely applied for the production of charcoal, but over the last decades it has gained interest as adapted process conditions can tune the yields of the different product streams (*25*). Today, different 'modes' of pyrolysis can be distinguished: (i) near-complete gasification, obtained at very high temperatures (700-900 °C), (ii) fast pyrolysis characterized by short residence times (1 s), high heating rates and moderate temperatures (500 °C), typically reaching highest bio-oil yields up to 75%, and (iii) torrefaction, with a residence time of several minutes (10-60 minutes) at lower temperatures (290 °C) mainly yielding a solid charcoal fraction (up to 80%) (*25*). Since the obtained bio-oil after fast pyrolysis has a high oxygen content, it is often unstable, immiscible with petroleum-derived fuels and moreover, it has low energy content, and thus catalytic upgrading (deoxygenation) is required (*24*). Alternatively, fast pyrolysis can be executed in the presence of catalysts, referred to as 'catalytic fast pyrolysis' (CFP), hereby circumventing multiple operational steps. As summarized by Schutyser *et al.* various aluminates, silicates, oxides and metal on oxides serve as catalysts in CFP, ultimately boosting the monomer yield and selectivity to deoxygenated products (*24*).

As an alternative to landfill or incineration, the pyrolysis of waste materials is gaining interest, however, most research is not performed on true waste fractions, in the way they are collected or disposed of (*vide supra*). Generally, the impact of one, well-defined type of waste material (typically petro-derived plastics being PE, PP, and PET), blended with virgin lignocellulose (*e.g.* bamboo, poplar, sugarcane) is assessed. The high H:C ratio of these plastics, results in reduced char formation and high hydrocarbon yields (*26–29*). Pyrolysis of true PCW (*e.g.* construction and demolition wood) performed by Shonnard *et al.* (*30*) and Kawamoto *et al.* (*31*) showed similar results, *i.e.* reduced char formation and increased hydrocarbon yields, since these type of plastics, high in H:C ratio, were present in the collected C&D wood to high extent. However, next to PE, PP and PET, post-consumer wood can contain chlorinated compounds (*e.g.* PVC), causing exhaust of toxic polychlorinated dibenzofurans and dibenzo-*p*-dioxins (PCDF & PCDD) during incineration or gasification (*32*). This is the main incentive why such waste streams are incinerated in WID compliant installations (*vide supra*). Alternatively, torrefaction of PCW renders a solid char fraction with a diminished chlorine content as chlorine species, such as PCDF's and PCDD's, migrate to the gas phase. The extent by which this interphase transport occurs is however dependent on their molecular weights (*8, 33, 34*). Moreover, inorganic metal contaminants largely remain in the solid char fraction and in turn catalyze the formation of PCDF's and PCDD's upon incineration (*34*).

Comparable to virgin lignocellulose pyrolysis, the use of silica/alumina (MCM-41, SBA-16, Na_2CO_3/γ-alumina) or zeolites (H-ZSM5), *i.e.* catalytic fast pyrolysis, boosted the yield of hydrocarbons and deoxygenated products (*26, 27, 29*). As the waste materials (PE, PP, PET) are free of chlorine and inorganic metals, coke formation on the aluminosilicate material is the main cause of deactivation in catalytic fast pyrolyisis (*29, 35–37*). The impact of organic (chlorine) and inorganic contaminants on catalyst deactivation in CFP of post-consumer wood, to our knowledge, remains underexplored.

Even with catalytic upgrading or direct use of catalysts, the selectivity of pyrolysis remains low, partially attributed to the heterogeneity in the lignocellulose structure, consisting of an aromatic and carbohydrate biopolymer *(vide supra)*. Biorefinery processes therefore focus on separating the lignin from the carbohydrate fraction. Most recurring strategies are harsh pulping processes (*e.g.* Kraft- or sulfite pulping), yielding a lignin free carbohydrate pulp. The lignin fraction is isolated and depolymerized at high temperatures (140-170 °C) using alkaline reaction mixtures, *i.e.* pulping liquor, containing Na_2S or sulfite salts, these strongly nucleophilic species assist in the lignin depolymerization which occurs *via* ether bond cleavage. The depolymerized lignin fragments are prone to C-C bond formation, resulting in a highly degraded lignin precipitate, with thiol or sulfonate groups. Alternatively, aqueous alkaline pretreatment avoids the use of Na_2S and sulfite salts, resulting in a sulphur free lignin fraction, but often requires anthraquinone additives or subsequent chlorine bleaching of the carbohydrate pulp. Next to alkaline fractionation, 'organosolv processes' make use of (volatile) organic solvents (*e.g.* short alcohols, THF, dioxane), often in combination with water and/or mineral or organic acids. The lignin is more soluble in these organic solvents and the solvent facilitates the penetration of the lignocellulose matrix. Moreover, the use of low boiling alcohols enables solvent recovery from the lignin precipitate. American value-added pulping (AVAP) a fractionation process using a SO_2-ethanol-water mixture, combines the benefits of a nucleophile (SO_2) assisting in lignin depolymerization and an organic solvent (EtOH) *i.e.* enhanced lignin solubility and easy solvent recovery (*38*). Regarding PCW, AVAP fractionation of A and B waste wood grades was performed by Retsina *et al* (*17*), yielding high pulp yields (87-99%) with high enzymatic digestibility for cellulose (> 90%) and hemicellulose (near 100%). Alkaline pulping of MDF, C grade waste wood and D grade waste wood towards fiberboard pulps and cellulose nanocrystals has been demonstrated by Asikainen *et al.* (*18*) and Cathala *et al.* (*39*)

Direct enzymatic hydrolysis of lignocellulose avoids using harsh fractionation methods, but struggles to overcome the lignocellulose recalcitrance. The lignin yield therefore is low and the lignin residue is undervalued in such biorefinery strategies, mostly focusing on obtaining monomeric sugars. Nevertheless, enzymatic hydrolysis of post-consumer wood has been successfully demonstrated. Enzymatic hydrolysis of MSW, C grade waste wood, D grade waste wood and OSB isolates a carbohydrate fraction readily available for enzymatic fermentation towards drop-in chemicals such as bioethanol (*17*, *40–43*), lactic acid (*44*), methane, VFA (*45*). Prior to enzymatic hydrolysis, the biomass is pretreated to make the lignocellulose fibers more accessible, *e.g.* (diluted) acid pretreatment (*40*, *41*), water wash (*42*), steaming (*43*) or ionic liquid pretreatment (*46*, *47*). Additionally, these methods remove (in)organic contaminants resulting in a clean cellulose pulp with lowered metal concentrations below enzyme specific minimal inhibition concentrations (*43*).

High pulp and product yields demonstrate the potential of PCW in biorefinery strategies, as depicted in Figure 1. However, the lignin fraction is undervalued in these fractionation processes, as the lignin fraction is precipitated and consists of degraded, high Mw phenolic oligomers and is merely suited for energy recovery. Envisioning a sustainable use of PCW, the valorization of lignin towards value-added materials, fuels and chemicals should be considered. Key to this lignin valorization is the preservation of unique functional moieties in the lignin biopolymer.

Catalyst Deactivation in Reductive Catalytic Fractionation Processes

As biomass recalcitrance often hinders the valorization of both the lignin and carbohydrate fraction, several protection strategies have been developed to suppress condensation reactions in lignocellulose biorefining (*48*). Note that the physical separation of reactive intermediates from the reaction mixture (and the carbohydrate pulp) *e.g.* by using flow-through setups (*vide infra*), is one way of lowering the occurrence of condensation reactions, but will not completely eliminate them (*49*). Besides, such setups generate diluted products streams which are partly explaining less occurrence of condensation reactions with the reactive intermediate given higher reaction orders for such reactions. Chemical stabilization strategies prevent condensation reactions *via* (ir)reversibly stabilizing the functional groups in the reactive intermediates or the native lignin structure, so condensation reactions can no longer occur. One example is the aldehyde assisted stabilization of lignin *via* the formation of a cyclic acetal from α,γ-diol structures in the lignin side-chain (*50*). Hereby, dehydration of the α-C atom and subsequent cleavage of β-O-4 linkages is inhibited preserving these reactive moieties during lignin fractionation. These acetals can also be formed later in the fractionation process, stabilizing unsaturated lignin fragments resulting from ether bond cleavage, making aldehyde assisted fractionation of lignocellulose biomass an interesting stabilizing strategy (*51*). Other stabilizing agents targeting functional groups in reactive lignin fragments are diols, again forming an acetal structure, or alcohols stabilizing the α-C carbocation via an nucleophilic addition (*50*). Alternatively, the unsaturated lignin fragments can be irreversibly stabilized *via* hydrogenation of the double bonds, thus avoiding the use of an additional organic reagent. This biorefinery strategy is usually referred to as reductive catalytic fractionation (RCF) (*52–55*).

RCF yields a low Mw lignin oil fraction (Figure 2), consisting of monomeric, dimeric and oligomeric biophenolics, together with a solid carbohydrate pulp, readily available for production of materials or conversion towards fuels (*56–58*), polyols (*53*, *59*, *60*) or bioethanol (*54*). Usually performed at elevated temperatures (180-250 °C) using a heterogeneous hydrogenation catalyst and an alcohol (*e.g.* MeOH, EtOH, *i*-PrOH *n*-BuOH) or ether as a solvent, possibly with water as a co-solvent (*e.g.* EtOH/H_2O, *i*-PrOH/H_2O *n*-BuOH/H_2O), lignin is isolated from the lignocellulose matrix. Lignin depolymerization occurs *via* ether bond cleavage, yielding reactive lignin fragments which are prone to subsequent repolymerization, forming stable C-C bonds. Stabilization of these reactive fragments, *i.e.* the hydrogenation of double bonds but also hydrogenolysis of C-O bonds, prevents new C-C bond formation, ensuring a low Mw, soluble and functional lignin oil. Crucial is the presence of a redox catalyst (*e.g.* Pd/C, Rh/C, Ru/C) and a hydrogen source, present as pressurized hydrogen gas or hydrogen donors such as additives (*61*), (co)solvent (*62–65*), or *in-situ* generated soluble carbohydrates from the lignocellulose substrate (*66–68*). The majority of RCF research is performed in stirred batch reactors, where the solvolytic lignin extraction, lignin depolymerization and reductive stabilization occur simultaneously (*52*). Recently, also flow-through reactors have been proposed. In this semi-continuous setup, a solvent flow is maintained over separated biomass and catalyst beds, physically separating the solvolytic lignin extraction and (partial) lignin depolymerization from the reductive catalytic depolymerization and stabilization of reactive lignin fragments (*68–71*). The benefits and drawbacks of batch and flow-through setups for both research and scale-up purposes, were reviewed earlier and outside the scope of this manuscript (*19, 52*).

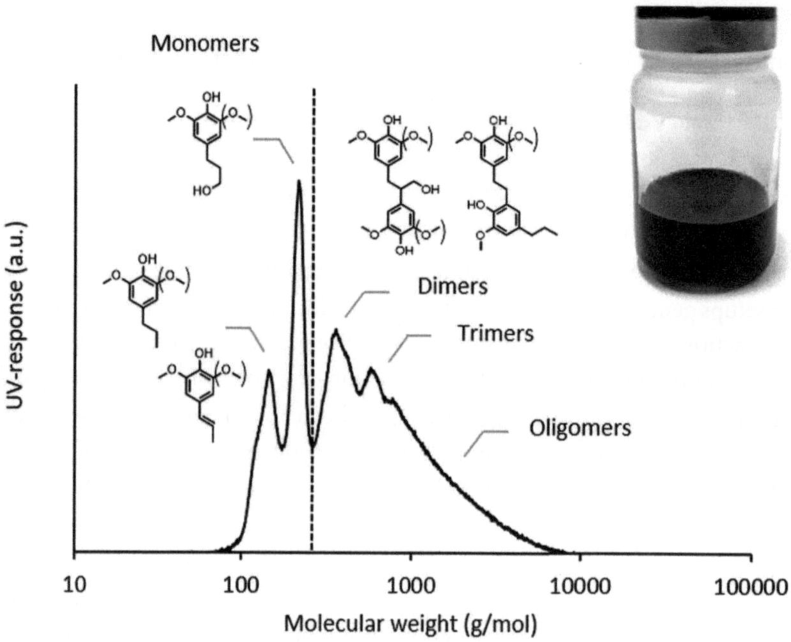

Figure 2. Gel permeation chromatogram, illustrating the Mw distribution of a softwood-derived RCF lignin oil, comprising monomers (+-30 wt%), dimers, trimers and oligomers. Adapted with permission from reference (52). Copyright 2019 Elsevier.

Success of the RCF biorefinery largely depends on the redox catalyst, which plays a threefold role: (i) assisting in ether bond cleavage of solvolytically extracted lignin fragments, (ii) hydrogenation of double bonds, thereby stabilizing the lignin fragments and (iii) directing high product selectivity (72). Moreover, since the catalyst cost impacts the economic feasibility of biorefinery processes, a lot of recent research is targeting stable, selective and cheap non-noble metal catalysts (*e.g.* MoS_2 (73), CuPMO (74), Raney nickel (75), Ni/C (*69*), Ni/Al_2O_3 (*54*)). Nevertheless, according to a recent study, even noble metal catalysts in RCF can be profitable, if the catalyst lifetime exceeds at least 2 years (*76*). Prolonging the catalyst lifetime is thus crucial to the economic viability of the RCF biorefinery. In this perspective, thorough insight in the possible deactivation mechanisms of the redox catalysis is of utmost importance.

The reuse of the redox catalyst in the RCF process requires either (i) physical separation during the reaction, *e.g.* a catalyst basket in batch or catalyst bed in flow-through setup, or (ii) downstream separation of catalyst and pulp, *e.g.* through magnetic separation, biphasic solvent extraction (*77*). In most recent studies, this recuperated catalyst is used in consecutive runs, thereby assessing its deactivation by monitoring the decrease in monomer yield and selectivity. Often, this is complemented by analysis of the recuperated catalyst *via* analytical techniques such as microscopy, physi- and chemisorption, ICP and XRD (*54, 69, 73, 78–81*).

Regardless of the reactor setup and separation techniques, several deactivation mechanisms were found to occur in RCF. For example, heterogeneous redox catalysts are prone to metal sintering, *i.e.* the (irreversible) agglomeration of metallic species under influence of heat, given the high reaction temperatures needed for lignin solvolysis or reductive stabilization (*82*). Both supported nickel and ruthenium catalyst have shown to be susceptible to sintering after repeated catalyst reuse under RCF conditions (*69, 79, 80*). The highest amount of sintering has been reported in a work of Román-Leshkov *et al.* where a 50% particle increase (14 to 21 nm) of Ni/C catalyst in MeOH using an RCF flow-through reactor was observed (*69*). In other studies, if reported, the growth of particle size was limited to 11% (*79, 80*). Next to sintering, metal species are also prone to oxidation, but given the reductive process conditions, it is seldom observed in RCF, and if so, it is not regarded to be a main cause of catalyst deactivation (*78*). Not only physicochemical alteration, but also the loss of catalyst metal by leaching is a well-known deactivation mechanism in biorefinery strategies. Nickel catalysts are reported to be more prone to leaching compared to ruthenium catalysts (*83–85*). Especially in acidic circumstances it is long known that nickel leaches, mainly due to NiO solubilization, if the reductive circumstances are mild. Noble metal catalysts (Rh/C, Ru/C) indeed show limited leaching in RCF, even when exposed to aqueous/acidic solvent mixtures (*86, 87*). Nevertheless, also copper and nickel-based catalysts, CuPMO, Ni/Al_2O_3 and $Ni/Zn(methylimidazole)_2$-metal organic framework have proven to be stable in MeOH solvent RCF by few authors (*54, 74, 79*), whereas nickel leaching from a Ni/C catalyst, did occur in a flow-through reactor using MeOH, the nickel content decreased from 13.5 wt% to 9.8 wt% after four consecutive runs. Leaching was here attributed to the presence of soluble sugar and lignin components, as leaching from solvent and hydrogen flow alone was found negligible (*69*). Increased leaching from Raney Ni, induced by soluble sugars was also reported by Rinaldi *et al.* (*88*) Lastly, catalyst fouling is most often considered the main cause of deactivation in RCF research. Soluble sugar components are reported to alter catalyst activity and selectivity in RCF of biomass or conversion of lignin model compounds (*88, 89*) and, together with lignin oligomers, contribute to the fouling of the catalyst surface. Organic components block the catalyst surface and pores, leading to lower surface area, pore volume and metal dispersion, ultimately deactivating the redox catalyst (*54, 69, 81*). To our best knowledge, in contrast with sintering and leaching, deactivation through fouling can be reversed by desorption of these organic components. This was successfully demonstrated by Song *et al.* and Sels *et al.*, the activity of the spent catalyst was restored *via* H_2-treatment at elevated temperatures (400 °C) (*54, 81*).

Unfortunately, catalyst poisoning due to inorganic compounds (*e.g.* alloying) has not been sufficiently studied. This mode of catalyst deactivation could be highly relevant for processing post-consumer wood and residual biomass, which typically show a high inorganic content (*vide supra*) (*90*), but the possible accumulation and impact on the heterogeneous catalyst in RCF remains largely uninvestigated. Ma *et al.* stated that increased inorganic sulphur content in biomass not necessarily leads to catalyst poisoning and subsequent lower product yields. Only biogenic sulphur species (methionine and tryptophan amino acids) drastically decreased product yields, while inorganic sulphur species had negligible effect. Biogenic sulphur poisoning mainly impacted product yields from (hemi)cellulose, while lignin product yields were impacted to a lower extent (*86*). Still, chlorine (*91, 92*) Sulphur (*92–94*) and phosphor (*95*) species are known catalyst poisons for non-noble metal (Cu, Fe, Zn) hydrogenation catalysts in fatty acid and fatty ester hydrogenation (*96*). Thorough research on catalyst poisoning by (in)organic contaminants is a remaining hurdle for unrolling large scale RCF as a diverse and performant biorefinery strategy. Not only catalyst deactivation will play a role, but also monitoring and removing toxic metals should be a point of attention. The potential of waste wood grade B, *i.e.* wood containing increased amounts of non-wood materials and (in)organic contaminants (*vide supra*), as feedstock in RCF biorefineries has already been demonstrated *via* techno-economical assessment, based on experimental lab-scale data. An important conclusion here was the need to valorize the entire lignin oil product, implying valorization both the lignin-derived monomers and oligomers (*97*).

Functional Phenolic Oligomers for Attractive Value-Added Applications

As most RCF research primarily focuses on obtaining biophenolic monomers, various parameters (*e.g.* catalyst, solvent, lignocellulose biomass, operational conditions) have been optimized in order to maximize lignin monomer yields (*98*). In parallel, valorization routes for these phenolic monomers towards (i) drop-ins (*e.g.* phenol and propylene (*76, 99–102*), and alkylcyclohexanones (*103*)), (ii) fuel additives (*78*) and (iii) polymer building blocks and –additives (*104–106*) such as bisphenol A, caprolactam and acrylic acid (*64*) were investigated. However, despite the optimization of process conditions and development of advanced catalysts, lignin depolymerization in RCF remains limited to cleavage of ether bonds, while the more stable C-C bonds remain intact. The monomer yield is therefore limited to a theoretical maximum, roughly proportional to the square of the fraction of ether bonds in the native lignin (*48*). Typically, the share of ether bonds ranges between 60 – 75% in case of hardwoods and 45 - 50% in case of softwoods (*62*), corresponding to a maximal monomer yield of approximately 55 and 30 wt%, respectively (*24*). To tackle this, genetically engineered lignins were introduced: the overexpression of F5H gene resulted in high syringyl content (98%) in poplar wood (*107*), reaching a monomer yield of 77% after formaldehyde assisted fractionation and subsequent hydrogenolysis (*51*). Other modified lignins have been developed, but their use in RCF biorefineries remains largely underexploited, and more research needs to be performed here (*52, 108, 109*). Nevertheless, in most cases and especially in softwood RCF, the majority of lignin products thus consists of dimers and oligomers (Figure 2). Since PCW is largely composed of softwoods (*17*), the lignin oil is expected to mainly consist of dimers and oligomers. Valorization routes for these intermediate Mw fractions are of utmost importance when enrolling RCF biorefinery strategies for PCW and will enhance the viability of RCF biorefineries in general. Prior to chemical valorization into polymer applications, lignin oligomers can be isolated from the bio-oil mixture *via* solvent extraction (*53, 76, 78, 110–112*), membrane separation (*113*) or distillation (*105*). Note that RCF may not only be an interesting biorefinery, as it has also demonstrated to be an interesting, complementary tool, besides the traditional ones, to analyze structures of native lignin, as has been recently also shown for more complex feedstock such as bark (*87, 114–116*).

Just as the early RCF research was primarily focused on obtaining high monomer yields, structural analysis of lignin products was mostly limited to the identification and quantification of these monomers *via* chromatographic techniques such as GC and GC-MS. The GC analysis of dimers was only made possible by increasing their volatility through derivatization, and showed structures related to the typical original lignin interlinkages and end groups, though in their modified forms due to the hydrogenation and hydrogenolysis chemistry (*53*, *54*, *112*). The molecular weight of the lignin oil (fractions) could be determined *via* gel permeation chromatography (GPC) (Figure 2), and this was one of the methods firstly used to indeed prove the lower molecular weight of typical RCF lignin oils compared to technical lignins, *e.g.* Kraft or organosolv lignins (*53*, *72*). But still, lack of insights in the physicochemical properties of the lignin-derived oligomers was impeding the possibility towards targeted chemistry and the exploration of value-added applications. Regarding the analysis of near native lignin structures, *e.g.* milled wood lignin (MWL) and cellulolytic enzyme lignin (CEL), ^{1}H-^{13}C 2D heteronuclear single quantum coherence spectroscopy (HSQC) NMR has been developed for the (semi-)quantification of inter-unit linkages in *in planta* lignin (*62*, *117*). This powerful tool also improved the understanding of functionalities and inter-unit linkages in technical lignins, *e.g.* Kraft or organosolv lignins (*118*). Driven by this success, several studies attempted to evaluate the occurrence of C-O or C-C linkages in the lignin oil *via* 2D HSQC NMR, but despite these efforts, detailed insight in the nature of these C-C linkages and there native protolignin analogues was not fully established (*53*, *54*, *69*, *72*). Recently, a study from Van Aelst *et al.* thoroughly elucidated the molecular structures in pine-derived RCF lignin oil, appending chromatographic (GC, GC-MS, GPC, column chromatography separation) and spectroscopic (^{1}H-^{13}C 2D HSQC NMR, ^{31}P NMR) techniques, and over 80% of linkages and end groups in the RCF lignin oil, including all molecules from monomers to large oligomers, were identified (*112*). This analytical toolbox will enable research on the physicochemical properties of these lignin oligomers (oils) and will provide direct insight for further valorization purposes.

The oligomeric fraction can indeed be considered as an interesting product stream, due to unique chemical properties, *i.e.* an intermediate Mw, high content of both phenolic and aliphatic hydroxyl groups (the latter is tunable through catalyst selection (*72*)), and aromatic moieties providing strength and rigidity (*52*, *78*, *119*). Therefore intermediate Mw oligomers, might be an attractive starting material for the production of high Mw polymers. The unique combination of phenolic, and aliphatic moieties makes lignin-derived biophenolics most suited for the production of epoxy thermosets. Today, 90% of all epoxy thermoset polymers are synthesized from bisphenol A diglycidyl ether (BADGE), a prepolymer formed by base catalyzed coupling of bisphenol A and epichlorihydrin (*120*). Since bisphenol A (BPA) is produced *via* petrochemical routes and more importantly is an endocrine disruptor (*121*), sustainable BPA replacements, with lower oestrogenic activity are being considered. In addition to the conversion of monomers from RCF (*e.g.* cresol, vanillyl alcohol and alkyl-guaiacols) into novel bisphenols with much lower oestrogenic activity (*104*, *105*, *122*, *123*) (attributed to the presence of *ortho*-methoxy groups on the phenolic ring (*124*)) also lignin oligomers may serve as replacer. Initially, residual lignin precipitates from harsh fractionation processes were appended for BPA replacement, but low solubility and reactivity of these lignins often result in poor epoxy polymer properties or required chemical pretreatment such as phenolization or hydroxypropylation (*110*, *120*). Alternatively, Torr *et al.* isolated lignin-derived oligomers from pine wood RCF, separated *via* diethyl ether extraction, which were blended in BADGE prepolymers (up to 75% replacement), resulting in increased flexural strength and stiffness, in the final epoxy resin (*110*, *111*).

High hydroxyl content of RCF lignin oil also allows (partial) replacement of current petro-based polyols in polyurethane, synthesized from a polyol and isocyanate component and for example widely used as thermal insulation material because of its low density and low thermal conductivity (*78, 125*). The use of bio-based polyols has been investigated, as a replacement for petroleum-based chemicals, leading to a reduced environmental impact and a higher biodegradability of the resulting polyurethane materials (*126*). Although residual lignin is used as such in polyurethane coatings (*127, 128*), adequate polyol functionalities for rigid polyurethane foams often require oxypropylation (*129–131*). As demonstrated by Tan *et al.* rigid polyurethane foams were synthesized from high Mw oligomers, isolated as residue after hexane extraction on RCF lignin oil from apple wood (*78*). Up to 50% of the polyether polyol was replaced with hexane residue prior to polymerization with polyphenylmethane polyisocyanate (PAPI), resulting in a rigid polyurethane foam with uniform cell structure, density and improved water resistance and flame resistance (*78*).

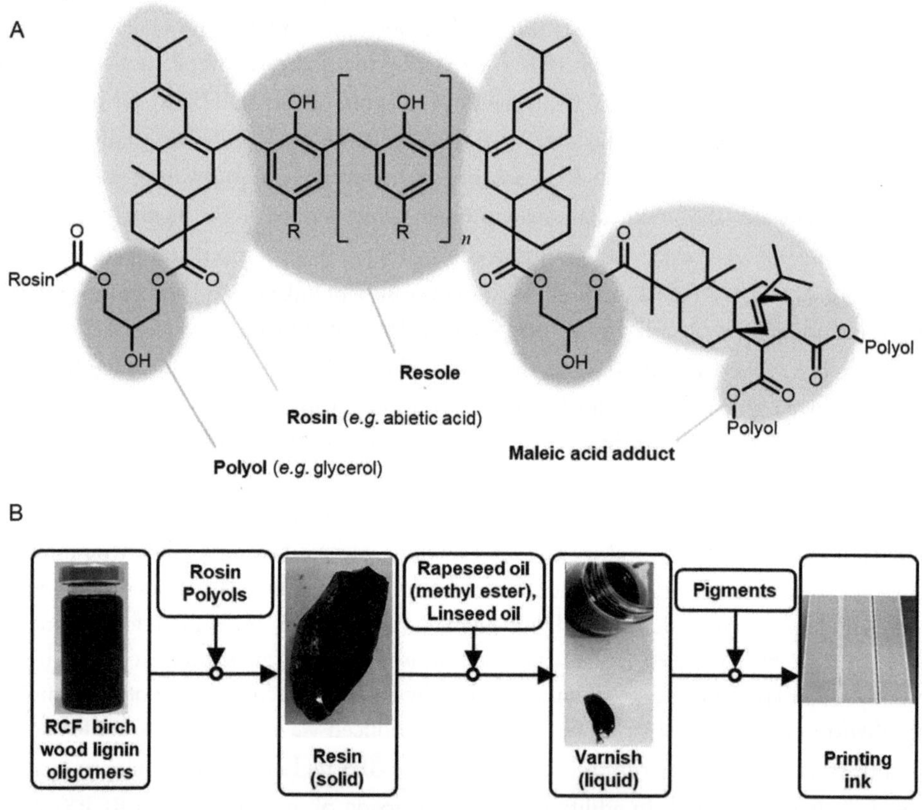

Figure 3. A) Structure of rosin-modified resin, the resole part is synthesized from cardanol, formaldehyde and p-nonylphenol, which was replaced by birch wood lignin oligomers. B) Stepwise synthesis of printing ink from RCF birch wood lignin oligomers. Reprinted with permission from reference (76). Copyright (2020) The American Association for the Advancement of Science.

Another application of lignin oligomers from RCF can be found in printing ink formulation, the resole part in printing ink is normally synthesized from cardanol, formaldehyde and the endocrine disruptive *p*-nonylphenol. The *p*-nonylphenol could be replaced by birch wood lignin oligomers, obtained after hexane extraction of the low Mw fraction of RCF lignin oil, as demonstrated by Sels *et al.* The resin (with a total biorenewable content of 85%), next to the resole component also consisting of a polyol (*e.g.* glycerol), maleic acid and rosin (*e.g.* abietic acid), was by addition of linseed oil, rapeseed oil and pigments, successfully converted to lithographic printing ink with high quality (Figure 3) (*76*).

Conclusions

The urge to find sustainable bio-based alternatives for energy, materials and chemicals has resulted in the development of various biorefinery strategies, converting lignocellulose biomass into materials, fuels and chemicals. However, both for economical and sustainable motives, post-consumer wood (PCW), of which 67% is directly landfilled or incinerated in the EU, is ought to be regarded as suitable alternative for freshly harvested biomass. International frameworks for the collection, sorting, pretreatment and trade of PCW can help to overcome the different technical, economical and governmental barriers that still exist and impede its implementation in biorefinery strategies. Valorization of the carbohydrate fraction in PCW has proven to be successful, but such fractionation processes disregard the lignin fraction, which is merely suited for energy recovery. A more holistic use of wood materials implies the valorization of both the carbohydrate and lignin fraction towards value-added applications. Reductive catalytic fractionation (RCF) can provide this holistic approach, as it enables the production of a solid (hemi)cellulose fraction, and a functional, low Mw bio-oil, consisting of lignin monomers, dimers and oligomers. Crucial in RCF is the redox catalyst, which has a threefold role: (i) assisting in ether bond cleavage of solvolytic extracted lignin fragments, (ii) hydrogenation of double bonds in lignin intermediates and (iii) directing the product selectivity. Moreover, the lifetime of the catalyst can significantly impact the economic viability of RCF biorefineries. Recent research showed that supported noble metal and non-noble metal catalysts show decent stability in RCF process conditions, the main cause of deactivation likely being catalyst fouling by soluble lignin and carbohydrate species. Nevertheless, increased insight in deactivation mechanisms, most importantly the impact of (in)organic impurities present in PCW is of utmost importance in order to further develop RCF towards a sustainable, economically viable biorefining strategy for PCW. Finally, early RCF research mostly focused on the production of monomeric products and their applications, while the intermediate Mw dimers and oligomers were often overlooked, in part due to complexity of the involved analytics. However, the paradigm is shifting and lignin-derived oligomers, showing unique chemical properties, *viz.* intermediate Mw and high content of phenolic and aliphatic hydroxyl groups, have been successfully appended in applications such as polymer synthesis and printing ink formulation.

Abbreviations

Al_2O_3, alumina

AVAP, American value-added pulping

As, arsenic

BPA, bisphenol A

BADGE, bisphenol A diglycidyl ether

n-BuOH, *n*-butanol

Cd, cadmium

CFP, catalytic fast pyrolysis

CEL, cellulolytic enzyme lignin

Cr, chromium

C&D, construction and demolition wood

C&I, commercial and industrial wood
Cu, copper
CCA, copper chrome arsenic
EtOH, ethanol
F5H, ferulate 5-hydroxylase
GC, gas chromatography
GC-MS, gas chromatography-mass spectroscopy
GPC, gel permeation chromatography
HSQC, heteronuclear single quantum coherence
ICP, inductive coupled plasma
Pb, lead
MDF, medium density fiberboard
Hg, mercury
MeOH, methanol
MWL, milled wood lignin
MSW, municipal solid waste
Ni, nickel
NMR, nuclear magnetic resonance
OSB, oriented strand board
PCDF, polychlorinated dibenzofuran
PCDD, polychlorinated dibenzo-*p*-dioxin
PE, polyethylene
PET, polyethylene terephthalate
PAPI, polyphenylmethane polyisocyanate
PP, polypropylene
PVC, polyvinyl chloride
PMO, porous metal oxides
PCW, post-consumer wood
i-PrOH, *iso*-propanol
RCF, reductive catalytic fractionation
Rh, rhodium
Ru, ruthenium
Na_2CO_3, sodium carbonate
SO_2 , sulphurdioxide
THF, tetrahydrofuran
VFA, volatile fatty acids
WID, waste incineration directive
XRD, X-ray diffraction
Zn, zinc

Acknowledgments

G.V.d.B. acknowledges funding through catalisti-SBO project NIBCON. T.V. acknowledges the Research Foundation Flanders (FWO Vlaanderen) for his doctoral fellowship. K. V. A. acknowledges funding through EoS project BIOFACT, FWO-SBO project BioWood and Catalisti-SBO project NIBCON. L.T. acknowledges funding through FWO-SBO project BioWood. K.N. acknowledges funding from VITO. S.V.d.B. acknowledges Flanders Innovation & Entrepreneurship (innovation mandate – postdoc), as well as the FWO-SBO project BioWood. W. A. and B. F. S. acknowledges funding from internal funds of KULeuven (ICN project – FFASSD).

References

1. Brunet-Navarro, P.; Jochheim, H.; Kroiher, F.; Muys, B. Effect of cascade use on the carbon balance of the German and European wood sectors. *Journal of Cleaner Production* **2018**, *170*, 137–146.
2. Brunet-Navarro, P.; Jochheim, H.; Muys, B. The effect of increasing lifespan and recycling rate on carbon storage in wood products from theoretical model to application for the European wood sector. *Mitigation and Adaptation Strategies for Global Change* **2017**, *22* (8), 1193–1205.
3. de Carvalho Araújo, C. K.; Salvador, R.; Moro Piekarski, C.; Sokulski, C. C.; de Francisco, A. C.; de Carvalho Araújo Camargo, S. K. Circular Economy Practices on Wood Panels: A Bibliographic Analysis. *Sustainability* **2019**, *11* (4), 1057.
4. Rivela, B.; Hospido, A.; Moreira, T.; Feijoo, G. Life Cycle Inventory of Particleboard: A Case Study in the Wood Sector (8 pp). *The International Journal of Life Cycle Assessment* **2006**, *11* (2), 106–113.
5. Vis, M.; Mantau, U.; Allen, B.; Essel, R.; Reichenbach, J., *CASCADES, Study on the optimised cascading use of wood*. 2016.
6. Löser, C.; Ulbricht, H.; Hoffmann, P.; Seidel, H. Composting of Wood Containing Polycyclic Aromatic Hydrocarbons (PAHs). *Compost Science & Utilization* **1999**, *7* (3), 16–32.
7. Abb, M.; Breuer, J. V.; Zeitz, C.; Lorenz, W. Analysis of pesticides and PCBs in waste wood and house dust. *Chemosphere* **2010**, *81* (4), 488–493.
8. Gao, Q.; Edo, M.; Larsson, S. H.; Collina, E.; Rudolfsson, M.; Gallina, M.; Oluwoye, I.; Altarawneh, M.; Dlugogorski, B. Z.; Jansson, S. Formation of PCDDs and PCDFs in the torrefaction of biomass with different chemical composition. *Journal of Analytical and Applied Pyrolysis* **2017**, *123*, 126–133.
9. Krook, J.; Mårtensson, A.; Eklund, M. Metal contamination in recovered waste wood used as energy source in Sweden. *Resources, Conservation and Recycling* **2004**, *41* (1), 1–14.
10. Krook, J.; Mårtensson, A.; Eklund, M. Sources of heavy metal contamination in Swedish wood waste used for combustion. *Waste Management* **2006**, *26* (2), 158–166.
11. Edo, M.; Björn, E.; Persson, P.-E.; Jansson, S. Assessment of chemical and material contamination in waste wood fuels – A case study ranging over nine years. *Waste Management* **2016**, *49*, 311–319.
12. Faraca, G.; Boldrin, A.; Astrup, T. Resource quality of wood waste: The importance of physical and chemical impurities in wood waste for recycling. *Waste Management* **2019**, *87*, 135–147.
13. Jonsson, R., Jonsson, R. *2010: Econometric modelling*. pp 33-45 in: Mantau, U. 2010: EUwood - Real potential for changes in growth and use of EU forests. Methodology report. Hamburg/Germany, June 2010. 165 p. 2010; pp 33-45.
14. Marques, A.; Cunha, J.; De Meyer, A.; Navare, K. Contribution Towards a Comprehensive Methodology for Wood-Based Biomass Material Flow Analysis in a Circular Economy Setting. *Forests* **2020**, *11* (1), 106.

15. Höglmeier, K.; Weber-Blaschke, G.; Richter, K. Potentials for cascading of recovered wood from building deconstruction—A case study for south-east Germany. *Resources, Conservation and Recycling* **2013**, *78*, 81–91.
16. Greenhalf, M. B., Miles, *The buisiness case for wood waste collection hubs*. 2012.
17. Iakovlev, M.; Survase, S.; Segers, P.; Sideri, S.; Rouzinou, S.; Pylkkanen, V.; Retsina, T. Sulfur dioxide-ethanol-water fractionation platform for conversion of recycled wood to sugars, lignin and lignosulfonates. *Bioresource Technology* **2020**, *300*, 122652.
18. Rautkoski, H.; Vähä-Nissi, M.; Kataja, K.; Gestranius, M.; Liukkonen, S.; Määttänen, M.; Liukkonen, J.; Kouko, J.; Asikainen, S. Recycling of Contaminated Construction and Demolition Wood Waste. *Waste and Biomass Valorization* **2016**, *7* (3), 615–624.
19. Cooreman, E.; Vangeel, T.; Van Aelst, K.; Van Aelst, J.; Lauwaert, J.; Thybaut, J. W.; Van den Bosch, S.; Sels, B. F. A perspective on overcoming scale-up hurdles for the reductive catalytic fractionation of lignocellulose biomass. *Industrial & Engineering Chemistry Research* **2020**.
20. DeMartini, J. D.; Pattathil, S.; Miller, J. S.; Li, H.; Hahn, M. G.; Wyman, C. E. Investigating plant cell wall components that affect biomass recalcitrance in poplar and switchgrass. *Energy & Environmental Science* **2013**, *6* (3), 898–909.
21. Himmel, M. E.; Ding, S.-Y.; Johnson, D. K.; Adney, W. S.; Nimlos, M. R.; Brady, J. W.; Foust, T. D. Biomass Recalcitrance: Engineering Plants and Enzymes for Biofuels Production. *Science* **2007**, *315* (5813), 804–807.
22. Van den Bosch, S.; Koelewijn, S. F.; Renders, T.; Van den Bossche, G.; Vangeel, T.; Schutyser, W.; Sels, B. F. Catalytic Strategies Towards Lignin-Derived Chemicals. *Topics in Current Chemistry* **2018**, *376* (5), 36.
23. Dusselier, M.; Mascal, M.; Sels, B. F., Top Chemical Opportunities from Carbohydrate Biomass: A Chemist's View of the Biorefinery. In *Selective Catalysis for Renewable Feedstocks and Chemicals*, Nicholas, K. M. , Ed. Springer International Publishing: Cham, 2014; pp 1-40.
24. Schutyser, W.; Renders, T.; Van den Bosch, S.; Koelewijn, S. F.; Beckham, G. T.; Sels, B. F. Chemicals from lignin: an interplay of lignocellulose fractionation, depolymerisation, and upgrading. *Chemical Society Reviews* **2018**, *47* (3), 852–908.
25. Bridgwater, A. V. Review of fast pyrolysis of biomass and product upgrading. *Biomass and Bioenergy* **2012**, *38*, 68–94.
26. Kim, Y.-M.; Lee, H. W.; Choi, S. J.; Jeon, J.-K.; Park, S. H.; Jung, S.-C.; Kim, S. C.; Park, Y.-K. Catalytic co-pyrolysis of polypropylene and Laminaria japonica over zeolitic materials. *International Journal of Hydrogen Energy* **2017**, *42* (29), 18434–18441.
27. Kim, Y.-M.; Jae, J.; Kim, B.-S.; Hong, Y.; Jung, S.-C.; Park, Y.-K. Catalytic co-pyrolysis of torrefied yellow poplar and high-density polyethylene using microporous HZSM-5 and mesoporous Al-MCM-41 catalysts. *Energy Conversion and Management* **2017**, *149*, 966–973.
28. Zheng, X.; Ying, Z.; Wang, B.; Chen, C. Effect of Calcium Oxide Addition on Tar Formation During the Pyrolysis of Key Municipal Solid Waste (MSW) Components. *Waste and Biomass Valorization* **2018**, *10*.
29. Ghorbannezhad, P.; Park, S.; Onwudili, J. A. Co-pyrolysis of biomass and plastic waste over zeolite- and sodium-based catalysts for enhanced yields of hydrocarbon products. *Waste Management* **2020**, *102*, 909–918.

30. Klemetsrud, B.; Ukaew, S.; Thompson, V. S.; Thompson, D. N.; Klinger, J.; Li, L.; Eatherton, D.; Puengprasert, P.; Shonnard, D. Characterization of Products from Fast Micropyrolysis of Municipal Solid Waste Biomass. *ACS Sustainable Chemistry & Engineering* **2016**, *4* (10), 5415–5423.
31. Hwang, I.-H.; Kobayashi, J.; Kawamoto, K. Characterization of products obtained from pyrolysis and steam gasification of wood waste, RDF, and RPF. *Waste Management* **2014**, *34* (2), 402–410.
32. Nzihou, A.; Stanmore, B. The fate of heavy metals during combustion and gasification of contaminated biomass—A brief review. *Journal of Hazardous Materials* **2013**, *256-257*, 56–66.
33. Yuan, H.; Wang, Y.; Kobayashi, N.; Zhao, D.; Xing, S. Study of Fuel Properties of Torrefied Municipal Solid Waste. *Energy & Fuels* **2015**, *29* (8), 4976–4980.
34. Edo, M.; Skoglund, N.; Gao, Q.; Persson, P.-E.; Jansson, S. Fate of metals and emissions of organic pollutants from torrefaction of waste wood, MSW, and RDF. *Waste Management* **2017**, *68*, 646–652.
35. Dorado, C.; Mullen, C.; Boateng, A. Origin of carbon in aromatic and olefin products derived from HZSM-5 catalyzed co-pyrolysis of cellulose and plastics via isotopic labeling. *Applied Catalysis B: Environmental* **2015**, *162*, 338–345.
36. Li, X.; Li, J.; Zhou, G.; Feng, Y.; Wang, Y.; Yu, G.; Deng, S.; Huang, J.; Wang, B. Enhancing the production of renewable petrochemicals by co-feeding of biomass with plastics in catalytic fast pyrolysis with ZSM-5 zeolites. *Applied Catalysis A: General* **2014**, *481*, 173–182.
37. Li, X.; Zhang, H.; Li, J.; Su, L.; Zuo, J.; Komarneni, S.; Wang, Y. Improving the aromatic production in catalytic fast pyrolysis of cellulose by co-feeding low-density polyethylene. *Applied Catalysis A: General* **2013**, *455*, 114–121.
38. Nelson, K.; Retsina, T.; Iakovlev, M.; van Heiningen, A.; Deng, Y.; Shatkin, J. A.; Mulyadi, A., American Process: Production of Low Cost Nanocellulose for Renewable, Advanced Materials Applications. In *Materials Research for Manufacturing: An Industrial Perspective of Turning Materials into New Products*, Madsen, L. D.; Svedberg, E. B., Eds. Springer International Publishing: Cham, 2016; pp 267-302.
39. Couret, L.; Irle, M.; Belloncle, C.; Cathala, B. Extraction and characterization of cellulose nanocrystals from post-consumer wood fiberboard waste. *Cellulose* **2017**, *24* (5), 2125–2137.
40. Silva, N. L. C.; Betancur, G. J. V.; Vasquez, M. P.; de Barros Gomes, E.; Pereira, N. Ethanol Production from Residual Wood Chips of Cellulose Industry: Acid Pretreatment Investigation, Hemicellulosic Hydrolysate Fermentation, and Remaining Solid Fraction Fermentation by SSF Process. *Applied Biochemistry and Biotechnology* **2011**, *163* (7), 928–936.
41. Jafari, V.; Labafzadeh, S. R.; Jeihanipour, A.; Karimi, K.; Taherzadeh, M. J. Construction and demolition lignocellulosic wastes to bioethanol. *Renewable Energy* **2011**, *36* (11), 2771–2775.
42. Puri, D. J.; Heaven, S.; Banks, C. J. Improving the performance of enzymes in hydrolysis of high solids paper pulp derived from MSW. *Biotechnology for Biofuels* **2013**, *6* (1), 107.
43. Dornau, A.; Robson, J. F.; Thomas, G. H.; McQueen-Mason, S. J. Robust microorganisms for biofuel and chemical production from municipal solid waste. *Microbial Cell Factories* **2020**, *19* (1), 68.

44. Dreschke, G.; Probst, M.; Walter, A.; Pümpel, T.; Walde, J.; Insam, H. Lactic acid and methane: Improved exploitation of biowaste potential. *Bioresource Technology* **2015**, *176*, 47–55.
45. Beccari, M.; Longo, G.; Majone, M.; Rolle, E.; Scarinci, A. Modeling of Pretreatment and Acidogenic Fermentation of the Organic Fraction of Municipal Solid Wastes. *Water Science and Technology* **1993**, *27* (2), 193–200.
46. Sun, N.; Xu, F.; Sathitsuksanoh, N.; Thompson, V.; Cafferty, K.; Li, C.; Tanjore, D.; Narani, A.; Pray, T.; Simmons, B.; Singh, S. Blending Municipal Solid Waste with Corn Stover for Sugar Production Using Ionic Liquid Process. *Bioresource Technology* **2015**, *186*.
47. Li, C.; Liang, L.; Sun, N.; Thompson, V. S.; Xu, F.; Narani, A.; He, Q.; Tanjore, D.; Pray, T. R.; Simmons, B. A.; Singh, S. Scale-up and process integration of sugar production by acidolysis of municipal solid waste/corn stover blends in ionic liquids. *Biotechnology for Biofuels* **2017**, *10* (1), 13.
48. Abu-Omar, M. M.; Barta, K.; Beckham, G. T.; Luterbacher, J. S.; Ralph, J.; Rinaldi, R.; Román-Leshkov, Y.; Samec, J. S. M.; Sels, B. F.; Wang, F. Guidelines for performing lignin-first biorefining. *Energy & Environmental Science* **2020**.
49. Luo, X.; Li, Y.; Gupta, N. K.; Sels, B.; Ralph, J.; Shuai, L. Protection Strategies Enable Selective Conversion of Biomass. *Angewandte Chemie International Edition* **2020**, *59* (29), 11704–11716.
50. Questell-Santiago, Y. M.; Galkin, M. V.; Barta, K.; Luterbacher, J. S. Stabilization strategies in biomass depolymerization using chemical functionalization. *Nature Reviews Chemistry* **2020**, *4* (6), 311–330.
51. Shuai, L.; Amiri, M. T.; Questell-Santiago, Y. M.; Héroguel, F.; Li, Y.; Kim, H.; Meilan, R.; Chapple, C.; Ralph, J.; Luterbacher, J. S. Formaldehyde stabilization facilitates lignin monomer production during biomass depolymerization. *Science* **2016**, *354* (6310), 329–333.
52. Renders, T.; Van den Bossche, G.; Vangeel, T.; Van Aelst, K.; Sels, B. Reductive catalytic fractionation: state of the art of the lignin-first biorefinery. *Current Opinion in Biotechnology* **2019**, *56*, 193–201.
53. Van den Bosch, S.; Schutyser, W.; Vanholme, R.; Driessen, T.; Koelewijn, S. F.; Renders, T.; De Meester, B.; Huijgen, W. J. J.; Dehaen, W.; Courtin, C. M.; Lagrain, B.; Boerjan, W.; Sels, B. F. Reductive lignocellulose fractionation into soluble lignin-derived phenolic monomers and dimers and processable carbohydrate pulps. *Energy & Environmental Science* **2015**, *8* (6), 1748–1763.
54. Van den Bosch, S.; Renders, T.; Kennis, S.; Koelewijn, S. F.; Van den Bossche, G.; Vangeel, T.; Deneyer, A.; Depuydt, D.; Courtin, C. M.; Thevelein, J. M.; Schutyser, W.; Sels, B. F. Integrating lignin valorization and bio-ethanol production: on the role of Ni-Al2O3 catalyst pellets during lignin-first fractionation. *Green Chemistry* **2017**, *19* (14), 3313–3326.
55. Vangeel, T.; Schutyser, W.; Renders, T.; Sels, B. F. Perspective on Lignin Oxidation: Advances, Challenges, and Future Directions. *Topics in Current Chemistry* **2018**, *376* (4), 30.
56. Deneyer, A.; Ennaert, T.; Cavents, G.; Dijkmans, J.; Vanneste, J.; Courtin, C. M.; Dusselier, M.; Sels, B. F. Compositional and structural feedstock requirements of a liquid phase cellulose-to-naphtha process in a carbon- and hydrogen-neutral biorefinery context. *Green Chemistry* **2016**, *18* (20), 5594–5606.

57. Deneyer, A.; Peeters, E.; Renders, T.; Van den Bosch, S.; Van Oeckel, N.; Ennaert, T.; Szarvas, T.; Korányi, T. I.; Dusselier, M.; Sels, B. F. Direct upstream integration of biogasoline production into current light straight run naphtha petrorefinery processes. *Nature Energy* **2018**, 3 (11), 969–977.
58. Deneyer, A.; Tlatli, S.; Dusselier, M.; Sels, B. F. Branching-First: Synthesizing C–C Skeletal Branched Biobased Chemicals from Sugars. *ACS Sustainable Chemistry & Engineering* **2018**, *6* (6), 7940–7950.
59. Geboers, J.; Van de Vyver, S.; Carpentier, K.; de Blochouse, K.; Jacobs, P.; Sels, B. Efficient catalytic conversion of concentrated cellulose feeds to hexitols with heteropoly acids and Ru on carbon. *Chemical Communications* **2010**, *46* (20), 3577–3579.
60. Ennaert, T.; Op de Beeck, B.; Vanneste, J.; Smit, A. T.; Huijgen, W. J. J.; Vanhulsel, A.; Jacobs, P. A.; Sels, B. F. The importance of pretreatment and feedstock purity in the reductive splitting of (ligno)cellulose by metal supported USY zeolite. *Green Chemistry* **2016**, *18* (7), 2095–2105.
61. Rautiainen, S.; Di Francesco, D.; Katea, S. N.; Westin, G.; Tungasmita, D. N.; Samec, J. S. M. Lignin Valorization by Cobalt-Catalyzed Fractionation of Lignocellulose to Yield Monophenolic Compounds. *ChemSusChem* **2019**, *12* (2), 404–408.
62. Rinaldi, R.; Jastrzebski, R.; Clough, M. T.; Ralph, J.; Kennema, M.; Bruijnincx, P. C. A.; Weckhuysen, B. M. Paving the Way for Lignin Valorisation: Recent Advances in Bioengineering, Biorefining and Catalysis. *Angewandte Chemie International Edition* **2016**, *55* (29), 8164–8215.
63. Song, Q.; Wang, F.; Cai, J.; Wang, Y.; Zhang, J.; Yu, W.; Xu, J. Lignin depolymerization (LDP) in alcohol over nickel-based catalysts via a fragmentation-hydrogenolysis process. *Energy & Environmental Science* **2013**, *6* (3), 994–1007.
64. Paola, F.; Roberto, R. Catalytic Biorefining of Plant Biomass to Non-Pyrolytic Lignin Bio-Oil and Carbohydrates through Hydrogen Transfer Reactions. *Angewandte Chemie International Edition* **2014**, *53* (33), 8634–8639.
65. Klein, I.; Saha, B.; Abu-Omar, M. M. Lignin depolymerization over Ni/C catalyst in methanol, a continuation: effect of substrate and catalyst loading. *Catalysis Science & Technology* **2015**, *5* (6), 3242–3245.
66. V., G. M.; M., S. J. S. Selective Route to 2-Propenyl Aryls Directly from Wood by a Tandem Organosolv and Palladium-Catalysed Transfer Hydrogenolysis. *ChemSusChem* **2014**, 7 (8), 2154–2158.
67. V., G. M.; T., S. A.; Elena, S.; A., A. K.; Jonas, B.; J., H. W. J.; M., S. J. S. Hydrogen-free catalytic fractionation of woody biomass. *ChemSusChem* **2016**, *9* (23), 3280–3287.
68. Kumaniaev, I.; Subbotina, E.; Savmarker, J.; Larhed, M.; Galkin, M. V.; Samec, J. S. M. Lignin depolymerization to monophenolic compounds in a flow-through system. *Green Chemistry* **2017**, *19* (24), 5767–5771.
69. Anderson, E. M.; Stone, M. L.; Katahira, R.; Reed, M.; Beckham, G. T.; Román-Leshkov, Y. Flowthrough Reductive Catalytic Fractionation of Biomass. *Joule* **2017**, *1* (3), 613–622.
70. Anderson, E. M.; Stone, M. L.; Katahira, R.; Reed, M.; Muchero, W.; Ramirez, K. J.; Beckham, G. T.; Román-Leshkov, Y. Differences in S/G ratio in natural poplar variants do not predict catalytic depolymerization monomer yields. *Nature Communications* **2019**, *10* (1), 2033.

71. Anderson, E. M.; Stone, M. L.; Hülsey, M. J.; Beckham, G. T.; Román-Leshkov, Y. Kinetic Studies of Lignin Solvolysis and Reduction by Reductive Catalytic Fractionation Decoupled in Flow-Through Reactors. *ACS Sustainable Chemistry & Engineering* **2018**.
72. Van den Bosch, S.; Schutyser, W.; Koelewijn, S. F.; Renders, T.; Courtin, C. M.; Sels, B. F. Tuning the lignin oil OH-content with Ru and Pd catalysts during lignin hydrogenolysis on birch wood. *Chemical Communications* **2015**, *51* (67), 13158–13161.
73. Li, S.; Li, W.; Zhang, Q.; Shu, R.; Wang, H.; Xin, H.; Ma, L. Lignin-first depolymerization of native corn stover with an unsupported MoS2 catalyst. *RSC Advances* **2018**, *8* (3), 1361–1370.
74. Sun, Z.; Bottari, G.; Afanasenko, A.; Stuart, M. C. A.; Deuss, P. J.; Fridrich, B.; Barta, K. Complete lignocellulose conversion with integrated catalyst recycling yielding valuable aromatics and fuels. *Nature Catalysis* **2018**, *1* (1), 82–92.
75. Cao, Z.; Dierks, M.; Clough, M. T.; Daltro de Castro, I. B.; Rinaldi, R. A Convergent Approach for a Deep Converting Lignin-First Biorefinery Rendering High-Energy-Density Drop-in Fuels. *Joule* **2018**.
76. Liao, Y.; Koelewijn, S.-F.; Van den Bossche, G.; Van Aelst, J.; Van den Bosch, S.; Renders, T.; Navare, K.; Nicolaï, T.; Van Aelst, K.; Maesen, M.; Matsushima, H.; Thevelein, J. M.; Van Acker, K.; Lagrain, B.; Verboekend, D.; Sels, B. F. A sustainable wood biorefinery for low–carbon footprint chemicals production. *Science* **2020**, *367* (6484), 1385–1390.
77. Sun, Z.; Fridrich, B.; de Santi, A.; Elangovan, S.; Barta, K. Bright Side of Lignin Depolymerization: Toward New Platform Chemicals. *Chemical Reviews* **2018**, *118* (2), 614–678.
78. Huang, Y.; Duan, Y.; Qiu, S.; Wang, M.; Ju, C.; Cao, H.; Fang, Y.; Tan, T. Lignin-first biorefinery: a reusable catalyst for lignin depolymerization and application of lignin oil to jet fuel aromatics and polyurethane feedstock. *Sustainable Energy & Fuels* **2018**, *2* (3), 637–647.
79. Liu, X.; Li, H.; Xiao, L.-P.; Sun, R.-C.; Song, G. Chemodivergent hydrogenolysis of eucalyptus lignin with Ni@ZIF-8 catalyst. *Green Chemistry* **2019**, *21* (6), 1498–1504.
80. Li, X.; Guo, T.; Xia, Q.; Liu, X.; Wang, Y. One-Pot Catalytic Transformation of Lignocellulosic Biomass into Alkylcyclohexanes and Polyols. *ACS Sustainable Chemistry & Engineering* **2018**, *6* (3), 4390–4399.
81. Sun, J.; Li, H.; Xiao, L.-P.; Guo, X.; Fang, Y.; Sun, R.-C.; Song, G. Fragmentation of Woody Lignocellulose into Primary Monolignols and Their Derivatives. *ACS Sustainable Chemistry & Engineering* **2019**, *7* (5), 4666–4674.
82. Poels, E. K.; Dekker, J. G.; Van Leeuwen, W. A., Hydrothermal Sintering of the Active Phase in Alumina Supported Fixed Bed Nickel Catalysts During Reduction. In *Studies in Surface Science and Catalysis*, Poncelet, G.; Jacobs, P. A.; Grange, P.; Delmon, B. , Eds. Elsevier: 1991; Vol. 63, pp 205-214.
83. van Gorp, K.; Boerman, E.; Cavenaghi, C. V.; Berben, P. H. Catalytic hydrogenation of fine chemicals: sorbitol production. *Catalysis Today* **1999**, *52* (2), 349–361.
84. Hoffer, B. W.; Crezee, E.; Mooijman, P. R. M.; van Langeveld, A. D.; Kapteijn, F.; Moulijn, J. A. Carbon supported Ru catalysts as promising alternative for Raney-type Ni in the selective hydrogenation of d-glucose. *Catalysis Today* **2003**, *79-80*, 35–41.
85. Gallezot, P.; Nicolaus, N.; Flèche, G.; Fuertes, P.; Perrard, A. Glucose Hydrogenation on Ruthenium Catalysts in a Trickle-Bed Reactor. *Journal of Catalysis* **1998**, *180* (1), 51–55.

86. Liu, Y.; Chen, L.; Wang, T.; Zhang, Q.; Wang, C.; Yan, J.; Ma, L. One-Pot Catalytic Conversion of Raw Lignocellulosic Biomass into Gasoline Alkanes and Chemicals over LiTaMoO6 and Ru/C in Aqueous Phosphoric Acid. *ACS Sustainable Chemistry & Engineering* **2015**, 3 (8), 1745–1755.
87. Garrett, M. D.; Bennett, S. C.; Hardacre, C.; Patrick, R.; Sheldrake, G. N. New methods in biomass depolymerisation: catalytic hydrogenolysis of barks. *RSC Advances* **2013**, 3 (44), 21552–21557.
88. Chesi, C.; de Castro, I. B. D.; Clough, M. T.; Ferrini, P.; Rinaldi, R. The Influence of Hemicellulose Sugars on Product Distribution of Early-Stage Conversion of Lignin Oligomers Catalysed by Raney Nickel. *ChemCatChem* **2016**, *8* (12), 2079–2088.
89. Dwiatmoko, A. A.; Lee, S.; Ham, H. C.; Choi, J.-W.; Suh, D. J.; Ha, J.-M. Effects of Carbohydrates on the Hydrodeoxygenation of Lignin-Derived Phenolic Compounds. *ACS Catalysis* **2015**, *5* (1), 433–437.
90. Chakrabarti, R.; Tupy, S. A.; Schmidt, L. D. Effects of Biomass Inorganics on Rhodium Catalysts: II. Autothermal Reforming of Ethanol. *Energy & Fuels* **2011**, *25* (10), 4763–4769.
91. Huang, H.; Wang, S.; Wang, S.; Cao, G. Deactivation Mechanism of Cu/Zn Catalyst Poisoned by Organic Chlorides in Hydrogenation of Fatty Methyl Ester to Fatty Alcohol. *Catalysis Letters* **2010**, *134* (3-4), 351–357.
92. Twigg, M. V.; Spencer, M. S. Deactivation of supported copper metal catalysts for hydrogenation reactions. *Applied Catalysis A: General* **2001**, *212* (1), 161–174.
93. Huang, H.; Cao, G.; Wang, S. An evaluation of alkylthiols and dialkyl disulfides on deactivation of Cu/Zn catalyst in hydrogenation of dodecyl methyl ester to dodecanol. *Journal of Industrial and Engineering Chemistry* **2014**, *20* (3), 988–993.
94. Brands, D. S.; U-A-Sai, G.; Poels, E. K.; Bliek, A. Sulfur Deactivation of Fatty Ester Hydrogenolysis Catalysts. *Journal of Catalysis* **1999**, *186* (1), 169–180.
95. Huang, H.; Cao, G.; Wang, S. An evaluation of trimethyl phosphate on deactivation of Cu/Zn catalyst in hydrogenation of dodecyl methyl ester. *Korean Journal of Chemical Engineering* **2013**, *30* (9), 1710–1715.
96. Thakur, D. S.; Kundu, A. Catalysts for Fatty Alcohol Production from Renewable Resources. *JAOCS, Journal of the American Oil Chemists' Society* **2016**, *93* (12), 1575–1593.
97. Tschulkow, M.; Compernolle, T.; Van den Bosch, S.; Van Aelst, J.; Storms, I.; Van Dael, M.; Van den Bossche, G.; Sels, B.; Van Passel, S. Integrated techno-economic assessment of a biorefinery process: The high-end valorization of the lignocellulosic fraction in wood streams. *Journal of Cleaner Production* **2020**, *266*, 122022.
98. Sudarsanam, P.; Ruijten, D.; Liao, Y.; Renders, T.; Koelewijn, S.-F.; Sels, B. F. Towards Lignin-Derived Chemicals Using Atom-Efficient Catalytic Routes. *Trends in Chemistry* **2020**, *2* (10), 898–913.
99. Liao, Y.; d'Halluin, M.; Makshina, E.; Verboekend, D.; Sels, B. F. Shape selectivity vapor-phase conversion of lignin-derived 4-ethylphenol to phenol and ethylene over acidic aluminosilicates: Impact of acid properties and pore constraint. *Applied Catalysis B: Environmental* **2018**, *234*, 117–129.

100. Liao, Y.; Zhong, R.; d'Halluin, M.; Verboekend, D.; Sels, B. F. Aromatics Production from Lignocellulosic Biomass: Shape Selective Dealkylation of Lignin-Derived Phenolics over Hierarchical ZSM-5. *ACS Sustainable Chemistry & Engineering* **2020**, *8* (23), 8713–8722.
101. Liao, Y.; Zhong, R.; Makshina, E.; d'Halluin, M.; van Limbergen, Y.; Verboekend, D.; Sels, B. F. Propylphenol to Phenol and Propylene over Acidic Zeolites: Role of Shape Selectivity and Presence of Steam. *ACS Catalysis* **2018**, *8* (9), 7861–7878.
102. Verboekend, D.; Liao, Y.; Schutyser, W.; Sels, B. F. Alkylphenols to phenol and olefins by zeolite catalysis: a pathway to valorize raw and fossilized lignocellulose. *Green Chemistry* **2016**, *18* (1), 297–306.
103. Schutyser, W.; Van den Bosch, S.; Dijkmans, J.; Turner, S.; Meledina, M.; Van Tendeloo, G.; Debecker, D. P.; Sels, B. F. Selective Nickel-Catalyzed Conversion of Model and Lignin-Derived Phenolic Compounds to Cyclohexanone-Based Polymer Building Blocks. *ChemSusChem* **2015**, *8* (10), 1805–1818.
104. Koelewijn, S. F.; Van den Bosch, S.; Renders, T.; Schutyser, W.; Lagrain, B.; Smet, M.; Thomas, J.; Dehaen, W.; Van Puyvelde, P.; Witters, H.; Sels, B. F. Sustainable bisphenols from renewable softwood lignin feedstock for polycarbonates and cyanate ester resins. *Green Chemistry* **2017**, *19* (11), 2561–2570.
105. Koelewijn, S. F.; Cooreman, C.; Renders, T.; Andecochea Saiz, C.; Van den Bosch, S.; Schutyser, W.; De Leger, W.; Smet, M.; Van Puyvelde, P.; Witters, H.; Van der Bruggen, B.; Sels, B. F. Promising bulk production of a potentially benign bisphenol A replacement from a hardwood lignin platform. *Green Chemistry* **2018**, *20* (5), 1050–1058.
106. Zhao, S.; Abu-Omar, M. M. Biobased Epoxy Nanocomposites Derived from Lignin-Based Monomers. *Biomacromolecules* **2015**, *16* (7), 2025–2031.
107. Stewart, J. J.; Akiyama, T.; Chapple, C.; Ralph, J.; Mansfield, S. D. The Effects on Lignin Structure of Overexpression of Ferulate 5-Hydroxylase in Hybrid Poplar¹. *Plant Physiology* **2009**, *150* (2), 621–635.
108. Mottiar, Y.; Vanholme, R.; Boerjan, W.; Ralph, J.; Mansfield, S. D. Designer lignins: harnessing the plasticity of lignification. *Current Opinion in Biotechnology* **2016**, *37*, 190–200.
109. Vanholme, R.; Sundin, L.; Seetso, K. C.; Kim, H.; Liu, X.; Li, J.; De Meester, B.; Hoengenaert, L.; Goeminne, G.; Morreel, K.; Haustraete, J.; Tsai, H.-H.; Schmidt, W.; Vanholme, B.; Ralph, J.; Boerjan, W. COSY catalyses trans–cis isomerization and lactonization in the biosynthesis of coumarins. *Nature Plants* **2019**, *5* (10), 1066–1075.
110. van de Pas, D. J.; Torr, K. M. Biobased Epoxy Resins from Deconstructed Native Softwood Lignin. *Biomacromolecules* **2017**, *18* (8), 2640–2648.
111. Feghali, E.; van de Pas, D. J.; Torr, K. M. Toward Bio-Based Epoxy Thermoset Polymers from Depolymerized Native Lignins Produced at the Pilot Scale. *Biomacromolecules* **2020**, *21* (4), 1548–1559.
112. Van Aelst, K.; Van Sinay, E.; Vangeel, T.; Cooreman, E.; Van den Bossche, G.; Renders, T.; Van Aelst, J.; Van den Bosch, S.; Sels, B. F. Reductive catalytic fractionation of pine wood: elucidating and quantifying the molecular structures in the lignin oil. *Chemical Science* **2020**, *11* (42), 11498–11508.

113. Sultan, Z.; Graça, I.; Li, Y.; Lima, S.; Peeva, L. G.; Kim, D.; Ebrahim, M. A.; Rinaldi, R.; Livingston, A. G. Membrane Fractionation of Liquors from Lignin-First Biorefining. *ChemSusChem* **2019**, *12* (6), 1203–1212.

114. McCallum, C. S.; Strachan, N.; Bennett, S. C.; Forsythe, W. G.; Garrett, M. D.; Hardacre, C.; Morgan, K.; Sheldrake, G. N. Catalytic depolymerisation of suberin rich biomass with precious metal catalysts. *Green Chemistry* **2018**, *20* (12), 2702–2705.

115. Kumaniaev, I.; Samec, J. S. M. Valorization of Quercus suber Bark toward Hydrocarbon Bio-Oil and 4-Ethylguaiacol. *ACS Sustainable Chemistry & Engineering* **2018**, *6* (5), 5737–5742.

116. Vangeel, T.; Renders, T.; Van Aelst, K.; Cooreman, E.; Van den Bosch, S.; Van den Bossche, G.; Koelewijn, S. F.; Courtin, C. M.; Sels, B. F. Reductive catalytic fractionation of black locust bark. *Green Chemistry* **2019**, *21* (21), 5841–5851.

117. Constant, S.; Wienk, H. L. J.; Frissen, A. E.; Peinder, P. d.; Boelens, R.; van Es, D. S.; Grisel, R. J. H.; Weckhuysen, B. M.; Huijgen, W. J. J.; Gosselink, R. J. A.; Bruijnincx, P. C. A. New insights into the structure and composition of technical lignins: a comparative characterisation study. *Green Chemistry* **2016**, *18* (9), 2651–2665.

118. Lancefield, C. S.; Wienk, Hans L. J.; Boelens, R.; Weckhuysen, B. M.; Bruijnincx, P. C. A. Identification of a diagnostic structural motif reveals a new reaction intermediate and condensation pathway in kraft lignin formation. *Chemical Science* **2018**, *9* (30), 6348–6360.

119. Torr, K. M.; van de Pas, D. J.; Cazeils, E.; Suckling, I. D. Mild hydrogenolysis of in-situ and isolated Pinus radiata lignins. *Bioresource Technology* **2011**, *102* (16), 7608–7611.

120. Ma, S.; Li, T.; Liu, X.; Zhu, J. Research progress on bio-based thermosetting resins. *Polymer International* **2016**, *65* (2), 164–173.

121. Dodds, E. C.; Lawson, W. Synthetic strogenic Agents without the Phenanthrene Nucleus. *Nature* **1936**, *137* (3476), 996–996.

122. Nicastro, K. H.; Kloxin, C. J.; Epps, T. H. Potential Lignin-Derived Alternatives to Bisphenol A in Diamine-Hardened Epoxy Resins. *ACS Sustainable Chemistry & Engineering* **2018**, *6* (11), 14812–14819.

123. Hernandez, E. D.; Bassett, A. W.; Sadler, J. M.; La Scala, J. J.; Stanzione, J. F. Synthesis and Characterization of Bio-based Epoxy Resins Derived from Vanillyl Alcohol. *ACS Sustainable Chemistry & Engineering* **2016**, *4* (8), 4328–4339.

124. Koelewijn, S.-F.; Ruijten, D.; Trullemans, L.; Renders, T.; Van Puyvelde, P.; Witters, H.; Sels, B. F. Regioselective synthesis, isomerisation, in vitro oestrogenic activity, and copolymerisation of bisguaiacol F (BGF) isomers. *Green Chemistry* **2019**, *21* (24), 6622–6633.

125. Hu, Y. H.; Gao, Y.; Wang, D. N.; Hu, C. P.; Zu, S.; Vanoverloop, L.; Randall, D. Rigid polyurethane foam prepared from a rape seed oil based polyol. *Journal of Applied Polymer Science* **2002**, *84* (3), 591–597.

126. Hu, S.; Luo, X.; Li, Y. Polyols and polyurethanes from the liquefaction of lignocellulosic biomass. *ChemSusChem* **2014**, *7* (1), 66–72.

127. Griffini, G.; Passoni, V.; Suriano, R.; Levi, M.; Turri, S. Polyurethane Coatings Based on Chemically Unmodified Fractionated Lignin. *ACS Sustainable Chemistry & Engineering* **2015**, *3* (6), 1145–1154.

128. Jia, Z.; Lu, C.; Zhou, P.; Wang, L. Preparation and characterization of high boiling solvent lignin-based polyurethane film with lignin as the only hydroxyl group provider. *RSC Advances* **2015**, *5* (66), 53949–53955.

129. Ahvazi, B.; Wojciechowicz, O.; Ton-That, T.-M.; Hawari, J. Preparation of Lignopolyols from Wheat Straw Soda Lignin. *Journal of Agricultural and Food Chemistry* **2011**, *59* (19), 10505–10516.

130. Cateto, C. A.; Barreiro, M. F.; Rodrigues, A. E.; Belgacem, M. N. Optimization Study of Lignin Oxypropylation in View of the Preparation of Polyurethane Rigid Foams. *Industrial & Engineering Chemistry Research* **2009**, *48* (5), 2583–2589.

131. Xue, B.-L.; Wen, J.-L.; Xu, F.; Sun, R.-C. Polyols production by chemical modification of autocatalyzed ethanol-water lignin from Betula alnoides. *Journal of Applied Polymer Science* **2013**, *129* (1), 434–442.

Chapter 4

Proteomic Approaches for Advancing the Understanding and Application of Oleaginous Bacteria for Bioconversion of Lignin to Lipids

Xiaolu Li,[a] Zhangyang Xu,[a] Austin Gluth,[a] Wei-Jun Qian,[b] and Bin Yang[a,b,*]

[a]Department of Biological Systems Engineering and the Bioproduct Sciences and Engineering Laboratory, Washington State University, Richland, Washington 99354, United States

[b]Biological Sciences Division, Pacific Northwest National Laboratory, Richland, Washington 99352, United States

[*]Email: bin.yang@wsu.edu.

Bioconversion of cellulosic biorefinery lignin waste is important for complete valorization of all biomass carbons by biofuels and biochemical production. However, the deconstruction of lignin's complex polymeric framework into intermediates amenable to assimilation and subsequent lipid production for biofuels has proven challenging. Recent systems biology advances have shown promise in facilitating the bioconversion of lignin to lipids. Proteomics-guided synthetic biology approaches have proven to be effective tools for defining and designing new pathways. These pathways may help achieve optimal carbon conversion efficiency of biomass. Nevertheless, substantial challenges remain for the economic conversion of lignin to fuel. Herein, we review salient problems in the bacterial conversion of lignin to lipids with an emphasis on lignin depolymerization activities, *in vivo* aromatic degradation pathways, and recent advances in lipid production by oleaginous bacteria. Several mass spectrometry-based proteomics technologies and their applications to oleaginous bacteria are highlighted in this review. Together, these studies will help establish a system-level understanding of lignin bioconversion and drive efforts to build a platform for the maximal conversion of lignocellulosic biomass to economical next generation biorefinery.

Introduction

The development and commercialization of affordable green energy alternatives to fossil fuels is a global urgency. Sustainable 2nd generation biofuels are a promising alternative. These can be produced from a wide array of non-food feedstocks including agricultural and forest residues as well as energy crops. The price of waste lignocellulosic biomass is significantly lower than that of edible feedstocks like corn and soybean used for 1st generation biofuels production (*1*). However, converting the polymers in waste lignocellulosic biomass to value-added products is considerably more complicated. Many conversion processes employ a harsh physicochemical pretreatment requiring high energy inputs and producing toxic waste streams. In contrast, biological approaches are lower cost, eco-friendly, and do not produce toxic inhibitors (*2*).

To date, a majority of research in lignocellulose conversion has focused on separating the cellulose and hemicellulose fractions from that of lignin. After pretreatment, cellulose and hemicellulose are enzymatically hydrolyzed to hexose and pentose, respectively, and used as substrates for bioethanol and lipids fermentations (*3–6*). By design these carbohydrate-based biorefineries generate large quantities of lignin waste. Lignin is the fourth most abundant polymer in the world; though, it is often burnt for energy recovery because it is difficult to convert to valuable products (*7*). Lignin valorization has become a focal point for maximizing the carbon efficiency of biorefinery processes and improving commercial viability. Lignin's recalcitrant attributes are due to its complex heterogenous structure. Depolymerization of lignin results in complex mixture of aromatic molecules. A number of microorganisms with natural or engineered lignin degrading capabilities are being explored for the bioconversion of lignin to energy. Promising candidates are defined as metabolizing a wide variety of substrates, quickly adapting to and tolerating toxic environments, as well as synthesizing appreciable titers of lipids.

Oleaginous bacteria are capable of intracellularly accumulating more than 20% of cell dry weight (CDW) as neutral lipids. Notable examples of oleaginous Gram-positive bacteria are *Actinomycetes* and include *Gordonia sp.*, *Rhodococcus sp.*, *Streptomyces sp.*, *Mycobacterium sp.*, etc. A notable example of Gram-negative bacteria is *Acinetobacter sp.* (*8, 9–11*). Triacylglycerol (TAG) is the most abundant type of neutral lipid stored in oleaginous Gram-positive bacteria, whereas wax ester (WE) is the main lipid accumulated in some Gram-negative bacteria (*12*). They are suitable precursors for production of biodiesel, lubricants, and other commodities (*13*). TAGs and WEs are produced by transferring activated fatty acids to glycerol and long-chain fatty alcohol, respectively. Fatty acids vary in chain length and degree of unsaturation according to bacteria species and growth conditions (*14*). Neutral lipid is stored in lipid droplets (LD), an important and conserved cellular organelle (*15, 16*). LD functionality is highly dynamic and essential for growth, division, energy homeostasis, and stress response (e.g. nutrient depletion, oxidative stress, toxins, etc.) (*17*). Nitrogen limitation stress coupled to supply of excess carbon is a commonly used strategy to increase lipid accumulation. For example, *Rhodococcus opacus* PD630, one of the most studied oleaginous bacteria, accumulated up to 76% of CDW as lipid when cultured in nitrogen-limited medium with 1% (w/v) gluconate as the carbon source (*18*). Processes driving lipid accumulation using monomeric and/or dimeric sugars (e.g. glucose, xylose, cellobiose) from hemicellulose and cellulose have been studied extensively in oleaginous bacteria (*19–22*). In contrast, lipid accumulation using lignin and lignin-derived aromatics is still challenging as questions in lignin depolymerization, aromatics catabolism, and lipid metabolism remain. Nevertheless, the metabolic events occurring during the transition from growth to lipid accumulation are gaining interest as this shift alters carbon flux, reducing power requirements, and cellular redox state (*23, 24*). Specifically, how these reprogramming events affect the routing of excess carbon as aromatics to fatty acids is understudied. Thus, it is of great interest to identify mechanisms that regulate carbon flux, cofactor availability, and redox state during aerobic aromatics catabolism, and how these networks interact with those governing fatty acid synthesis and neutral lipid accumulation. Comprehending these complicated regulatory networks in oleaginous bacteria is critical for overcoming fortified barriers to lignin valorization.

High-throughput omics technologies including genomics, transcriptomics, proteomics, and metabolomics enable system-level characterization of complex biological samples. Mass spectrometry (MS)-based proteomics facilitates the study of integrated cellular processes by directly analyzing the functional units, proteins (*25*). This technology also facilitates protein localization studies by identifying proteins associated with isolated organelles (e.g. LD, cell membrane). Functional proteomics technologies are emerging as robust ways to characterize protein regulatory networks by identifying post-translational modifications (PTM) and protein-protein interactions. Proteomics has been used to investigate lipid accumulation, lignin depolymerization, and stress response (e.g. nitrogen limitation) in many microorganisms (*16, 26, 27*). MS-based proteomics has driven enzyme and pathway discovery. Results from proteomics studies help guide strain selection and engineering efforts for improving carbon flux to lipids. This is crucial for developing a bioprocess to convert lignin to lipids using oleaginous bacteria.

This review gives an overview of recent research in lignin depolymerization to aromatics, their subsequent conversion to lipid by oleaginous bacteria, and a focus on recent progress in augmenting bacterial lipid production. Innovations in MS-based proteomics technologies and highlighted advances in understanding conversion of lignin to lipid are summarized. Challenges and prospects are also discussed.

Lignin Structure

Lignocellulosic biomass consists of tightly intersected cellulose, hemicellulose, and lignin polymers. The lignin content in lignocellulosic biomass varies from 17 to 33% depending on the source (*28, 29*). Lignin is a three-dimensional phenolic heteropolymer primarily comprised of three basic subunits: p-hydroxyphenyl (H), guaiacyl (G) and syringyl (S). It is biosynthesized by oxidative coupling reactions employing three hydroxycinnamyl alcohols called monolignols (Figure 1) (*30*). These subunits are predominantly linked by β-O-4 motifs (up to ~80%) among others: 5-5, β-5, β-β, β-1, 4-O-5 (*31–33*). Compared to C–C bonds, ether bonds (e.g. β-O-4) are more reactive. β-O-4 linkages are primary targets in lignin depolymerization.

A variety of fractionation and extraction methods have been developed to separate lignin from cellulose and hemicellulose. Most methods can be summarized in several catagories (*34, 35*): (1) acid hydrolysis of biomass; (2) alkaline hydrolysis of biomass; (3) enzymatic hydrolysis of biomass; (4) reductive or oxidative fractionation of biomass, and (5) other fractionation techniques that employ mechanical forces (e.g. milling), ionic liquids, microwave, etc (*34*). Depending on the type of biomass, fractionation methods and conditions, the resulting lignin varies in structure, molecular size (and weight), solubility, purity, and, thus, reactivity (*36*). It is noteworthy to mention that in various fractionation processes, lignin depolymerization and repolymerization occur to different extents, leading to lower β-O-4 content and the formation of C–C bonds (i.e. lignin condensation), which stymies subsequent conversion. Generally, less condensed lignin with low molecular weight, high purity, and aqueous solubility is more readily converted by microorganisms.

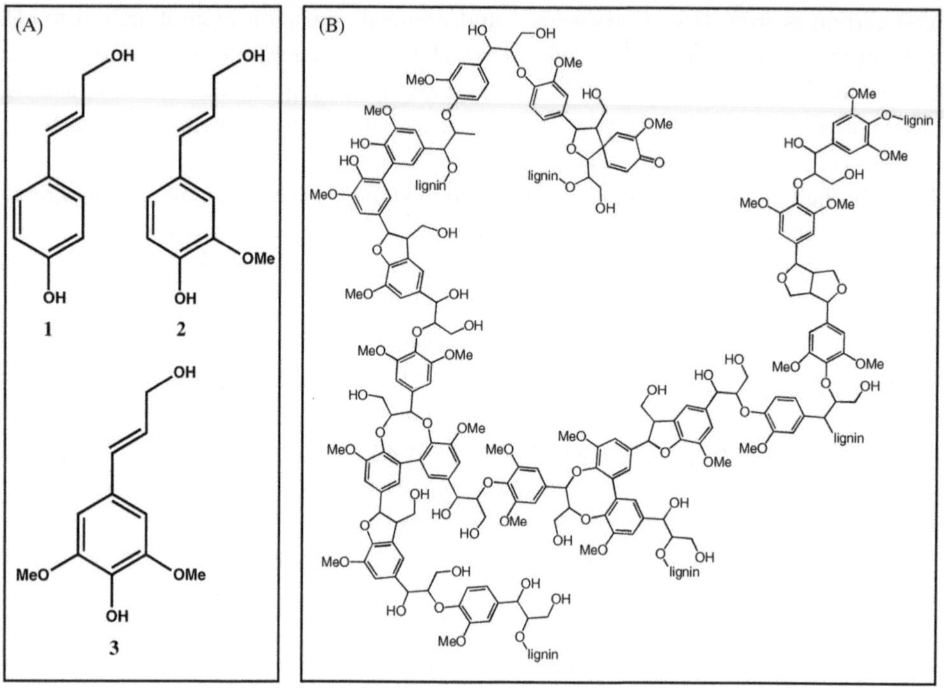

Figure 1. Monolignols, including p-coumaryl alcohol (1), coniferyl alcohol (2), and sinapyl alcohol (3), involved in lignin biosynthesis (A) and generic structure of lignin (e.g. hardwood and grass lignin) (B). Reproduced with permission from reference (29). Copyright 2013 John Wiley & Sons, Ltd.

Bacterial Bioconversion of Lignin to Lipid

In general, bioconversion of lignin to lipid consists of three major steps: a) depolymerization of lignin to low-molecular-weight (LMW) lignin (e.g. aromatic monomers, dimers, and oligomers) (*37, 38*); b) funneling (*39*), O-demethylation (*40*), and aromatics ring-cleavage (*41*) to generate central metabolites; c) the use of central metabolites as precursors for fatty acid and lipid synthesis. β-O-4 ether, di-aryl ether and biphenyl are three representative lignin depolymerization fragments and thus become the carbon source for subsequent degradation pathways (*42–45*). Two oxidative pathways for aromatics degradation (ortho- and meta-cleavage) have been extensively studied in soil bacteria. Lignin-derived aromatics ultimately enter the tricarboxylic acid cycle as succinyl-CoA, acetyl-CoA, pyruvate, etc. (*46*).

Depolymerization of Lignin

Though it is difficult, many microorganisms are capable of depolymerizing lignin to low molecular weight compounds (mono- and oligomers) which are funneled to useful metabolites (*47*). The most well characterized lignin depolymerizing enzymes are those secreted by white-rot fungi and include laccases and several types of heme-containing peroxidases (*48*). Laccases are cofactor-independent enzymes with low redox potential which can only oxidize phenolic lignin structure and generate water (*42, 49*). The substrate of laccases can be extended to non-phenolic lignin in the presence of a mediator which also improves the oxidation ability and prevents the repolymerization of phenolic groups. Peroxidases show higher redox potential than laccases and attack both phenolic and non-phenolic lignin by free radical mechanisms (e.g. veratryl alcohol cation radicals) and/or

Mn^{2+} as a diffusible oxidant (*50*). The peroxidases found in fungi include lignin peroxidases (LiP), manganese peroxidases (MnP), versatile peroxidases (VP) and dye-decolorizing peroxidases (DyP) (*51*). Various accessory enzymes are secreted by fungi to produce hydrogen peroxide needed for peroxidase function (*52*). Despite their ligninolytic potential, fungi grow slowly, and many produce harmful secondary metabolites complicating processes that employ them (*53–55*).

The ligninolytic capabilities of bacteria are rigorously being studied, for most bacteria grow faster than fungi and are amenable to genetic engineering (*56*). However, to date, neither VPs nor putative homologs have been identified in ligninolytic bacteria. Several studies report LiP activity in *Streptomyces viridosporus* T7A, and the MnPs in *Bacillus pumilus* and *Paenibacillus sp.* (*57, 58*). Conversely, DyPs have been annotated in many bacteria. Bacterial DyPs can be divided into 3 types, A, B and C, which overall share similar catalytic properties with fungal peroxidases (*59*). A-type DyPs are secreted by the Tat-system in bacteria and fungi, while the secretion signals are potentially lacking for B- and C-type DyPs suggesting that these two types of DyPs are located intracellularly or other secretion systems are involved (*28*). Laccases are secreted by the Tat-system in bacteria as well. Genomics has facilitated identification of numerous bacterial laccases. Bacterial laccases are comprised of two (so-called small laccases) or three domains. It is believed that bacterial laccases have great potential as ligninolytic biocatalysts due their cofactor-free mechanisms, which may be overexpressed and purified from *E. coli* for industrial use (*60*). Other proposed lignin-degrading enzymes include β-etherases, superoxide dismutases, and catalase-peroxidases (*61–63*).

In contrast to enzymatic lignin depolymerization, reactions that use Fenton chemistry are another proposed mechanism. The redox cycle is completed by quinones produced extracellularly which reduce Fe^{3+} to Fe^{2+}. Subsequently, the Fe^{2+} reacts with H_2O_2 to generate hydroxyl radicals to attack lignin (*42*). Some bacteria possess multiple lignin-degrading enzymes and harness Fenton chemistry (*38, 64, 65*). This may explain the strong lignin-degrading capacity of some strains and is an appealing option for lignin depolymerization. This justifies further exploration of bacterial secretomes for novel ligninolytic enzymes and synergism between different oxidative enzymes (*30*).

A large body of research aims to identify mechanisms by which lignin biaryl model compounds are catalyzed: these include dehydrodiconiferyl alcohol (DCA), guaiacylglycerol-β-guaiacyl ether (GGE), 5,5′-dehydrodivanillate (DDVA), 1,2-bis(4-hydroxy-3-methoxyphehyl)-1,3-propanediol (HMPPD), pinoresinol, etc. In Figure 2, five model lignin depolymerization pathways are shown (*40, 46*). These pathways were established primarily using *Sphingobium sp.* SYK-6, which is a model for bacterial lignin depolymerization. The five lignin model compounds including DCA (containing β-5 linkage), GGE (containing β-O-4 linkage), DDVA (containing 5-5′ biphenyl linkage), HMPPD (containing β-1 linkage) and pinoresinol (containing β-β biphenyl linkage) represent five common linkage motifs among p-hydroxyphenyl, guaiacyl, and syringyl units. These depolymerization routes yield vanillate, ferulate, and syringate in *Sphingobium sp.* SYK-6 (*39, 66–72*). The pinoresinol route is putatively catalyzed by *Sphingobium sp.* SYK-6. The model lignin derived aromatics vanillate, ferulate, and syringate are thus frequently used to study catabolic pathways.

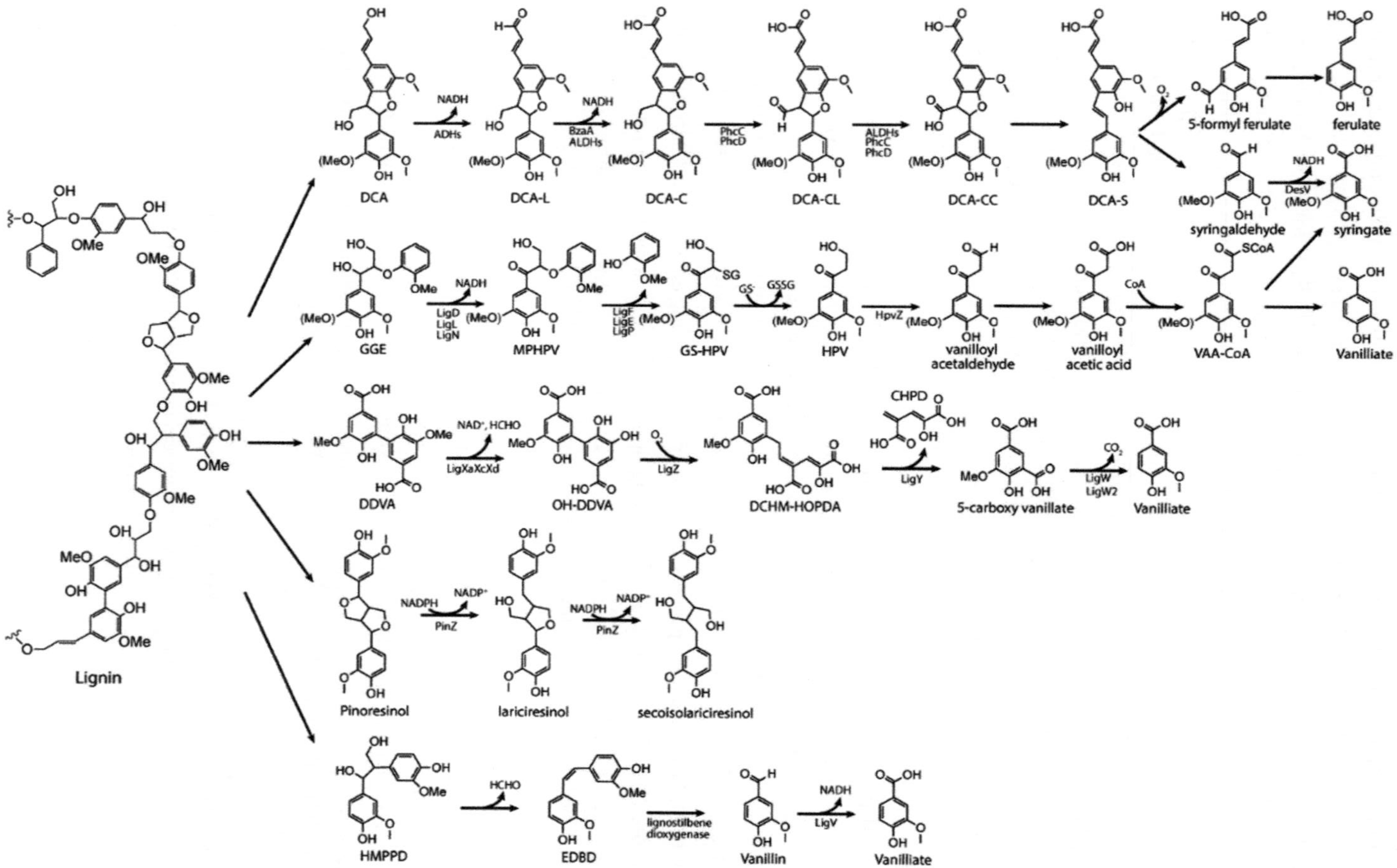

Figure 2. Model lignin depolymerization pathways in Sphingobium sp. SYK-6. Abbreviations: ADH, alcohol dehydrogenase; ALDH, aldehyde dehydrogenase; BzaA, Benzaldehyde-derivatives dehydrogenase; DesV, Benzaldehyde-derivatives dehydrogenase; HpvZ, Choline dehydrogenase; LigD, Cα-dehydrogenase for αR-isomers of β-aryl ether; LigE, β-Etherase for βR-isomers of Cα-keto β-aryl ether; LigF, β-Etherase for βS-isomers of Cα-keto β-aryl ether; LigG, Glutathione removing enzyme for

(βR)-GS-HPV; LigL, Cα-dehydrogenase for αS-isomers of β-aryl ether; LigN, Cα-dehydrogenase for αS-isomers of β-aryl ether; LigP, β-Etherase for βR-isomers of Cα-keto β-aryl ether; LigW, 5-Carboxyvanillate decarboxylase; LigW2, 5-Carboxyvanillate decarboxylase; LigXa, DDVA O-demethylase oxygenase component; LigXc, DDVA O-demethylase ferredoxin component; LigXd, DDVA O-demethylase reductase component; LigY, OH-DDVA meta-cleavage compound hydrolase; LigZ, OH-DDVA dioxygenase; PhcC, (+)-DCA-C oxidase; PhcD, (−)-DCA-C oxidase; PinZ, Pinoresinol/lariciresinol reductase; DCA, dehydrodiconiferyl alcohol; DCA-L, 3-(2-(4-hydroxy-3-methoxyphenyl)-3-(hydroxymethyl)-7-methoxy-2,3-dihydrobenzofuran-5-yl)acrylaldehyde; DCA-C, 3-(2-(4-hydroxy-3-methoxyphenyl)-3-(hydroxymethyl)-7-methoxy-2,3-dihydrobenzofuran-5-yl)acrylate; DCA-CL, 3-(3-formyl-2-(4-hydroxy-3-methoxyphenyl)-7-methoxy-2,3-dihydrobenzofuran-5-yl)acrylate; DCA-CC, 5-(2-carboxyvinyl)-2-(4-hydroxy-3-methoxyphenyl)-7-methoxy-2,3-dihydrobenzofuran-3-carboxylate; DCA-S, 3-(4-hydroxy-3-(4-hydroxy-3-methoxystyryl)-5-methoxyphenyl)acrylate; CoA, coenzyme A; HMPHP-CoA, 4-hydroxy-3-methoxyphenyl-β-hydroxypropionyl-CoA; GGE, guaiacylglycerol-β-guaiacyl ether; MPHPV, α-(2-methoxyphenoxy)-β-hydroxypropiovanillone; GS–, reduced glutathione; GSSG, oxidized glutathione; GS-HPV, α-glutathionyl-β-hydroxypropiovanillone; HPV, β-hydroxypropiovanillone; VAA-CoA, 4-hydroxy-3-methoxyphenyl-β-ketopropionyl-CoA; DDVA, 5,5′-dehydrodivanillate; OH-DDVA, 2,2′,3-trihydroxy-3′-methoxy-5,5′-dicarboxybiphenyl; DCHM-HOPDA, 4,11-dicarboxy-8-hydroxy-9-methoxy-2-hydroxy-6-oxo-6-phenylhexa-2,4-dienoate; CHPD, 4-carboxy-2-hydroxypenta-2,4-dienoate; HMPPD, 1,2-bis(4-hydroxy-3-methoxyphehyl)-1,3-propanediol; H_4folate, tetrahydrofolate.

Catabolism of Lignin-Derived Aromatic Monomers to Central Metabolites

Unlike most fungi which infiltrate lignin to access hemi- and cellulosic sugars, bacteria assimilate and utilize aromatics for biomass and energy accumulation; examples include *Streptomyces*, *Cupriavidus*, *Psuedomonas*, *Rhodococcus*, etc. (*48*, *73*, *74*). Due to the heterogeneous structure of lignin, depolymerization results in a complex mixture of monomers and oligomers of substituted phenols and propyl phenols (*48*). Thus, fundamental studies of bacterial aromatic degradation pathways have focused on model lignin-derived monomers or their analogs such as benzoate, ferulic acid, p-coumaric acid, vanillin, vanillic acid, 4-hydroxybenzoic acid, etc. (*14*, *38*, *75*, *76*). It is generally accepted that peripheral pathways funnel various aromatics into central aromatic intermediates catechol and protocatechuate. These intermediates undergo ring cleavage flowing through central pathways to ultimately join the TCA cycle (*77*). Diverse aerobic and anaerobic strategies are employed by bacteria to catabolize aromatics; however, for lignin-derived single-ring aromatics, aerobic strategies are generally used. One of aerobic approach employs various oxygenases and/or dioxygenases to incorporate hydroxyl groups into the aromatic ring, which yields central intermediates (e.g. catechol, protocatechuate). This activation is followed by ring cleavage between (intradiol, aka. ortho-cleavage) or adjacent to (extradiol, aka. meta-cleavage) the hydroxyl groups (*78*, *79*). An example common to species of *Pseudomonas*, *Rhodococcus*, *Acinetobacter*, *Agrobacterium*, and *Burkholderia* is the catabolism of vanillate by vanillate O-demethylase oxygenase. Vanillate is demethylated producing protocatechuate which enters an ortho-cleavage pathway, the β-ketoadipate pathway, ultimately yielding acetyl-CoA and succinyl-CoA (*80–83*). 4,5-dioxygenase (encoded by *LigAB*) from *S. paucimobilis* SYK-6 is an example of an oxygenase used in the meta-cleavage pathway for protocatechuate catabolism (*84*). Among currently reported lignin-degrading microorganisms, ortho-cleavage activity is more prevalent than that of meta-cleavage. The β-ketoadipate pathway remains the most intensively studied central aromatic pathway (*42*, *48*).

The phenylacetic acid (PAA) pathway is the other strategy used for aerobic aromatic degradation. This approach requires co-enzyme A (CoA) to activate aromatic acids and prime the energy intensive non-oxygenolytic ring fission. This pathway has been studied in *Rhodococci* (*79*, *81*, *84*, *85*) and is proposedly the central aromatics degradation pathway during lignin fermentation (*26*). Other aromatics catabolism pathways are recognized in bacteria including the gentisate pathway and hydroxquinol pathway; though, it is unknown if they are involved in lignin degradation. Some bacteria harbor multiple catabolic pathways for aromatics catabolism and are likely promising candidates for efficient conversion of depolymerized lignin (*79*). It is worth highlighting that not all types of lignin and lignin-derived aromatics can be degraded and/or used by bacteria especially because high concentrations of some compounds inhibit growth and/or product formation (*4*, *80*). Nevertheless, several oleaginous bacteria have the capacity to tolerate toxic aromatics and are attuned to using catabolic pathways as a response (*86*, *87*).

Lignin Degradation Capacity of Oleaginous Bacteria

The degradation of lignin or lignin-derived model compounds has been reported in a number of oleaginous bacteria strains (Table 1).

Several oleaginous *Rhodococci* are capable of metabolizing low molecular weight lignin and lignin-derived aromatics: examples are *R. jostii* RHA1, *R. opacus* PD630, *R. opacus* DSM 1069, *R. erythropolis*, *R. fascians*, *R. rhodochrous* (*88*). It was reported that *R. erythropolis* and *R. jostii* RHA1 are able to degrade milled wood lignin extracted from wheat straw, pine and miscanthus (*89*, *90*).

R. fascians and *R. rhodochrous* accumulate TAG when fed vanillate or 4-hydroxybenzoic acid as the sole carbon source (*91*, *92*). Notably, the potential of *R. jostii* RHA1, *R. opacus* PD630, and *R. opacus* DSM 1069 to convert lignin or lignin-rich hydrolysates to lipid has been addressed (*26*, *93–95*). The dye-decolorizing peroxidase DypB of *R. jostii* RHA1 was the first well characterized bacterial lignin-degrading enzyme. It shows hydrogen peroxide-dependent activity against the β-O-4 ether lignin model compound GGE (*96*). Differing from *Sphingobium sp.* SYK-6 (Figure 2), the cleavage of Cα – Cβ bond in GGE by *R. jostii* RHA1 results in vanillin, guaiacol, and oxalic acid as products (*90*, *96*). Consistent with this study, vanillin and its analogs were detected by GC-MS in the fermentation supernatant of *Rhodococcus* strains that were fed on alkali corn stover lignin (*86*). It is proposed that DypB may be secreted by an encapsulation mechanism in *R. jos*tii RHA1: in an *in vitro* experiment, it was determined that encapsulin packs DypB by binding its terminal targeting peptide. The precise mechanism by which the DypB-encapsulin complex functions in secretion is unknown (*97–99*). On the other hand, the ligninolytic capacity of *R. jostii* RHA1 may not depend solely on extracellular peroxidases; instead, other extracellular liginolytic enzymes such as O_2-consuming laccases may assume a dominant role. This is hypothesized because it is unknown how *R. jostii* RHA1 produces H_2O_2 to enable peroxidase activity (*26*, *89*, *100*). Though the activities of laccases, MnP or other Mn^{2+}-independent peroxidases were detected in culture supernatants of *R. jostii* RHA1 fed lignin-rich alkaline pretreated liquor (APL), it is still necessary to identify the functional ligninolytic enzymes (*38*). In summary, oleaginous *Rhodococcus spp.* have strong aromatics degradation capabilities, which include eight central aromatic pathways (one of which is the β-ketoadipate pathway) and dozens of peripheral pathways (*79*).

Other oleaginous bacteria have been studied for their lignin degradation capabilities, though not to the degree of *Rhodococci*. *A. baylyi* ADP1 is a model Gram-negative bacteria known to accumulate up to 27% of CDW as WE (*10*, *11*, *107*). *A. baylyi* ADP1 displays ligninolytic activities when fed APL (*38*). Several studies report conversion of lignin-derived model compounds (including ferulate, coumarate, and caffeate) to WE or alkanes via native or engineered pathways in *A. baylyi* ADP1 (*108*, *109*). The β-ketoadipate pathway has been observed in *A. baylyi* ADP1, though currently no report has identified ligninolytic enzymes in this strain (*110*). In contrast, several ligninolytic enzymes have been reported in *Streptomyces sp.* among which four strains (*S. lividans*, *S. coelicolor* A3(2), *S. griseus*, *S. albus*) demonstrated lipid accumulation capability (up to 150 mg/L) in early studies (*19*, *105*). Laccases were identified in studies of *S. coelicolor* A3(2) and *S. griseus*. Among them, a small laccase (SLAC) from *S. coelicolor* A3(2) was characterized to determine thermostability, optimized pH, substrate range, and the role of residues in a T1 copper site (*104*, *111–114*). A lignin peroxidase was reported in *S. lividans* (*115*). However, *Streptomyces sp.* are frequently studied for production of antibiotics and other chemicals rather than their lipid accumulation capacity (*116*). *Amycolatopsis sp.* 75iv2 (previously known as *Streptomyces setonii* and *Streptomyces griseus* 75iv2) expresses a C-type DyP2 and a small laccase (AMLAC). It is capable of producing 13mg/L fatty acids when fed on APL under nitrogen-limiting conditions (*38*, *104*, *117*). This is relatively low compared to that of *R. jostii* RHA1 (288 mg/L); nonetheless, the ligninolytic capacity of *Amycolatopsis sp.* is notable.

Table 1. Oleaginous Bacteria Reported to Degrade Lignin

Strains	*Storage Compound*	*Ligninolytic Enzyme*	*Substrate*	*Lignin Degradation Rate*	*Ref.*
Acinetobacter baylyi adp1	TAG/ TFA[a]/ WE	Activity of laccase, MnP and Mn^{2+}-independnet peroxidase was detected	Lignin-rich alkaline pretreated liquor	17.5%	(*38*)
Amycolatopsis sp. 75iv2	TFA	DyP2, a small laccase (AMLAC)	Lignin-rich alkaline pretreated liquor	16%	(*38*)
Mycobacterium smegmatis LZ-K2	TAG	Activity of laccase-like, LiP-like and MnP-like enzymes was detected, combined with Fenton reaction proposed based on measured Fe^{3+}-reducing activity and H_2O_2 production	Corn straw	51.17%	(*65*)
Rhodococcus erythropolis	TAG	NA	Milled wood lignin	NA	(*46, 89*)
Rhodococcus jostii RHA1	TAG	DypB; Activity of laccase, MnP and Mn^{2+}-independent peroxidase was detected	Lignin-rich alkaline pretreated liquor	20%	(*38*)
			Alkali corn stover lignin	23.20%	(*26*)
Rhodococcus opacus PD630	TAG	NA	Kraft lignin	15%	(*101*)
			Ethanol organosolv lignin (sugar content 1.43%)	36%	(*93*)
Rhodococcus opacus DSM 1069	TAG	NA	Oxygen-pretreated kraft lignin	28.00%	(*95*)
			Ethanol organosolv lignin (sugar content 1.43%)	23.00%	(*93*)
Co-culture (R. opacus PD630 and R. jostii RHA1 VanA–)	TAG	DypB; Activity of laccase, MnP and Mn^{2+}-independent peroxidase was detected in RHA1	Alkali corn stover Lignin	39.60%	(*86*)

Table 1. (Continued). Oleaginous Bacteria Reported to Degrade Lignin

Strains	*Storage Compound*	*Ligninolytic Enzyme*	*Substrate*	*Lignin Degradation Rate*	*Ref.*
Co-culture (R. opacus PD630, R. jostii RHA1 and R. jostii RHA1 VanA−)	TAG	DypB; Activity of laccase, MnP and Mn^{2+}-independent peroxidase was detected in RHA1	Alkali corn stover Lignin	33.60%	(*26*)
Rhodococcus pyridinivorans	TAG	NA	Alkali lignin	30.20%	(*102*)
Rhodococcus opacus NRRL B-3311	TAG	NA	Ammonia fiber expansion (AFEX) corn stover lignin	20.00%	(*103*)
Streptomyces coelicolor A3(2)	TAG	Small laccase (SLAC/ SCLAC)	Acid-precipitated polymeric lignin	NA	(*104, 105*)
Streptomyces lividans	TAG	Peroxidase	Extracted corn stalk lignocellulose	NA	(*19, 106*)

[a] TFA: total fatty acid.

Challenges and Recent Advances in Bioconverting Lignin to Lipid

Central metabolites of the TCA cycle are generated from lignin degradation, and some of these metabolites enter pathways for carbon storage. Many studies have reported lipid production by oleaginous bacteria using lignin model compounds, extracted lignin, and lignin-rich streams (e.g. alkaline pretreated liquor). A yield of 0.004 g/ g substrate and 4.08% CDW as lipid was reported in *R. opacus* utilizing ultrasonicated ethanol organosolv lignin as the sole carbon source (*93*). Lignin-rich dilute-acid pretreatment residue yielded 0.015 g/L lipid by *R. opacus* PD630 or DSM 1069 (*94*). These yields and titers are substantially lower than those reported using lignin model compounds (e.g. vanillic acid, 4-hydroxybenzoic acid) or favorable carbohydrates (e.g. glucose) as the carbon source (*118*). The plausible reasons include: a) the ligninolytic capability of the investigated bacteria is not adequate to provide sufficient carbon sources; b) some lignin-derived aromatics cannot be utilized by the strains and even inhibit growth; c) lignin and its derivatives induce oxidative stress resulting in lipid degradation; and/or d) other regulatory events (e.g. differentiated carbon routing) affected by carbon sources (*26, 80, 119*).

To improve lignin conversion, some efforts have added exogenous ligninolytic enzymes or chemical pretreatment reactions to modify lignin structure and depolmyerization lignin to low molecular weight fragments. A laccase supplemented *R. opacus* PD630 system for lignin conversion resulted in a 17-fold increase in lipid production (*101*). Oxygen-treated kraft lignin enabled the degradation by *R. opacus* DSM1069 (*95*). Using ammonia fiber expansion (AFEX) pretreated improved assimilation of lignin derivatives and lipid production in *R. opacus* NRRL B-3311 (*103*). The laccase-mediator (1-hydroxybenzotriazole) treatment resulted solubilized lignin as superior substrate for bioconversion by *R. opacus* PD630 (*120*). Combinatorial integration of multiple strategies (pretreatment, addition of laccase, detoxification of potential inhibitors) increased lipid yield to 1.83 g/L: this may be due to an increase in aromatic monomers and a decrease in volatile inhibitors (*121*). Recently, Xie et al. engineered *R. opacus* PD630 to heterologously express and secrete a laccase. Guided by proteomics data, they also improved lipid accumulation. The strain engineered with laccase secretion module and lipid production module showed superior growth and lipid production from kraft lignin and, separately, AFEX lignin (containing 5% glucose) (*27*).

Some studies suggest that lignin depolymerization and/or lipid yield can be increased by combining lignin with nutrient-rich substrates. This combination may help supply central metabolites, reducing power, and ATP to obtain a higher cell density and more catalytic activity for lignin depolymerization, lignin-derived aromatics catabolism, and product accumulation (*26, 38, 122*). However, in the presence of glucose and/or other preferred carbon sources, carbon catabolite repression (CCR) may occur reducing the overall assimilation of aromatics and increasing fermentation times for lipid production (*123*). The observation and mechanism of CCR differs among microbes depending on global and operon-specific regulations. CCR has been studied extensively in model bacteria *E. coli* and *Bacillus subtilis* as well as in *Pseudonomas putida*, which is a promising candidate in the field of lignin to bioplastic (*123*). The presence of succinate significantly suppressed the transcription of *pheA2A1* and *pheR* involved in phenol catabolism in both *R. erythropolis* and *R. jostii*. However, the addition of glucose or glycerol resulted in partial repression in *R. jostii* but not in *R. erythropolis* (*124*). A comprehensive analysis of CCR's interaction with the β-ketoadipate pathway at the nodes succinate and acetate revealed a global regulator protein termed Crc (catabolite repression control) in *A. baylyi* (*125*). This Crc protein was also identified in *P. putida*. The conversion of aromatic lignin monomers to muconate was reportedly enhanced by eliminating of Crc protein in *P. putida* while glucose was used as a co-fed carbon source for promoting cell growth (*126*). Additional steps like separation inevitably increase bioprocess operational expenses; thus, it is possible that mixed carbon sources such as biorefinery lignin-rich residues rather than individual platform sugars will be used in carbon efficient conversion of lignocellulose to lipids. Therefore, altering CCR in some oleaginous bacteria is an intriguing approach to possibly improve mixed carbon lignin utilization.

In addition to reductionist approaches, systems biology approaches have been employed to understand how oleaginous bacteria can act as cellular refineries for lignin valorization. A combination of these approaches will uncover answers to questions about, for example, CCR. These approaches include proteomics, transcriptomics, C13-marked metabolic analysis etc., which are often integrated and used in combination (*14, 127–129*). Systems biology research aims to holistically detail intricate cellular processes by building models using data from system-wide experiments (generally 'omics studies): this is essential for metabolically engineering oleaginous bacteria to improve lignin depolymerization and carbon flux towards lipids. Mass spectrometry-based proteomics technologies provide system-wide characterization of proteins in global processes. Their applications to studies of oleaginous bacteria are reviewed in the following section.

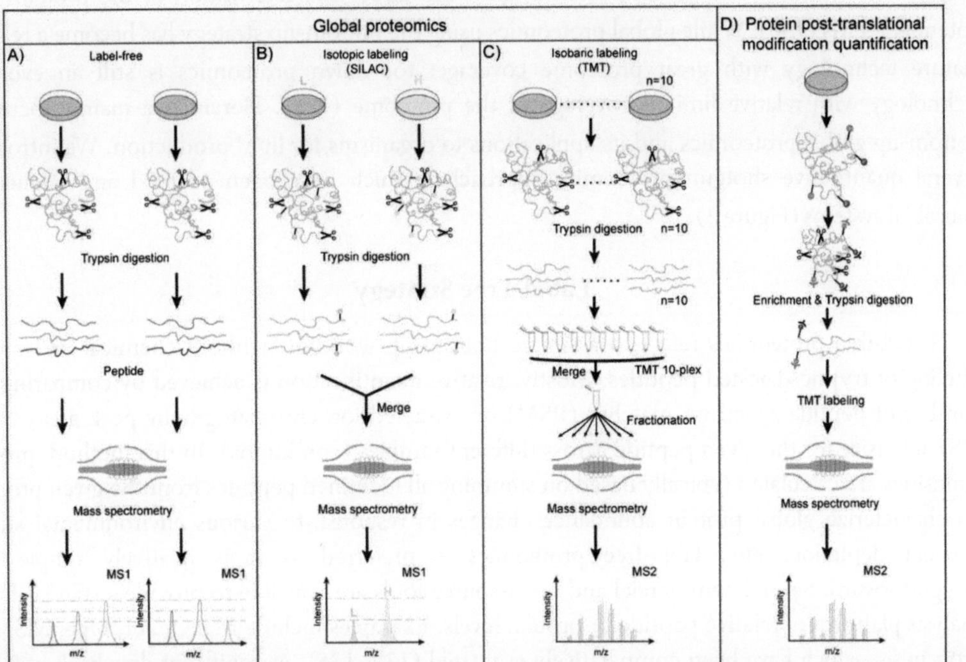

Figure 3. Overview of selected shotgun global quantitative proteomics technologies. A) Label-free strategy: proteins are extracted and digested by trypsin. The resulting peptides are subjected to LC-MS/MS. Peptide spectrum matches or extracted ion chromatography intensity of a given peptide is used to compare the peptide abundance across conditions. B) Isotopic labeling strategy (e.g. stable isotope labeling with amino acids in cell culture): isotopic labels are tagged to proteins during culture. Tagged proteins are extracted, digested, and pooled together for MS analysis. Ratio of paired (heavy and light isotopic labeled) peak intensity in MS1 spectrum is used to determine peptide abundance. C) Isobaric labeling (e.g. tandem mass tags): after proteins digestion, the resulting peptides are labeled by isobaric tags and pooled together for MS analysis. In MS^2, isobaric tags are fragmented to release reporter ions. Ratio of reporter ion intensity in MS2 spectrum is used to compare peptide abundance. D) Protein post-translational modification (PTM) quantification: after/during protein extraction, PTM-containing proteins or peptides are enriched prior to MS analysis.

MS-based proteomics technologies are mainly developed to provide global proteome-wide relative quantification across experimental conditions. Typically, quantitative proteomics strategies can be divided into two groups: bottom-up and top-down strategies. In the bottom-up proteomics (which is also called shotgun proteomics), proteins extracted from cell lysates, tissues or other sources are enzymatically digested into a mixture of peptides with high complexity, which is subjected to analysis by liquid chromatography coupled with tandem mass spectrometry (LC-MS/MS). Peptides are separated by liquid chromatography (LC) which is followed by electrospray ionization of peptides to enter MS in gas phase. The mass-to-charge (m/z) of intact peptides in MS1 spectrum at any given time can be used to distinguish peptides. Then, selected peptide ions with adequate signals are subjected to fragmentation to generate b (containing N terminus), y (containing C terminus) ions which can be detected in the tandem MS. The m/z values of b and y ions MS/MS (or MS^2) spectrum allow confident identification of a given peptide sequence through proteome database searching algorithm against a proteome database. Top-down proteomics is designed to introduce intact proteins into MS instead of peptides. In this strategy, the intact proteins and their fragment ions masses can be measured to allow the direct characterization of the proteins and potential PTMs (*130*). While global proteomics using the bottom-up strategy has become a relative mature technology with great proteome coverage, top-down proteomics is still an evolving technology with relative limited coverage of the proteome (*131*). Herein, we mainly focus on bottom-up global proteomics and its applications to organisms for lipid production. We introduce several quantitative shotgun proteomics approaches which have been adapted and applied to microbial systems (Figure 3).

Label-Free Strategy

Label-free proteomics refers to a simple bottom-up workflow without chemical or isotopic labeling of tryptic-digested peptides. Mostly, relative quantification is achieved by comparing the number of peptide spectrum matches (PSM) or extracted ion chromatography peak areas based MS1 intensity for the given peptide across different samples (conditions). In this method, protein intensities are calculated typically based on summing all identified peptides from the given protein. To characterize global protein abundance changes in response to various environmental stress, nutrient depletion, etc., label-free proteomics is preferred as it is relatively simple and straightforward. Several commercial and open-source tools are available to provide a standard data analysis platform of relative peptide or protein levels. Examples include MaxQuant, OpenMS, and Progenesis, which have been comparatively evaluated (*132–135*). A significant drawback of label-free proteomics is the potential missing data problem: some peptides are not detected in every sample even replicates (*136–138*). Some algorithms like those included in the "matches between run" feature of MaxQuant were designed to mitigate the problem.

Isotopic Labeling Strategy

Compared to label-free approaches, quantification strategies involving isotopic labeling provide higher precision of measurements. For example, the stable isotope labeling with amino acids in cell culture (SILAC) approach allows proteins to be assembled with heavy-isotope-containing amino acids during cell culture (*139*). Samples labeled with light or heavy isotopes are combined before injection into a mass spectrometer allowing co-analysis in one run. In MS1 spectra, the light and heavy peptides will form pairs. Relative quantification can be achieved by determining the ratio of ion intensities at the MS1 level between the parallel peaks of the same peptide.

Multiplex Isobaric Labeling Strategy

A common limitation of both label-free and isotopic labeling strategies is limited sample throughput given that only one or two samples can be analyzed per LC-MS/MS run. Multiplex isobaric labeling has become a promising strategy for high throughput proteomics because it employs isobaric mass tags with different distributions of heavy isotopes to allow 4-plex up to 16-plex labeling. An isobaric mass tag consists of a reporter group (containing different number of heavy isotopes), a mass normalizing group (to balance the mass of the whole molecule to be identical) and a peptide reactive (e.g. amine-reactive) group. As all labeled samples are pooled together and subjected to MS, signals of identical peptides but with different reporter tags are merged into one peak in MS1 spectrum due to the same mass. At MS^2 level, the isobaric tags are fragmented along with peptide backbone to release the specific reporter ions for each channel. Commonly used isobaric labels include the tandem mass tag (TMT) and the isobaric tag for relative and absolute quantitation (iTRAQ). TMT 6-16 plex and iTRAQ 8-plex are commercially available, which means up to 16 samples can be co-analyzed in one single experiment and ensures higher reproducibility and throughput. Combined with sample fractionation technologies, more comprehensive coverage of peptide identification can be achieved. The main issue with multiplex isobaric labeling is ratio compression due to co-eluting peptides along with the identified peptide (*137*). To alleviate this issue, MS^3 mode (further fragmentation of selected ions from MS2 spectrum) has been employed (*140, 141*).

Protein Post-Translational Modification (PTM) Quantification

Besides protein abundance measurements, a unique aspect of MS-based proteomics is its ability to probe dynamic functional information about proteins at a systems-level. This functional information includes various types of PTMs and protein-protein interactions. These characterizations are essential for elucidating global cellular signaling networks, regulations, protein complexing, etc.

Protein PTMs refer to the reversible or irreversible modification of protein amino acid residues through biochemical reactions, thus altering protein conformation, activity and function beyond genetic and transcriptional regulations. PTMs such as phosphorylation, ubiquitination, acetylation, proteolytic processing, thiol-based redox modification, etc. are global events that are labile, substoichiometric, and potentially in low abundance (*142, 143*). This requires the introduction of additional enrichment step(s) for modified peptides, a delicate workflow design, and cautious sample preparation to avoid artificial results. The enrichment strategies mainly include 1) immunoprecipitation by PTM-specific antibodies and 2) chemical affinity capture based on unique chemical groups of PTMs (*144, 145*). The modification on peptides can be detected by MS based on the specific mass differences of the modifications (e.g., acetyl groups, 42 Da; phosphate groups, 80 Da) (*142*). Isotopic or multiplex isobaric labeling technologies are widely applied to PTM identification to obtain high-throughput and good reproducibility.

The identification of PTMs in bacterial systems is a burgeoning field with novel bacterial PTMs and corresponding protein targets being continuously discovered. The role of phosphorylation in signal transduction and regulatory networks responsible for altering cellular metabolic programs is well established for stress response (e.g. nitrogen limitation, oxidative stress, etc.), nutrient availability (e.g. carbohydrates, aromatics, nitrogen sources, etc.), and other processes. Recently, PTMs of redox sensitive residues have drawn attention as the redox state of a cell likely changes according to stress and nutrient availability. Herein, we emphasize phosphorylation and thiol-based redox modifications.

Phosphorylation

Protein phosphorylation is one of the most characterized PTMs in bacteria and has been detected on side chains of serine, threonine, tyrosine (in the form of phosphoesters), histidine, arginine, lysine (as phosphoamidates), aspartate (as mixed phosphoacyl anhydrides), and cysteine (as phosphothiolates) (*142, 146*). Phosphorylation and dephosphorylation are catalyzed by kinases and phosphatases, respectively, which are often residue specific. Phosphorylation of different residues has multifarious effects on protein targets: how these events affect protein(s) functionalit(ies) is an area of continued research. Nevertheless, the role of phosphorylation in signal transduction in eukaryotes is widely accepted and emphasizes the value of exploring bacterial signaling pathways. Currently most bacterial phosphoproteomics studies focus on serine/threonine/tyrosine phosphorylation due to available assays (e.g. phosphopeptide enrichment with TiO_2 affinity chromatography) adapted from eukaryote analyses. Few studies describe histidine phosphoproteomes (*147*). Many bacterial phosphoproteomics studies are conducted in the medical microbiology field and aim to elucidate pathogen survival mechanisms and virulence (*148–150*).

Thiol-Based Redox Modification

Profiling protein thiol-based redox modifications and understanding their physiological roles is an emerging area of research. In eukaryotes, protein thiol modifications are recognized as essential components in regulatory networks that involve a number of essential processes such as cellular growth, differentiation, inflammation, oxidative stress response, mitochondrial biogenesis, etc. (*151, 152*). These regulatory networks are responsible for monitoring and addressing changes to cellular redox homeostasis. Protein thiol modification is a PTM that generally occurs on cysteine residues: examples include S-glutathionylation (SSG), S-nitrosylation (SNO), S-sulfhydration (aka. persulfidation, SSH), and S-sulfenylation (SOH). These redox modifications can be reversibly modified by small molecules (e.g. reactive oxygen species, reactive nitrogen species, low molecular weight thiols like glutathione) or enzymes (e.g. glutathione reductase, thioredoxin system) (*143*). Among the various types of modifications, total oxidation (a sum of all types) and S-glutathionylation are frequently investigated mostly because of their prevalence and the existence of relevant assays. It was reported that S-glutathionylation can modulate protein activities as well as protect proteins from oxidative stress and store glutathione (GSH) (*153*). In bacteria, S-glutathionylation is generally found in Gram-negative bacteria. S-mycothiolation and S-bacillithiolation occurs in the Gram-positive *Actinomycetes* and some *Firmicutes* (e.g. *Bacillus sp.* and *Staphylococcus sp.*), respectively. These modifications use mycothiol (MSH) or bacillithiol (BSH) as the LMW thiols (*154, 155*). Compared to S-glutathionylation, only non-quantitative assays are available to identify protein S-mycothiolation or S-bacillithiolation by MS based on their mass increase. These assays do not specifically enrich and quantify MSH or BSH modified peptides/proteins to provide corresponding dynamic information (*156, 157*).

Systematic characterization of biological processes will direct vast efforts in synthetic biology, which is a field that applies engineering principles to design organisms for useful tasks. Global quantitative proteomics approaches aim to elucidate pathways and processes involved in conversion of various substrates to final products. Results of these studies can be applied to those in synthetic biology to improve lignin depolymerization, aromatics metabolism, and lipid production by oleaginous bacteria.

Application of Proteomics to Oleaginous Bacteria

Many proteomics studies have investigated lipid accumulation in oleaginous bacteria under nitrogen limited conditions with excess carbon for lipogenesis supplied as sugars (*3, 9, 158*). An integrative metabolic model of bacterial lipogenesis was proposed by Costa et al. based on results from a label-free proteomics study of *R. jostii* RHA1 (*159*). Protein abundances were compared between nitrogen-limiting and nitrogen-rich conditions. Proteins in the Entner-Doudoroff pathway and pentose phosphate shunt were up-regulated in the nitrogen-limiting condition suggesting the importance of NADPH generation for fatty acid synthesis. Many processes associated with the metabolic rearrangement were activated such as glyceroneogenesis and amino acid catabolism to provide glycerol 3-phosphate, acetyl-CoA, propionyl-CoA and NADPH to fully support TAG accumulation. Digging into the global proteomics data generated by Costa et al., a switch from NAD^+ to $NADP^+$ dependent oxidation of glyceraldehyde 3-phosphate was identified by observing relative abundance change of related proteins, highlighting a key metabolic adjustment to TAG accumulation phase. Furthermore, observations that select proteins are modified at their reactive thiols on cysteine residues during TAG accumulation in the presence of a pro-oxidant suggests potential redox regulation of lipogenesis.

Lipid droplets (LDs) are an important and conserved cellular organelle in which a majority of accumulated neutral lipids are stored. They are found in nearly all domains of life from bacteria to man (*15*). Combined with high-resolution imaging, proteomics has been used to characterize isolated LDs. These allow the profiling of LD associated proteins, many of which have roles in lipid metabolism and LD biogenesis (*16*). Several global as well as LD isolated bacterial proteomics studies have been performed (Table 2). As aforementioned, *R. opacus* PD630 is a promising oleaginous bacterium for lipid accumulation (up to 80% CDW), and its LD proteome has been characterized. 430 LD associated proteins were quantitatively identified, and their abundance changed over culture time. Besides the 231 proteins that were directly involved in lipid synthesis and degradation, many proteins with functions related to transportation, transcription and translation, stress response, cell division, etc. were identified implying dynamic regulatory networks that govern lipid metabolism (*160*). Guided by this proteomics result, a microorganism lipid droplet small (MLDS) protein was discovered, which not only affects LD size but also binds DNA and enhances survival under stress conditions (nitrogen limitation and ultraviolet exposure) (*161*, *162*). As mentioned above, nitrogen limitation is one of the most common strategies to induce lipid accumulation, though is less than favorable for biomass accumulation. Further investigation of MLDS and its underlying mechanistic role in stress response and/or protection may be useful for directing engineering towards biofuels production.

As discussed in previous sections, compared to sugars, lignin conversion to lipid by oleaginous bacteria presents a number of challenges: low depolymerization efficiency, toxicity and inhibitory effects, energy-consuming ring cleavage activity, inefficient carbon flux from aromatics to fatty acids, etc. (*26*, *80*, *120*, *121*). Global proteomics studies can be used to profile proteins and enrich biological processes involved in lignin degradation, aerobic aromatics catabolism, central metabolism, and lipid synthesis. In parallel, based on database annotation and sequence homology, potential ligninolytic enzymes and pathways can be uncovered. Overall, these tools will improve comprehensive understanding of bacterial conversion of lignin to lipid, help identify functional modules and metabolic bottlenecks causing low lipid yield, and thus, guide strain manipulation to optimize the lignin valorization process. Conversion of glucose to TAGs by *R. opacus* PD630 was investigated using global proteomics (*27*). The proteomics results suggested that expression of FASI is positively correlated with lipid yield, which guided the design of a functional module to enhance lipid production by overexpressing *fasI* operon and another gene *atf2* (encoding diacylglyceride acyl transferase in TAG synthesis pathway). Additionally, using secretome results, the authors were able to screen for functional promotor/ribosome binding sites and signaling peptides. Some were used to build secretion modules to express a small laccase from *Streptomyces coelicolor* in *R. opacus* PD630, enabling higher titer lipid production from lignin-enriched biorefinery waste. Here, proteome analyses have driven hypothesis formulation and module design to rewire *Rhodococcus opacus*. Proteomics guided synthetic biology studies of *Pseudomonas putida* for lignin to bioplastic conversion have also been conducted (*64*). Furthermore, proteomics can provide fast and robust characterization of co-cultures and even whole microbial communities, which is useful for probing the interactive and contributive effects of consortia in synthetic and natural processes. The consortia proteome of three *Rhodococci* strains co-cultured on corn stover alkali lignin was compared to that on glucose. This revealed the phenylacetic acid (PAA) pathway contributing to lignin-derived aromatics ring cleavage (Figure 4) (*26*). Analysis of up-regulated proteins led to a hypothesis of redox balance and reducing power competition between oxidative stress caused by lignin degradation and lipid synthesis. This defines a possible mechanism for low lipid yields from lignin. Secretome mapping supports the potential role of several peroxidases, accessory oxidases, and a laccase-like protein in extracellular lignin depolymerization, which pinpointed novel ligninolytic enzymes for further characterization (*26*).

Table 2. Selected Proteomics Application on Oleaginous Bacteria

Strains	*Conditions*	*Proteomics Approaches*	*Observations*	*Ref.*
Rhodococcus rhodochrous	Glucose, phenol mixed or individually fermented, 0 or 24 h	Label-free	Almost 500 proteins were identified in each condition. Using phenol compared to glucose, proteins involved in phenol degradation pathway were up-regulated, whereas proteins involved in fatty acid metabolism and nitrogen metabolism were down-regulated. These results correlated with lipid yields in different conditions.	(*92*)
Rhodococcus jostii RHA1	Glucose, ammonium chloride 1 g/L or 0 g/L	Label-free	2200 proteins were identified. Abundances of 108 proteins were significantly changed in presence or absence of nitrogen source. Proteins involved in NADPH, precursor generation, and nitrogen metabolism were up-regulated during nitrogen starvation. Metabolic shift and carbon re-routing events to support TAG accumulation were proposed.	(*159*)
	Glucose, ammonium chloride 1 g/L or 0 g/L, supplemented with methyl viologen as oxidative stress	Thiol-based PTM quantification	The thiol total oxidation level of a group of proteins related to TAG accumulation was changed dependent on nitrogen-rich or limiting conditions in presence of oxidative stress. These proteins have been identified as redox-sensitive implicating redox regulation of lipogenesis.	(*159*)
	Glucose, ammonium chloride 1 g/L or 0 g/L, supplemented with methyl viologen as oxidative stress	Label-free	Antioxidants, and proteins involved in lipid and fatty acid degradation were significantly up-regulated in response to oxidative stress during TAG accumulation. Proteins involved in precursor generation, fatty acid synthesis, and TAG synthesis were down-regulated, whereas proteins involved in NADPH synthesis remained active.	(*163*)

Table 2. (Continued). Selected Proteomics Application on Oleaginous Bacteria

Strains	*Conditions*	*Proteomics Approaches*	*Observations*	*Ref.*
	Cholesterol, cholic acid, or pyruvate as carbon sources	Label-free; Protein localization	1703 proteins were identified. Abundances of 59 proteins were significantly changed (over two-fold). Up-regulation of proteins involved in steroid degradation, lipid metabolism, energy homeostasis, and secondary metabolism. It was determined that some steroid-degrading proteins are membrane associated.	(*164*)
	Lipid accumulatLipid accumulation phase LD isolated	Label-free	Identification of 228 proteins associated with LDs to study LD dynamics and function. A protein that regulates LD size was identified.	(*162*)
Rhodococcus opacus PD630	Lipid accumulation phase LD isolated	Label-free	Identification of 238 proteins in LDs. Ribosomal proteins, RNA polymerases, and transcriptional regulators were differentially expressed. These results lead to hypothesis that LDs have more biological functions than simply lipid accumulation and metabolism.	(*160*)
	Glucose or alkali lignin as sole carbon sources	Label-free	Secretome analysis identified promotors, signal peptides, membrane protein transporters as targets to test, which revealed the lack of efficient secretion system for heterologous expression of a laccase. Proteomics analysis identified a type I fatty acid synthase that was up-regulated and abundance of which was correlated with lipid accumulation. These results guided the design and integration of functional modules for laccase secretion and lipid production into R. opacus, enabling higher titer lipid production from lignin-enriched biorefinery waste.	(*27*)
Co-culture of Rhodococcus jostii RHA1, Rhodococcus jostii RHA1 vanA-, Rhodococcus opacus PD630	Glucose or corn-stover lignin as sole carbon sources	Label-free	2740 proteins were identified in lysate. 171 proteins were significantly up-regulated using lignin as sole carbon source compared to glucose. A major fraction is involved in aromatic degradation, fatty acid metabolism and oxidative stress response. Qualitative identification of 1800 proteins in the secretome identified possible lignin-degrading enzymes.	(*26*)

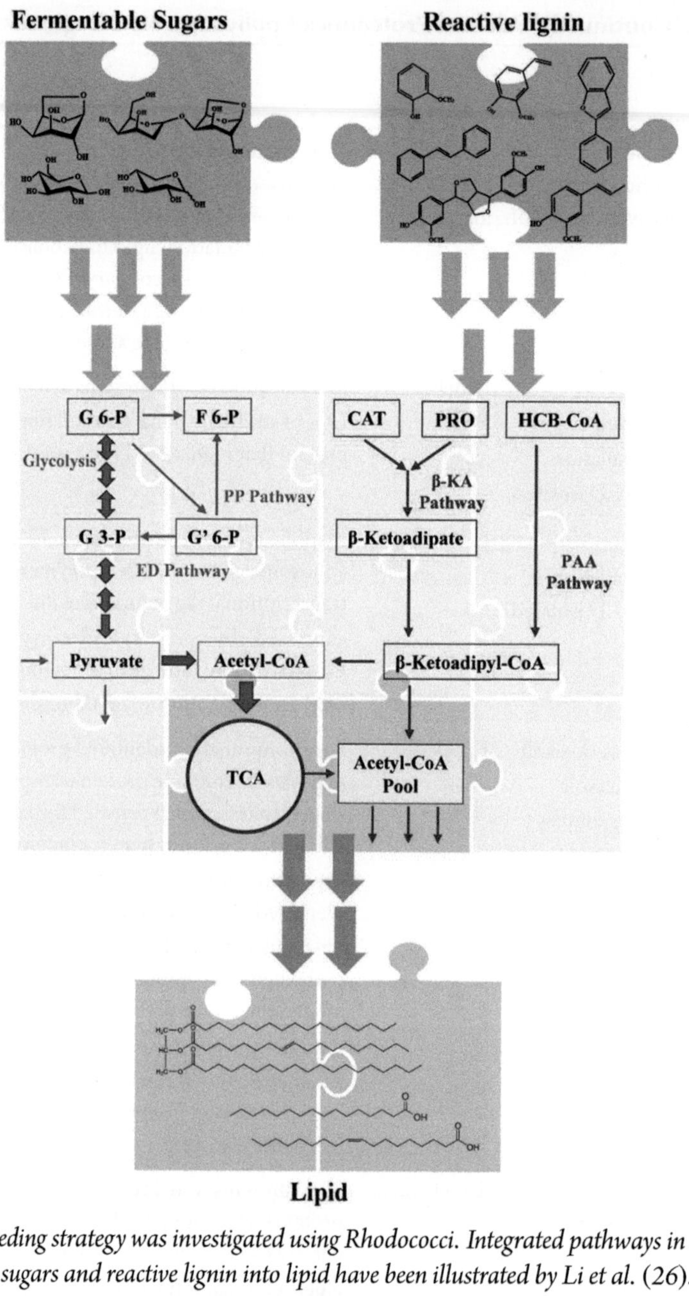

Figure 4. A co-feeding strategy was investigated using Rhodococci. Integrated pathways in Rhodococci for conversion of both sugars and reactive lignin into lipid have been illustrated by Li et al. (26). G 6-P, Glucose 6-P; F 6-P, Fructose 6-P; G 3-P, Glyceraldehyde 3-P; G′ 6-P, Gluconate 6-P; CAT, catechol; PRO, protocatechuate; HCB-CoA, 2-Hydroxy-4-carboxy-butanyl-CoA; PP Pathway, Pentose Phosphate Pathway; ED Pathway, Entner–Doudoroff Pathway; β-KA Pathway, β-Ketoadipate Pathway; PAA Pathway, Phenylacetic acid Pathway. Reproduced with permission from reference (26). Copyright 2019 Springer Nature.

Post-Translational Regulation in Oleaginous Bacteria

It is of great importance to unravel regulatory attributes to better understand the metabolic switch for transitioning from growth to lipid accumulation. Identification of global protein PTMs and their dynamics will provide novel insights into this metabolic switch. Few studies have explored the regulation at PTM level in oleaginous bacteria.

Phosphorylation

The most common form of protein phosphorylation in bacteria occurs on histidine. This is catalyzed by kinases belonging to the two-component systems (TCS). Phosphate from histidine is transferred to aspartate residue(s) of the respective response regulator, which is stimulated by specific environmental conditions (e.g. heat, salt, nutrient supply). The signals transmitted via TCS mostly affect target gene transcription by altering the DNA-binding activity of response regulator(s), thereby mediating a cellular response to environmental conditions (*165–167*). For oleaginous bacteria, nitrogen metabolism including uptake, assimilation, and regulation is of great significance as nitrogen depletion promotes lipid accumulation. The TCS response regulator GlnR is a global regulator playing a central role in transcriptionally regulating genes involved in nitrogen metabolism. This regulator has been reported in *Streptomyces* (*168, 169*), *Mycobacterium* (*170, 171*), *Saccharopolyspora* (*172*), and *Amicolatopsis* (*173, 174*). A transcriptional regulator NnaR, which can be activated by GlnR, is a co-activator associated with nitrate/nitrite assimilation. NnaR orthologues have been found in *R. jostii* and *R. opacus* and are named NlpR. NlpR exhibits functionality in modulating lipogenesis and lipid accumulation in addition to co-activator activity in response to ammonium limitation (*175*). This implies an important role of the GlnR-mediated system in lipid accumulation for oleaginous *Rhodococci*. Another TCS consisting of histidine kinase PrrB and response regulator PrrA was reported in *Mycobacterium smegmatis* and regulates expression of several genes involved in TAG and lipid biosynthesis pathways (*176*). It was recently reported that the activity of GlnR is regulated by serine/threonine phosphorylation which modulates GlnR binding to target DNA in *Streptomyces coelicolor* when excessive nitrogen source was available (*177, 178*). Application of phosphoproteomics to oleaginous bacteria under nitrogen limiting condition will map underlying signal transduction networks governing oleagenicity.

As in other microorganisms, phosphorylation may be an essential regulatory modification for the CCR program in oleaginous bacteria. Unlike the CCR mechanisms found in *Acinetobacter baylyi* and *Pseudomonas putida* which are governed by the global regulator Crc protein, CCR in *E. coli* and *Bacillus subtilis* involve multiple phosphorylation events which constitute the phosphoenolpyruvate–carbohydrate phosphotransferase system (PTS) (*179*). For example, the histidine protein (HPr) can be phosphorylated at Ser46 in the presence of preferred carbon sources (e.g. glucose) in *Bacillus subtilis*. The phosphorylated HPr subsequently binds to catabolite control protein A (CcpA), and the complex represses transcription of catabolic genes responsible for utilization of less preferred carbon sources. The phosphorylation of HPr is reversible and is catalyzed by the HPr kinase/phosphorylase (HPrK) system to reflect nutrient availability (*180*). In *Rhodococcus sp.*, putative cyclic AMP (cAMP) receptor protein (CRP) binding sites in the *pheR* promoter regions (involved in phenol catabolism) and the promoters of *thnA1* and *thnS* (involved in tetralin catabolism) were identified (*124, 181*). The role of cAMP-CRP complex has been elucidated: it binds and activates promoters of catabolic genes in *E. coli* and regulated by the phosphorylation status of the EIIA domain of the glucose transporter. This may serve as a reference to direct study of CCR mechanism in *Rhodococci* using phosphoproteomic approaches.

Thiol-Based Redox Modification

Several proteomics studies report the global identification of S-mycothiolated proteins in *Actinomycetes* under oxidative/hypochlorite stress. These proteins are involved in essential processes: energy metabolism, protein translation, antioxidant functions, as well as biosynthetic pathways of amino acid, nucleotide, and fatty acids (*156*, *182*, *183*). However, few studies of oleaginous bacteria have focused on redox regulation of lipogenesis. Total oxidation of proteins was identified in *R. jostii* RHA1 fed on gluconate when exposed to methyl viologen (MV), which promotes oxidative stress through superoxide generation. Proteins involved in glycolysis, gluconeogenesis, and fatty acid synthesis were oxidized to different extents and are perhaps prone to redox regulation (*159*). Reactive oxygen species (ROS) are important mediators of redox signaling and highlight the necessity to maintain redox balance (homeostasis) in cells. When ROS formation overcomes cellular antioxidant capacity, the resulting condition is called oxidative stress. High levels of cellular ROS are can cause damage to cellular components and may lead to cell death (*184*). The effects of oxidative stress caused by H_2O_2 and MV were also investigated in *R. opacus* PD630. The results of this study suggest a metabolic interaction between TAG biosynthesis and oxidative stress response: competition for the NADPH pool. H_2O_2 and MV had different oxidative effects, and the strain used different mechanisms for defense (*185*). A follow-up study showed up-regulation of proteins involved in lipolysis and β-oxidation of fatty acids. The authors discuss the significantly decreased TAG accumulation in the presence of MV at a proteome level. Moreover, they report an up-regulated mycothiol antioxidant system in addition to catalase and superoxide dismutase in response to oxidative stress as well, though no further profiling of S-mycothiolated proteins was conducted (*163*). On the other hand, it was proposed that low levels of ROS correlate with increased lipid accumulation, and that ROS signaling networks may be important for lipogenesis in many oleaginous microorganisms (*186–191*). However, direct evidence to the proposed underlying mechanisms and elucidation of specific signaling transduction pathways are largely lacking.

Oxidative stress caused by the toxicity of lignocellulose-derived phenolics and aldehydes in a variety of biorefinery microorganisms has been recognized (*192–195*). Though many lignin-degrading bacteria display excellent tolerance, adaption, and detoxification capacity against these compounds, their cellular redox homeostasis can still be impacted. It was reported that oxidative stress may occur during lignin fermentation as many antioxidants (e.g. catalases, members of the thioredoxin system) were observed unlike in glucose fermentation (*26*). Furthermore, the oxidative depolymerization process of lignin involving ligninolytic radicals (e.g. veratryl alcohol cation radical), hydroxyl radicals, quinones, H_2O_2, etc. is likely another potential source of increased ROS levels (*42*, *50*). A salient knowledge gap remains between redox homeostasis and its possible regulation of lipogenesis. This raises questions about the effect of lignin utilization on the cellular redox state of oleaginous bacteria. Mapping and quantitative analysis of thiol-modified proteins using functional proteomics assays will explicate the redox regulation network of lipogenesis during lignin depolymerization and/or lignin-derived aromatic catabolism. These results will potentially drive synthetic biology efforts to resolve the relatively low lipid yield from lignin as compared to sugars and other favorable carbon sources.

Conclusion

Oleaginous bacteria are potential cellular factories for the next generation of biofuel production from lignin. Unfortunately, such bioprocessing technologies have not yet been well established, at least partly because of the lack of fundamental knowledge about carbon, nitrogen, and redox metabolism. Important knowledge gaps remain in lignin depolymerization, aromatics catabolism, and lipid accumulation in oleaginous bacteria. Addressing these fundamental questions is required to achieve high lipid yields, titers, and rates which are vital for economic and carbon efficient biofuels production from biomass. Systems biology driven by advanced omics technologies like genomics, transcriptomics, proteomics, metabolomics, and lipidomics promises progress in bioconversion of biorefinery lignin waste. Here, we highlighted proteomics approaches and studies that were conducted to study lignin degradation, carbon and nitrogen metabolism, and lipid synthesis and storage in oleaginous bacteria. Further studies integrating these and other reductionist approaches are nonetheless important for developing synthetic biology approaches to address the issue of lignin waste valorization. We envision that the synthetic reconstruction and balanced modification of key regulators and enzymes in lignin depolymerization, aromatics catabolism, lipid biosynthesis, redox metabolism, and other relevant processes will enable bacteria to efficiently convert lignin to lipids.

Acknowledgments

We acknowledge the U.S. Department of Energy (DOE), the Office of Energy Efficiency & Renewable Energy (EERE) Awards (DE-EE0007104, DE-EE0006112, and DE-EE0008250), the USDA National Institute of Food and Agriculture, Hatch project 1014753, Hatch/Multi State project 1017904, and the Bioproducts, Science and Engineering Laboratory, Department of Biological Systems Engineering at Washington State University. A portion of the research was performed using EMSL (grid.436923.9), a DOE Office of Science user facility sponsored by the Biological and Environmental Research program. X.L. and Z, X. are grateful for support from the Pacific Northwest National Laboratory (PNNL)-Washington State University (WSU) Distinguished Graduate Research Program (DGRP) Fellowship.

References

1. Lee, R. A.; Lavoie, J.-M. From first- to third-generation biofuels: Challenges of producing a commodity from a biomass of increasing complexity. *Animal Frontiers* **2013**, 3 (2), 6–11.
2. Baruah, J.; Nath, B. K.; Sharma, R.; Kumar, S.; Deka, R. C.; Baruah, D. C.; Kalita, E. *Recent Trends in the Pretreatment of Lignocellulosic Biomass for Value-Added Products.* **2018**, *6* (141), 1–19.
3. Kurosawa, K.; Boccazzi, P.; de Almeida, N. M.; Sinskey, A. J. High-cell-density batch fermentation of Rhodococcus opacus PD630 using a high glucose concentration for triacylglycerol production. *J. Biotechnol.* **2010**, *147* (3), 212–218.
4. Wang, B.; Rezenom, Y. H.; Cho, K.-C.; Tran, J. L.; Lee, D. G.; Russell, D. H.; Gill, J. J.; Young, R.; Chu, K.-H. Cultivation of lipid-producing bacteria with lignocellulosic biomass: Effects of inhibitory compounds of lignocellulosic hydrolysates. *Bioresour. Technol.* **2014**, *161*, 162–170.
5. Senatham, S.; Chamduang, T.; Kaewchingduang, Y.; Thammasittirong, A.; Srisodsuk, M.; Elliston, A.; Roberts, I. N.; Waldron, K. W.; Thammasittirong, S. N.-R. Enhanced xylose

fermentation and hydrolysate inhibitor tolerance of Scheffersomyces shehatae. *Springerplus* **2016**, 5 (1), 1–8.

6. Yang, B.; Wyman, C. E. Pretreatment: the key to unlocking low-cost cellulosic ethanol. *Biofuels, Bioprod. Bioref.* **2008**, 2 (1), 26–40.
7. Humbird, D.; Davis, R.; Tao, L.; Kinchin, C.; Hsu, D.; Aden, A.; Schoen, P.; Lukas, J.; Olthof, B.; Worley, M. *Process design and economics for biochemical conversion of lignocellulosic biomass to ethanol: dilute-acid pretreatment and enzymatic hydrolysis of corn stover*; Golden, CO: National Renewable Energy Laboratory (NREL), 2011.
8. Alvarez, H.; Steinbüchel, A. Triacylglycerols in prokaryotic microorganisms. *Applied microbiology and biotechnology* **2002**, *60* (4), 367–376.
9. Gouda, M. K.; Omar, S. H.; Aouad, L. M. Single cell oil production by Gordonia sp. DG using agro-industrial wastes. *World Journal of Microbiology and Biotechnology* **2008**, *24* (9), 1703–1711.
10. Fixter, L. M.; Nagi, M. N.; Mccormack, J. G.; Fewson, C. A. Structure, Distribution and Function of Wax Esters in Acinetobacter calcoaceticus. *Microbiology* **1986**, *132* (11), 3147–3157.
11. Salcedo-Vite, K.; Sigala, J.-C.; Segura, D.; Gosset, G.; Martinez, A. Acinetobacter baylyi ADP1 growth performance and lipid accumulation on different carbon sources. *Applied Microbiology and Biotechnology* **2019**, *103* (15), 6217–6229.
12. Alvarez, H. M., Biotechnological Production and Significance of Triacylglycerols and Wax Esters. In *Handbook of Hydrocarbon and Lipid Microbiology*, Timmis, K. N., Ed. Springer Berlin Heidelberg: Berlin, Heidelberg, 2010; pp 2995-3002.
13. Madani, M.; Enshaeieh, M.; Abdoli, A. Single cell oil and its application for biodiesel production. *Process Safety and Environmental Protection* **2017**, *111*, 747–756.
14. Amara, S.; Seghezzi, N.; Otani, H.; Diaz-Salazar, C.; Liu, J.; Eltis, L. D. Characterization of key triacylglycerol biosynthesis processes in rhodococci. *Sci. Rep.* **2016**, *6*, 24985, 1–13.
15. Murphy, D. J. The biogenesis and functions of lipid bodies in animals, plants and microorganisms. *Progress in Lipid Research* **2001**, *40* (5), 325–438.
16. Yang, L.; Ding, Y.; Chen, Y.; Zhang, S.; Huo, C.; Wang, Y.; Yu, J.; Zhang, P.; Na, H.; Zhang, H.; Ma, Y.; Liu, P. The proteomics of lipid droplets: structure, dynamics, and functions of the organelle conserved from bacteria to humans. *Journal of Lipid Research* **2012**, 53 (7), 1245–1253.
17. Martin, S.; Parton, R. G. Lipid droplets: a unified view of a dynamic organelle. *Nature Reviews Molecular Cell Biology* **2006**, 7 (5), 373–378.
18. Wältermann, M.; Luftmann, H.; Baumeister, D.; Kalscheuer, R.; Steinbüchel, A. Rhodococcus opacus strain PD630 as a new source of high-value single-cell oil? Isolation and characterization of triacylglycerols and other storage lipids. *Microbiology* **2000**, *146* (5), 1143–1149.
19. Olukoshi, E. R.; Packter, N. M. Importance of stored triacylglycerols in Streptomyces: possible carbon source for antibiotics. *Microbiology* **1994**, *140* (4), 931–943.
20. Röttig, A.; Hauschild, P.; Madkour, M. H.; Al-Ansari, A. M.; Almakishah, N. H.; Steinbüchel, A. Analysis and optimization of triacylglycerol synthesis in novel oleaginous Rhodococcus and Streptomyces strains isolated from desert soil. *Journal of Biotechnology* **2016**, 225, 48–56.

21. Xiong, X.; Wang, X.; Chen, S. Engineering of a xylose metabolic pathway in Rhodococcus strains. *Appl. Environ. Microbiol.* **2012**, *78* (16), 5483–5491.

22. Xiong, X.; Wang, X.; Chen, S. Engineering of an L-arabinose metabolic pathway in Rhodococcus jostii RHA1 for biofuel production. *J. Ind. Microbiol. Biotechnol.* **2016**, *43* (7), 1017–1025.

23. Alvarez, H. M.; Herrero, O. M.; Silva, R. A.; Hernández, M. A.; Lanfranconi, M. P.; Villalba, M. S. Insights into the Metabolism of Oleaginous Rhodococcus spp. *Applied and environmental microbiology* **2019**, *85* (18), e00498-19, 1–12.

24. Hernández, M. A.; Alvarez, H. M. Increasing lipid production using an NADP+-dependent malic enzyme from Rhodococcus jostii. *Microbiology* **2019**, *165* (1), 4–14.

25. Angel, T. E.; Aryal, U. K.; Hengel, S. M.; Baker, E. S.; Kelly, R. T.; Robinson, E. W.; Smith, R. D. Mass spectrometry-based proteomics: existing capabilities and future directions. *Chem Soc Rev* **2012**, *41* (10), 3912–3928.

26. Li, X.; He, Y.; Zhang, L.; Xu, Z.; Ben, H.; Gaffrey, M. J.; Yang, Y.; Yang, S.; Yuan, J. S.; Qian, W.-J.; Yang, B. Discovery of potential pathways for biological conversion of poplar wood into lipids by co-fermentation of Rhodococci strains. *Biotechnol. Biofuels* **2019**, *12* (1), 60, 1–16.

27. Xie, S.; Sun, S.; Lin, F.; Li, M.; Pu, Y.; Cheng, Y.; Xu, B.; Liu, Z.; da Costa Sousa, L.; Dale, B. E.; Ragauskas, A. J.; Dai, S. Y.; Yuan, J. S. Mechanism-Guided Design of Highly Efficient Protein Secretion and Lipid Conversion for Biomanufacturing and Biorefining. *Advanced Science* **2019**, *6* (13), 1801980, 1–11.

28. de Gonzalo, G.; Colpa, D. I.; Habib, M. H.; Fraaije, M. W. Bacterial enzymes involved in lignin degradation. *Journal of Biotechnology* **2016**, *236*, 110–119.

29. Laskar, D. D.; Yang, B.; Wang, H.; Lee, J. Pathways for biomass-derived lignin to hydrocarbon fuels. *Biofuels, Bioproducts and Biorefining* **2013**, *7* (5), 602–626.

30. Liu, Z.-H.; Le, R. K.; Kosa, M.; Yang, B.; Yuan, J.; Ragauskas, A. J. Identifying and creating pathways to improve biological lignin valorization. *Renewable and Sustainable Energy Reviews* **2019**, *105*, 349–362.

31. Ralph, J.; Lundquist, K.; Brunow, G.; Lu, F.; Kim, H.; Schatz, P. F.; Marita, J. M.; Hatfield, R. D.; Ralph, S. A.; Christensen, J. H. Lignins: natural polymers from oxidative coupling of 4-hydroxyphenyl-propanoids. *Phytochemistry Reviews* **2004**, *3* (1-2), 29–60.

32. Ralph, J.; Akiyama, T.; Kim, H.; Lu, F.; Schatz, P. F.; Marita, J. M.; Ralph, S. A.; Reddy, M. S.; Chen, F.; Dixon, R. A. Effects of coumarate 3-hydroxylase down-regulation on lignin structure. *Journal of Biological Chemistry* **2006**, *281* (13), 8843–8853.

33. Brown, M. E.; Chang, M. C. Exploring bacterial lignin degradation. *Current opinion in chemical biology* **2014**, *19*, 1–7.

34. Wang, H.; Pu, Y.; Ragauskas, A.; Yang, B. From lignin to valuable products–strategies, challenges, and prospects. *Bioresource Technology* **2019**, *271*, 449–461.

35. Schutyser, W.; Renders, T.; Van den Bosch, S.; Koelewijn, S. F.; Beckham, G. T.; Sels, B. F. Chemicals from lignin: an interplay of lignocellulose fractionation, depolymerisation, and upgrading. *Chemical Society Reviews* **2018**, *47* (3), 852–908.

36. Watkins, D.; Nuruddin, M.; Hosur, M.; Tcherbi-Narteh, A.; Jeelani, S. Extraction and characterization of lignin from different biomass resources. *Journal of Materials Research and Technology* **2015**, *4* (1), 26–32.

37. Xu, R.; Zhang, K.; Liu, P.; Han, H.; Zhao, S.; Kakade, A.; Khan, A.; Du, D.; Li, X. Lignin depolymerization and utilization by bacteria. *Bioresource Technology* **2018**, *269*, 557–566.
38. Salvachua, D.; Karp, E. M.; Nimlos, C. T.; Vardon, D. R.; Beckham, G. T. Towards lignin consolidated bioprocessing: simultaneous lignin depolymerization and product generation by bacteria. *Green Chemistry* **2015**, *17* (11), 4951–4967.
39. Kamimura, N.; Sakamoto, S.; Mitsuda, N.; Masai, E.; Kajita, S. Advances in microbial lignin degradation and its applications. *Current Opinion in Biotechnology* **2019**, *56*, 179–186.
40. Naofumi, K.; Kenji, T.; Kosuke, M.; Takuma, A.; Masaya, F.; Yudai, H.; Eiji, M. Bacterial catabolism of lignin-derived aromatics: New findings in a recent decade: Update on bacterial lignin catabolism. *Environmental Microbiology Reports* **2017**, *9* (6), 679–705.
41. Xu, Z.; Lei, P.; Zhai, R.; Wen, Z.; Jin, M. Recent advances in lignin valorization with bacterial cultures: microorganisms, metabolic pathways, and bio-products. *Biotechnology for Biofuels* **2019**, *12* (1), 32, 1–19.
42. Bugg, T. D.; Ahmad, M.; Hardiman, E. M.; Rahmanpour, R. Pathways for degradation of lignin in bacteria and fungi. *Nat Prod Rep* **2011**, *28* (12), 1883–1896.
43. Zhu, D.; Liang, N.; Zhang, R.; Ahmad, F.; Zhang, W.; Yang, B.; Wu, J.; Geng, A.; Gabriel, M.; Sun, J. Insight into Depolymerization Mechanism of Bacterial Laccase for Lignin. *ACS Sustainable Chem. Eng.* **2020**, *8* (34), 12920–12933.
44. Zhu, D.; Si, H.; Zhang, P.; Geng, A.; Zhang, W.; Yang, B.; Qian, W.-j.; Gabriel, M.; Sun, J. Genomics and biochemistry investigation on the metabolic pathway of milled wood and alkali lignin-derived aromatic metabolites of Comamonas serinivorans SP-35. *Biotechnology for Biofuels* **2018**, *11*, 338, 1–15.
45. Zhu, D.; Zhang, P.; Xie, C.; Sun, J.; Zhu, D.; Zhang, W.; Qian, W.-J.; Yang, B. Biodegradation of alkaline lignin by Bacillus ligniniphilus L1. *Biotechnology for biofuels* **2017**, *10*, 44, 1–14.
46. Bugg, T. D. H.; Ahmad, M.; Hardiman, E. M.; Singh, R. The emerging role for bacteria in lignin degradation and bio-product formation. *Current Opinion in Biotechnology* **2011**, *22* (3), 394–400.
47. Nelsen, M. P.; DiMichele, W. A.; Peters, S. E.; Boyce, C. K. Delayed fungal evolution did not cause the Paleozoic peak in coal production. *Proceedings of the National Academy of Sciences* **2016**, *113* (9), 2442–2447.
48. Abdelaziz, O. Y.; Brink, D. P.; Prothmann, J.; Ravi, K.; Sun, M.; García-Hidalgo, J.; Sandahl, M.; Hulteberg, C. P.; Turner, C.; Lidén, G. Biological valorization of low molecular weight lignin. *Biotechnology Advances* **2016**, *34* (8), 1318–1346.
49. Sánchez, C. Lignocellulosic residues: biodegradation and bioconversion by fungi. *Biotechnology advances* **2009**, *27* (2), 185–194.
50. Hunt, C. G.; Houtman, C. J.; Jones, D. C.; Kitin, P.; Korripally, P.; Hammel, K. E. Spatial mapping of extracellular oxidant production by a white rot basidiomycete on wood reveals details of ligninolytic mechanism. *Environmental microbiology* **2013**, *15* (3), 956–966.
51. Lambertz, C.; Ece, S.; Fischer, R.; Commandeur, U. Progress and obstacles in the production and application of recombinant lignin-degrading peroxidases. *Bioengineered* **2016**, *7* (3), 145–154.

52. Ayyachamy, M.; Cliffe, F. E.; Coyne, J. M.; Collier, J.; Tuohy, M. G. Lignin: untapped biopolymers in biomass conversion technologies. *Biomass Conversion and Biorefinery* **2013**, *3* (3), 255–269.
53. Goyal, S.; Ramawat, K. G.; Mérillon, J.-M., Different Shades of Fungal Metabolites: An Overview. In *Fungal Metabolites*, Mérillon, J.-M.; Ramawat, K. G., Eds. Springer International Publishing: Cham, 2017; pp 1-29.
54. Plácido, J.; Capareda, S. Ligninolytic enzymes: a biotechnological alternative for bioethanol production. *Bioresources and Bioprocessing* **2015**, 2 (1), 23, 1–12.
55. Kumar, A.; Chandra, R. Ligninolytic enzymes and its mechanisms for degradation of lignocellulosic waste in environment. *Heliyon* **2020**, *6* (2), e03170, 1–18.
56. Billings, A. F.; Fortney, J. L.; Hazen, T. C.; Simmons, B.; Davenport, K. W.; Goodwin, L.; Ivanova, N.; Kyrpides, N. C.; Mavromatis, K.; Woyke, T.; DeAngelis, K. M. Genome sequence and description of the anaerobic lignin-degrading bacterium Tolumonas lignolytica sp. nov. *Stand Genomic Sci* **2015**, *10*, 106, 1–11.
57. Oliveira, P. L. d.; Duarte, M. C. T.; Ponezi, A. N.; Durrant, L. R. Purification and Partial characterization of manganese peroxidase from Bacillus pumilus AND Paenibacillus sp. *Brazilian Journal of Microbiology* **2009**, *40* (4), 818–826.
58. Ramachandra, M.; Crawford, D. L.; Hertel, G. Characterization of an extracellular lignin peroxidase of the lignocellulolytic actinomycete Streptomyces viridosporus. *Applied and Environmental Microbiology* **1988**, *54* (12), 3057–3063.
59. Liers, C.; Aranda, E.; Strittmatter, E.; Piontek, K.; Plattner, D. A.; Zorn, H.; Ullrich, R.; Hofrichter, M. Phenol oxidation by DyP-type peroxidases in comparison to fungal and plant peroxidases. *Journal of Molecular Catalysis B: Enzymatic* **2014**, *103*, 41–46.
60. Ihssen, J.; Reiss, R.; Luchsinger, R.; Thöny-Meyer, L.; Richter, M. Biochemical properties and yields of diverse bacterial laccase-like multicopper oxidases expressed in Escherichia coli. *Scientific reports* **2015**, *5*, 10465, 1–13.
61. Picart, P.; de María, P. D.; Schallmey, A. From gene to biorefinery: microbial β-etherases as promising biocatalysts for lignin valorization. *Frontiers in microbiology* **2015**, *6*, 916, 1–8.
62. Rashid, G. M.; Taylor, C. R.; Liu, Y.; Zhang, X.; Rea, D.; Fülöp, V.; Bugg, T. D. Identification of manganese superoxide dismutase from Sphingobacterium sp. T2 as a novel bacterial enzyme for lignin oxidation. *ACS chemical biology* **2015**, *10* (10), 2286–2294.
63. Brown, M. E.; Walker, M. C.; Nakashige, T. G.; Iavarone, A. T.; Chang, M. C. Discovery and characterization of heme enzymes from unsequenced bacteria: application to microbial lignin degradation. *Journal of the American Chemical Society* **2011**, *133* (45), 18006–18009.
64. Lin, L.; Cheng, Y.; Pu, Y.; Sun, S.; Li, X.; Jin, M.; Pierson, E. A.; Gross, D. C.; Dale, B. E.; Dai, S. Y.; Ragauskas, A. J.; Yuan, J. S. Systems biology-guided biodesign of consolidated lignin conversion. *Green Chem.* **2016**, *18* (20), 5536–5547.
65. Zhang, K.; Xu, R.; Abomohra, A. E.-F.; Xie, S.; Yu, Z.; Guo, Q.; Liu, P.; Peng, L.; Li, X. A sustainable approach for efficient conversion of lignin into biodiesel accompanied by biological pretreatment of corn straw. *Energy Conversion and Management* **2019**, *199*, 111928, 1–10.
66. Takahashi, K.; Kamimura, N.; Hishiyama, S.; Hara, H.; Kasai, D.; Katayama, Y.; Fukuda, M.; Kajita, S.; Masai, E. Characterization of the catabolic pathway for a phenylcoumaran-type lignin-derived biaryl in Sphingobium sp. strain SYK-6. *Biodegradation* **2014**, *25* (5), 735–745.

67. Takahashi, K.; Hirose, Y.; Kamimura, N.; Hishiyama, S.; Hara, H.; Araki, T.; Kasai, D.; Kajita, S.; Katayama, Y.; Fukuda, M.; Masai, E. Membrane-Associated Glucose-Methanol-Choline Oxidoreductase Family Enzymes PhcC and PhcD Are Essential for Enantioselective Catabolism of Dehydrodiconiferyl Alcohol. *Applied and Environmental Microbiology* **2015**, *81* (23), 8022–8036.
68. Zakzeski, J.; Bruijnincx, P. C. A.; Jongerius, A. L.; Weckhuysen, B. M. The Catalytic Valorization of Lignin for the Production of Renewable Chemicals. *Chem Rev* **2010**, *110* (6), 3552–3599.
69. Sato, Y.; Moriuchi, H.; Hishiyama, S.; Otsuka, Y.; Oshima, K.; Kasai, D.; Nakamura, M.; Ohara, S.; Katayama, Y.; Fukuda, M.; Masai, E. Identification of Three Alcohol Dehydrogenase Genes Involved in the Stereospecific Catabolism of Arylglycerol-β-Aryl Ether by Sphingobium sp. Strain SYK-6. *Applied and Environmental Microbiology* **2009**, *75* (16), 5195–5201.
70. Tanamura, K.; Abe, T.; Kamimura, N.; Kasai, D.; Hishiyama, S.; Otsuka, Y.; Nakamura, M.; Kajita, S.; Katayama, Y.; Fukuda, M.; Masai, E. Characterization of the third glutathione S-transferase gene involved in enantioselective cleavage of the beta-aryl ether by Sphingobium sp. strain SYK-6. *Biosci Biotechnol Biochem* **2011**, *75* (12), 2404–2407.
71. Yoshikata, T.; Suzuki, K.; Kamimura, N.; Namiki, M.; Hishiyama, S.; Araki, T.; Kasai, D.; Otsuka, Y.; Nakamura, M.; Fukuda, M.; Katayama, Y.; Masai, E. Three-Component Demethylase System Essential for Catabolism of a Lignin-Derived Biphenyl Compound in Sphingobium sp. Strain SYK-6. *Applied and Environmental Microbiology* **2014**, *80* (23), 7142–7153.
72. Masai, E.; Yamamoto, Y.; Inoue, T.; Takamura, K.; Hara, H.; Kasai, D.; Katayama, Y.; Fukuda, M. Characterization of ligV essential for catabolism of vanillin by Sphingomonas paucimobilis SYK-6. *Biosci Biotechnol Biochem* **2007**, *71* (10), 2487–2492.
73. Wang, W.; Yang, S.; Hunsinger, G. B.; Pienkos, P. T.; Johnson, D. K. Connecting lignin-degradation pathway with pre-treatment inhibitor sensitivity of Cupriavidus necator. *Frontiers in microbiology* **2014**, *5*, 247, 1–10.
74. Yam, K. C.; Okamoto, S.; Roberts, J. N.; Eltis, L. D. Adventures in Rhodococcus: From steroids to explosives. *Can. J. Microbiol.* **2011**, *57* (3), 155–168.
75. Salvachua, D.; Katahira, R.; Cleveland, N. S.; Khanna, P.; Resch, M. G.; Black, B. A.; Purvine, S. O.; Zink, E. M.; Prieto, A.; Martinez, M. J.; Martinez, A. T.; Simmons, B. A.; Gladden, J. M.; Beckham, G. T. Lignin depolymerization by fungal secretomes and a microbial sink. *Green Chemistry* **2016**, *18* (22), 6046–6062.
76. Varman, A. M.; He, L.; Follenfant, R.; Wu, W.; Wemmer, S.; Wrobel, S. A.; Tang, Y. J.; Singh, S. Decoding how a soil bacterium extracts building blocks and metabolic energy from ligninolysis provides road map for lignin valorization. *Proc Natl Acad Sci U S A* **2016**, *113* (40), 5802–5811.
77. Linger, J. G.; Vardon, D. R.; Guarnieri, M. T.; Karp, E. M.; Hunsinger, G. B.; Franden, M. A.; Johnson, C. W.; Chupka, G.; Strathmann, T. J.; Pienkos, P. T. Lignin valorization through integrated biological funneling and chemical catalysis. *Proceedings of the National Academy of Sciences* **2014**, *111* (33), 12013–12018.
78. Vaillancourt, F. H.; Bolin, J. T.; Eltis, L. D. The Ins and Outs of Ring-Cleaving Dioxygenases. *Critical Reviews in Biochemistry and Molecular Biology* **2006**, *41* (4), 241–267.

79. Alvarez, H. M., *Biology of Rhodococcus*. Springer Science & Business Media: 2010; Vol. 16.
80. Chen, H.-P.; Chow, M.; Liu, C.-C.; Lau, A.; Liu, J.; Eltis, L. D. Vanillin catabolism in Rhodococcus jostii RHA1. *Appl. Environ. Microbiol.* **2012**, *78* (2), 586–588.
81. Jiménez, J. I.; Miñambres, B.; García, J. L.; Díaz, E. Genomic analysis of the aromatic catabolic pathways from Pseudomonas putida KT2440. *Environmental Microbiology* **2002**, *4* (12), 824–841.
82. Harwood, C. S.; Parales, R. E. The β-ketoadipate pathway and the biology of self-identity. *Annu. Rev. Microbiol.* **1996**, *50* (1), 553–590.
83. Chang, H. K.; Zylstra, G. J. Novel organization of the genes for phthalate degradation from Burkholderia cepacia DBO1. *Journal of bacteriology* **1998**, *180* (24), 6529–6537.
84. Masai, E.; Katayama, Y.; Fukuda, M. Genetic and biochemical investigations on bacterial catabolic pathways for lignin-derived aromatic compounds. *Bioscience, biotechnology, and biochemistry* **2007**, *71* (1), 1–15.
85. Navarro-Llorens, J. M.; Patrauchan, M. A.; Stewart, G. R.; Davies, J. E.; Eltis, L. D.; Mohn, W. W. Phenylacetate catabolism in Rhodococcus sp. strain RHA1: a central pathway for degradation of aromatic compounds. *Journal of bacteriology* **2005**, *187* (13), 4497–4504.
86. He, Y.; Li, X.; Ben, H.; Xue, X.; Yang, B. Lipid Production from Dilute Alkali Corn Stover Lignin by Rhodococcus Strains. *ACS Sustainable Chem. Eng.* **2017**, *5* (3), 2302–2311.
87. He, Y.; Li, X.; Xue, X.; Swita, M. S.; Schmidt, A. J.; Yang, B. Biological conversion of the aqueous wastes from hydrothermal liquefaction of algae and pine wood by Rhodococci. *Bioresour. Technol.* **2017**, *224*, 457–464.
88. Alvarez, H. M.; Kalscheuer, R.; Steinbüchel, A. Accumulation of storage lipids in species of Rhodococcus and Nocardia and effect of inhibitors and polyethylene glycol. *Lipid/Fett* **1997**, *99* (7), 239–246.
89. Ahmad, M.; Taylor, C. R.; Pink, D.; Burton, K.; Eastwood, D.; Bending, G. D.; Bugg, T. D. Development of novel assays for lignin degradation: comparative analysis of bacterial and fungal lignin degraders. *Molecular Biosystems* **2010**, *6* (5), 815–821.
90. Taylor, C. R.; Hardiman, E.; Ahmad, M.; Sainsbury, P.; Norris, P.; Bugg, T. D. Isolation of bacterial strains able to metabolize lignin from screening of environmental samples. *Journal of applied microbiology* **2012**, *113* (3), 521–530.
91. Song, Y.-J. Characterization of aromatic hydrocarbon degrading bacteria isolated from pine litter. *Korean J Microbiol Biotechnol* **2009**, *37* (4), 333–339.
92. Shields-Menard, S. A.; AmirSadeghi, M.; Green, M.; Womack, E.; Sparks, D. L.; Blake, J.; Edelmann, M.; Ding, X.; Sukhbaatar, B.; Hernandez, R.; Donaldson, J. R.; French, T. The effects of model aromatic lignin compounds on growth and lipid accumulation of Rhodococcus rhodochrous. *International Biodeterioration & Biodegradation* **2017**, *121*, 79–90.
93. Kosa, M.; Ragauskas, A. J. Lignin to lipid bioconversion by oleaginous Rhodococci. *Green Chem.* **2013**, *15* (8), 2070–2074.
94. Le, R. K.; Das, P.; Mahan, K. M.; Anderson, S. A.; Wells, T.; Yuan, J. S.; Ragauskas, A. J. Utilization of simultaneous saccharification and fermentation residues as feedstock for lipid accumulation in Rhodococcus opacus. *AMB Express* **2017**, 7, 185, 1–10.

95. Wei, Z.; Zeng, G.; Huang, F.; Kosa, M.; Huang, D.; Ragauskas, A. J. Bioconversion of oxygen-pretreated Kraft lignin to microbial lipid with oleaginous Rhodococcus opacus DSM 1069. *Green Chemistry* **2015**, *17* (5), 2784–2789.
96. Ahmad, M.; Roberts, J. N.; Hardiman, E. M.; Singh, R.; Eltis, L. D.; Bugg, T. D. H. Identification of DypB from Rhodococcus jostii RHA1 as a lignin peroxidase. *Biochemistry* **2011**, *50* (23), 5096–5107.
97. Sutter, M.; Boehringer, D.; Gutmann, S.; Günther, S.; Prangishvili, D.; Loessner, M. J.; Stetter, K. O.; Weber-Ban, E.; Ban, N. Structural basis of enzyme encapsulation into a bacterial nanocompartment. *Nature structural & molecular biology* **2008**, *15* (9), 939–947.
98. Rahmanpour, R.; Bugg, T. D. H. Assembly in vitro of Rhodococcus jostii RHA1 encapsulin and peroxidase DypB to form a nanocompartment. *FEBS J.* **2013**, *280* (9), 2097–2104.
99. Contreras, H.; Joens, M. S.; McMath, L. M.; Le, V. P.; Tullius, M. V.; Kimmey, J. M.; Bionghi, N.; Horwitz, M. A.; Fitzpatrick, J. A.; Goulding, C. W. Characterization of a Mycobacterium tuberculosis nanocompartment and its potential cargo proteins. *Journal of Biological Chemistry* **2014**, *289* (26), 18279–18289.
100. Ausec, L.; Zakrzewski, M.; Goesmann, A.; Schlüter, A.; Mandic-Mulec, I. Bioinformatic analysis reveals high diversity of bacterial genes for laccase-like enzymes. *PLoS One* **2011**, *6* (10), e25724, 1–9.
101. Zhao, C.; Xie, S.; Pu, Y.; Zhang, R.; Huang, F.; Ragauskas, A. J.; Yuan, J. S. Synergistic enzymatic and microbial lignin conversion. *Green Chemistry* **2016**, *18* (5), 1306–1312.
102. Chong, G.-G.; Huang, X.-J.; Di, J.-H.; Xu, D.-Z.; He, Y.-C.; Pei, Y.-N.; Tang, Y.-J.; Ma, C.-L. Biodegradation of alkali lignin by a newly isolated Rhodococcus pyridinivorans CCZU-B16. *Bioprocess and Biosystems Engineering* **2018**, *41* (4), 501–510.
103. Wang, Z.; Li, N.; Pan, X. Transformation of Ammonia Fiber Expansion (AFEX) corn stover lignin into microbial lipids by Rhodococcus opacus. *Fuel* **2019**, *240*, 119–125.
104. Majumdar, S.; Lukk, T.; Solbiati, J. O.; Bauer, S.; Nair, S. K.; Cronan, J. E.; Gerlt, J. A. Roles of Small Laccases from Streptomyces in Lignin Degradation. *Biochemistry* **2014**, *53* (24), 4047–4058.
105. Shim, M.-s.; Kim, W.-s.; Kim, J.-h. Neutral lipids and lipase activity for actinorhodin biosynthesis of Streptomyces coelicolor A3 (2). *Biotechnology letters* **1997**, *19* (3), 221–224.
106. Wang, Z.; Bleakley, B. H.; Crawford, D. L.; Hertel, G.; Rafii, F. Cloning and expression of a lignin peroxidase gene from Streptomyces viridosporus in Streptomyces lividans. *Journal of biotechnology* **1990**, *13* (2-3), 131–144.
107. Luo, J.; Efimova, E.; Losoi, P.; Santala, V.; Santala, S. Wax ester production in nitrogen-rich conditions by metabolically engineered Acinetobacter baylyi ADP1. *Metabolic Engineering Communications* **2020**, *10*, e00128, 1–11.
108. Salmela, M.; Lehtinen, T.; Efimova, E.; Santala, S.; Santala, V. Alkane and wax ester production from lignin-related aromatic compounds. *Biotechnology and Bioengineering* **2019**, *116* (8), 1934–1945.
109. Salmela, M.; Lehtinen, T.; Efimova, E.; Santala, S.; Santala, V. Towards bioproduction of poly-α-olefins from lignocellulose. *Green Chemistry* **2020**, *22* (15), 5067–5076.
110. Fischer, R.; Bleichrodt, F. S.; Gerischer, U. C. Aromatic degradative pathways in Acinetobacter baylyi underlie carbon catabolite repression. *Microbiology* **2008**, *154* (10), 3095–3103.

111. Prins, A.; Kleinsmidt, L.; Khan, N.; Kirby, B.; Kudanga, T.; Vollmer, J.; Pleiss, J.; Burton, S.; Le Roes-Hill, M. The effect of mutations near the T1 copper site on the biochemical characteristics of the small laccase from Streptomyces coelicolor A3 (2). *Enzyme and microbial technology* **2015**, *68*, 23–32.
112. Yadav, D.; Ranjan, B.; Mchunu, N.; Le Roes-Hill, M.; Kudanga, T. Secretory expression of recombinant small laccase from Streptomyces coelicolor A3 (2) in Pichia pastoris. *International journal of biological macromolecules* **2018**, *108*, 642–649.
113. Endo, K.; Hayashi, Y.; Hibi, T.; Hosono, K.; Beppu, T.; Ueda, K. Enzymological Characterization of EpoA, a Laccase-Like Phenol Oxidase Produced by Streptomyces griseus. *The Journal of Biochemistry* **2003**, *133* (5), 671–677.
114. Dwivedi, U. N.; Singh, P.; Pandey, V. P.; Kumar, A. Structure–function relationship among bacterial, fungal and plant laccases. *Journal of Molecular Catalysis B: Enzymatic* **2011**, *68* (2), 117–128.
115. Yee, D. C.; Wood, T. K. Elicitation of lignin peroxidase in Streptomyces lividans. *Applied Biochemistry and Biotechnology* **1996**, *60* (2), 139–149.
116. Kashiwagi, N.; Ogino, C.; Kondo, A. Production of chemicals and proteins using biomass-derived substrates from a Streptomyces host. *Bioresource Technology* **2017**, *245*, 1655–1663.
117. Brown, M. E.; Barros, T.; Chang, M. C. Y. Identification and Characterization of a Multifunctional Dye Peroxidase from a Lignin-Reactive Bacterium. *ACS Chemical Biology* **2012**, 7 (12), 2074-2081.
118. Kosa, M.; Ragauskas, A. J. Bioconversion of lignin model compounds with oleaginous Rhodococci. *Appl. Microbiol. Biotechnol.* **2012**, *93* (2), 891–900.
119. Ravi, K.; Abdelaziz, O. Y.; Nöbel, M.; García-Hidalgo, J.; Gorwa-Grauslund, M. F.; Hulteberg, C. P.; Lidén, G. Bacterial conversion of depolymerized Kraft lignin. *Biotechnology for Biofuels* **2019**, *12* (1), 56, 1–14.
120. Xie, S.; Sun, Q.; Pu, Y.; Lin, F.; Sun, S.; Wang, X.; Ragauskas, A. J.; Yuan, J. S. Advanced chemical design for efficient lignin bioconversion. *ACS Sustainable Chemistry & Engineering* **2017**, *5* (3), 2215–2223.
121. Liu, Z.-H.; Xie, S.; Lin, F.; Jin, M.; Yuan, J. S. Combinatorial pretreatment and fermentation optimization enabled a record yield on lignin bioconversion. *Biotechnology for Biofuels* **2018**, *11* (1), 21, 1–20.
122. Vardon, D. R.; Franden, M. A.; Johnson, C. W.; Karp, E. M.; Guarnieri, M. T.; Linger, J. G.; Salm, M. J.; Strathmann, T. J.; Beckham, G. T. Adipic acid production from lignin. *Energy Environ. Sci.* **2015**, *8* (2), 617–628.
123. Görke, B.; Stülke, J. Carbon catabolite repression in bacteria: many ways to make the most out of nutrients. *Nature Reviews Microbiology* **2008**, *6* (8), 613–624.
124. Szőköl, J.; Rucká, L.; Šimčíková, M.; Halada, P.; Nešvera, J.; Pátek, M. Induction and carbon catabolite repression of phenol degradation genes in Rhodococcus erythropolis and Rhodococcus jostii. *Applied Microbiology and Biotechnology* **2014**, *98* (19), 8267–8279.
125. Bleichrodt, F. S.; Fischer, R.; Gerischer, U. C. The β-ketoadipate pathway of Acinetobacter baylyi undergoes carbon catabolite repression, cross-regulation and vertical regulation, and is affected by Crc. *Microbiology* **2010**, *156* (5), 1313–1322.

126. Johnson, C. W.; Abraham, P. E.; Linger, J. G.; Khanna, P.; Hettich, R. L.; Beckham, G. T. Eliminating a global regulator of carbon catabolite repression enhances the conversion of aromatic lignin monomers to muconate in Pseudomonas putida KT2440. *Metabolic engineering communications* **2017**, *5*, 19–25.

127. Yoneda, A.; Henson, W. R.; Goldner, N. K.; Park, K. J.; Forsberg, K. J.; Kim, S. J.; Pesesky, M. W.; Foston, M.; Dantas, G.; Moon, T. S. Comparative transcriptomics elucidates adaptive phenol tolerance and utilization in lipid-accumulating Rhodococcus opacus PD630. *Nucleic acids research* **2016**, *44* (5), 2240–2254.

128. Hara, H.; Onishi, Y.; Fukuda, M. Regulatory network analysis of actinomycetes based on genome information. *Kagaku to Seibutsu* **2011**, *49* (4), 234–242.

129. Tajparast, M.; Frigon, D. Genome-scale metabolic model of Rhodococcus jostii RHA1 (iMT1174) to study the accumulation of storage compounds during nitrogen-limited condition. *BMC Syst. Biol.* **2015**, *9*, 1–14.

130. Donnelly, D. P.; Rawlins, C. M.; DeHart, C. J.; Fornelli, L.; Schachner, L. F.; Lin, Z.; Lippens, J. L.; Aluri, K. C.; Sarin, R.; Chen, B.; Lantz, C.; Jung, W.; Johnson, K. R.; Koller, A.; Wolff, J. J.; Campuzano, I. D. G.; Auclair, J. R.; Ivanov, A. R.; Whitelegge, J. P.; Paša-Tolić, L.; Chamot-Rooke, J.; Danis, P. O.; Smith, L. M.; Tsybin, Y. O.; Loo, J. A.; Ge, Y.; Kelleher, N. L.; Agar, J. N. Best practices and benchmarks for intact protein analysis for top-down mass spectrometry. *Nature Methods* **2019**, *16* (7), 587–594.

131. Chen, B.; Brown, K. A.; Lin, Z.; Ge, Y. Top-Down Proteomics: Ready for Prime Time? *Analytical Chemistry* **2018**, *90* (1), 110–127.

132. Välikangas, T.; Suomi, T.; Elo, L. L. A comprehensive evaluation of popular proteomics software workflows for label-free proteome quantification and imputation. *Brief Bioinform* **2018**, *19* (6), 1344–1355.

133. Tyanova, S.; Temu, T.; Cox, J. The MaxQuant computational platform for mass spectrometry-based shotgun proteomics. *Nature Protocols* **2016**, *11* (12), 2301–2319.

134. Al Shweiki, M. H. D. R.; Mönchgesang, S.; Majovsky, P.; Thieme, D.; Trutschel, D.; Hoehenwarter, W. Assessment of Label-Free Quantification in Discovery Proteomics and Impact of Technological Factors and Natural Variability of Protein Abundance. *J Proteome Res* **2017**, *16* (4), 1410–1424.

135. Röst, H. L.; Sachsenberg, T.; Aiche, S.; Bielow, C.; Weisser, H.; Aicheler, F.; Andreotti, S.; Ehrlich, H.-C.; Gutenbrunner, P.; Kenar, E.; Liang, X.; Nahnsen, S.; Nilse, L.; Pfeuffer, J.; Rosenberger, G.; Rurik, M.; Schmitt, U.; Veit, J.; Walzer, M.; Wojnar, D.; Wolski, W. E.; Schilling, O.; Choudhary, J. S.; Malmström, L.; Aebersold, R.; Reinert, K.; Kohlbacher, O. OpenMS: a flexible open-source software platform for mass spectrometry data analysis. *Nature Methods* **2016**, *13* (9), 741–748.

136. Webb-Robertson, B.-J. M.; Wiberg, H. K.; Matzke, M. M.; Brown, J. N.; Wang, J.; McDermott, J. E.; Smith, R. D.; Rodland, K. D.; Metz, T. O.; Pounds, J. G.; Waters, K. M. Review, Evaluation, and Discussion of the Challenges of Missing Value Imputation for Mass Spectrometry-Based Label-Free Global Proteomics. *J Proteome Res* **2015**, *14* (5), 1993–2001.

137. Pappireddi, N.; Martin, L.; Wühr, M. A Review on Quantitative Multiplexed Proteomics. *ChemBioChem* **2019**, *20* (10), 1210–1224.

138. Cox, J.; Hein, M. Y.; Luber, C. A.; Paron, I.; Nagaraj, N.; Mann, M. Accurate proteome-wide label-free quantification by delayed normalization and maximal peptide ratio extraction, termed MaxLFQ. *Mol Cell Proteomics* **2014**, *13* (9), 2513–2526.

139. Ong, S.-E.; Blagoev, B.; Kratchmarova, I.; Kristensen, D. B.; Steen, H.; Pandey, A.; Mann, M. Stable isotope labeling by amino acids in cell culture, SILAC, as a simple and accurate approach to expression proteomics. *Molecular & cellular proteomics* **2002**, *1* (5), 376–386.

140. Ting, L.; Rad, R.; Gygi, S. P.; Haas, W. MS3 eliminates ratio distortion in isobaric multiplexed quantitative proteomics. *Nature Methods* **2011**, *8* (11), 937–940.

141. Wenger, C. D.; Lee, M. V.; Hebert, A. S.; McAlister, G. C.; Phanstiel, D. H.; Westphall, M. S.; Coon, J. J. Gas-phase purification enables accurate, multiplexed proteome quantification with isobaric tagging. *Nature Methods* **2011**, *8* (11), 933–935.

142. Macek, B.; Forchhammer, K.; Hardouin, J.; Weber-Ban, E.; Grangeasse, C.; Mijakovic, I. Protein post-translational modifications in bacteria. *Nature Reviews Microbiology* **2019**, *17* (11), 651–664.

143. Duan, J.; Gaffrey, M. J.; Qian, W.-J. Quantitative proteomic characterization of redox-dependent post-translational modifications on protein cysteines. *Molecular BioSystems* **2017**, *13* (5), 816–829.

144. Hattori, T.; Koide, S. Next-generation antibodies for post-translational modifications. *Current Opinion in Structural Biology* **2018**, *51*, 141–148.

145. Guo, J.; Gaffrey, M. J.; Su, D.; Liu, T.; Camp, D. G.; Smith, R. D.; Qian, W.-J. Resin-assisted enrichment of thiols as a general strategy for proteomic profiling of cysteine-based reversible modifications. *Nature Protocols* **2014**, *9* (1), 64–75.

146. Mijakovic, I.; Grangeasse, C.; Turgay, K. Exploring the diversity of protein modifications: special bacterial phosphorylation systems. *FEMS Microbiology Reviews* **2016**, *40* (3), 398–417.

147. Yagüe, P.; Gonzalez-Quiñonez, N.; Fernánez-García, G.; Alonso-Fernández, S.; Manteca, A. Goals and Challenges in Bacterial Phosphoproteomics. *Int J Mol Sci* **2019**, *20* (22), 5678, 1–14.

148. Ravichandran, A.; Sugiyama, N.; Tomita, M.; Swarup, S.; Ishihama, Y. Ser/Thr/Tyr phosphoproteome analysis of pathogenic and non-pathogenic Pseudomonas species. *PROTEOMICS* **2009**, *9* (10), 2764–2775.

149. Misra, S. K.; Milohanic, E.; Aké, F.; Mijakovic, I.; Deutscher, J.; Monnet, V.; Henry, C. Analysis of the serine/threonine/tyrosine phosphoproteome of the pathogenic bacterium Listeria monocytogenes reveals phosphorylated proteins related to virulence. *PROTEOMICS* **2011**, *11* (21), 4155–4165.

150. Ouidir, T.; Jarnier, F.; Cosette, P.; Jouenne, T.; Hardouin, J. Extracellular Ser/Thr/Tyr phosphorylated proteins of Pseudomonas aeruginosa PA14 strain. *PROTEOMICS* **2014**, *14* (17-18), 2017–2030.

151. Mittler, R. ROS Are Good. *Trends in Plant Science* **2017**, *22* (1), 11–19.

152. Handy, D. E.; Loscalzo, J. Redox Regulation of Mitochondrial Function. *Antioxidants & Redox Signaling* **2011**, *16* (11), 1323–1367.

153. Dalle-Donne, I.; Rossi, R.; Colombo, G.; Giustarini, D.; Milzani, A. Protein S-glutathionylation: a regulatory device from bacteria to humans. *Trends in Biochemical Sciences* **2009**, *34* (2), 85–96.

154. Imber, M.; Pietrzyk-Brzezinska, A. J.; Antelmann, H. Redox regulation by reversible protein S-thiolation in Gram-positive bacteria. *Redox Biol* **2019**, *20*, 130–145.
155. Newton, G. L.; Arnold, K.; Price, M. S.; Sherrill, C.; Delcardayre, S. B.; Aharonowitz, Y.; Cohen, G.; Davies, J.; Fahey, R. C.; Davis, C. Distribution of thiols in microorganisms: mycothiol is a major thiol in most actinomycetes. *Journal of Bacteriology* **1996**, *178* (7), 1990–1995.
156. Hillion, M.; Bernhardt, J.; Busche, T.; Rossius, M.; Maaß, S.; Becher, D.; Rawat, M.; Wirtz, M.; Hell, R.; Rückert, C.; Kalinowski, J.; Antelmann, H. Monitoring global protein thiol-oxidation and protein S-mycothiolation in Mycobacterium smegmatis under hypochlorite stress. *Sci Rep* **2017**, *7* (1), 1195, 1–20.
157. Imber, M.; Huyen, N. T. T.; Pietrzyk-Brzezinska, A. J.; Loi, V. V.; Hillion, M.; Bernhardt, J.; Thärichen, L.; Kolšek, K.; Saleh, M.; Hamilton, C. J.; Adrian, L.; Gräter, F.; Wahl, M. C.; Antelmann, H. Protein S-Bacillithiolation Functions in Thiol Protection and Redox Regulation of the Glyceraldehyde-3-Phosphate Dehydrogenase Gap in Staphylococcus aureus Under Hypochlorite Stress. *Antioxidants & redox signaling* **2018**, *28* (6), 410–430.
158. Voss, I.; Steinbüchel, A. High cell density cultivation of Rhodococcus opacus for lipid production at a pilot-plant scale. *Applied microbiology and biotechnology* **2001**, *55* (5), 547–555.
159. Costa, J. S. D.; Herrero, O. M.; Alvarez, H. M.; Leichert, L. Label-free and redox proteomic analyses of the triacylglycerol-accumulating Rhodococcus jostii RHA1. *Microbiology* **2015**, *161* (3), 593–610.
160. Chen, Y.; Ding, Y.; Yang, L.; Yu, J.; Liu, G.; Wang, X.; Zhang, S.; Yu, D.; Song, L.; Zhang, H. Integrated omics study delineates the dynamics of lipid droplets in Rhodococcus opacus PD630. *Nucleic acids research* **2013**, *42* (2), 1052–1064.
161. Zhang, C.; Yang, L.; Ding, Y.; Wang, Y.; Lan, L.; Ma, Q.; Chi, X.; Wei, P.; Zhao, Y.; Steinbüchel, A.; Zhang, H.; Liu, P. Bacterial lipid droplets bind to DNA via an intermediary protein that enhances survival under stress. *Nature Communications* **2017**, *8* (1), 15979, 1–15.
162. Ding, Y.; Yang, L.; Zhang, S.; Wang, Y.; Du, Y.; Pu, J.; Peng, G.; Chen, Y.; Zhang, H.; Yu, J.; Hang, H.; Wu, P.; Yang, F.; Yang, H.; Steinbüchel, A.; Liu, P. Identification of the major functional proteins of prokaryotic lipid droplets. *Journal of lipid research* **2012**, *53* (3), 399–411.
163. Costa, J. S. D.; Silva, R. A.; Leichert, L.; Alvarez, H. M. Proteome analysis reveals differential expression of proteins involved in triacylglycerol accumulation by Rhodococcus jostii RHA1 after addition of methyl viologen. *Microbiology* **2017**, *163* (3), 343–354.
164. Haussmann, U.; Wolters, D. A.; Fraenzel, B.; Eltis, L. D.; Poetsch, A. Physiological adaptation of the Rhodococcus jostii RHA1 membrane proteome to steroids as growth substrates. *J. Proteome Res.* **2013**, *12* (3), 1188–1198.
165. Gross, R.; Aricò, B.; Rappuoli, R. Families of bacterial signal-transducing proteins. *Molecular Microbiology* **1989**, *3* (11), 1661–1667.
166. Stock, A. M.; Robinson, V. L.; Goudreau, P. N. Two-Component Signal Transduction. *Annual Review of Biochemistry* **2000**, *69* (1), 183–215.
167. Dago, A. E.; Schug, A.; Procaccini, A.; Hoch, J. A.; Weigt, M.; Szurmant, H. Structural basis of histidine kinase autophosphorylation deduced by integrating genomics, molecular dynamics, and mutagenesis. *Proceedings of the National Academy of Sciences* **2012**, *109* (26), 1733–1742.

168. Tiffert, Y.; Supra, P.; Wurm, R.; Wohlleben, W.; Wagner, R.; Reuther, J. The Streptomyces coelicolor GlnR regulon: identification of new GlnR targets and evidence for a central role of GlnR in nitrogen metabolism in actinomycetes. *Molecular Microbiology* **2008**, *67* (4), 861–880.

169. Wang, J.; Zhao, G.-P. GlnR positively regulates nasA transcription in Streptomyces coelicolor. *Biochemical and Biophysical Research Communications* **2009**, *386* (1), 77–81.

170. Amon, J.; Bräu, T.; Grimrath, A.; Hänβler, E.; Hasselt, K.; Höller, M.; Jeβberger, N.; Ott, L.; Szököl, J.; Titgemeyer, F.; Burkovski, A. Nitrogen Control in Mycobacterium smegmatis: Nitrogen-Dependent Expression of Ammonium Transport and Assimilation Proteins Depends on the OmpR-Type Regulator GlnR. *Journal of Bacteriology* **2008**, *190* (21), 7108–7116.

171. Malm, S.; Tiffert, Y.; Micklinghoff, J.; Schultze, S.; Joost, I.; Weber, I.; Horst, S.; Ackermann, B.; Schmidt, M.; Wohlleben, W.; Ehlers, S.; Geffers, R.; Reuther, J.; Bange, F.-C. *The roles of the nitrate reductase NarGHJI, the nitrite reductase NirBD and the response regulator GlnR in nitrate assimilation of Mycobacterium tuberculosis.* **2009**, *155* (4), 1332–1339.

172. Yao, L.-L.; Liao, C.-H.; Huang, G.; Zhou, Y.; Rigali, S.; Zhang, B.; Ye, B.-C. GlnR-mediated regulation of nitrogen metabolism in the actinomycete Saccharopolyspora erythraea. *Applied Microbiology and Biotechnology* **2014**, *98* (18), 7935–7948.

173. Yu, H.; Yao, Y.; Liu, Y.; Jiao, R.; Jiang, W.; Zhao, G.-P. A complex role of Amycolatopsis mediterranei GlnR in nitrogen metabolism and related antibiotics production. *Archives of Microbiology* **2007**, *188* (1), 89–96.

174. Wang, Y.; Wang, J.-Z.; Shao, Z.-H.; Yuan, H.; Lu, Y.-H.; Jiang, W.-H.; Zhao, G.-P.; Wang, J. Three of Four GlnR Binding Sites Are Essential for GlnR-Mediated Activation of Transcription of the Amycolatopsis mediterranei nas Operon. *Journal of Bacteriology* **2013**, *195* (11), 2595–2602.

175. Hernández, M. A.; Gleixner, G.; Sachse, D.; Alvarez, H. M. *Carbon Allocation in Rhodococcus jostii RHA1 in Response to Disruption and Overexpression of nlpR Regulatory Gene, Based on 13C-labeling Analysis.* **2017**, *8* (1992), 1–11.

176. Maarsingh, J. D.; Haydel, S. E. *Mycobacterium smegmatis PrrAB two-component system influences triacylglycerol accumulation during ammonium stress.* **2018**, *164* (10), 1276–1288.

177. Amin, R.; Franz-Wachtel, M.; Tiffert, Y.; Heberer, M.; Meky, M.; Ahmed, Y.; Matthews, A.; Krysenko, S.; Jakobi, M.; Hinder, M.; Moore, J.; Okoniewski, N.; Maček, B.; Wohlleben, W.; Bera, A. Post-translational Serine/Threonine Phosphorylation and Lysine Acetylation: A Novel Regulatory Aspect of the Global Nitrogen Response Regulator GlnR in S. coelicolor M145. *Front Mol Biosci* **2016**, *3*, 38, 1–14.

178. Amin, R.; Reuther, J.; Bera, A.; Wohlleben, W.; Mast, Y. *A novel GlnR target gene, nnaR, is involved in nitrate/nitrite assimilation in Streptomyces coelicolor.* **2012**, *158* (5), 1172–1182.

179. Görke, B.; Stülke, J. J. N. R. M. *Carbon catabolite repression in bacteria: many ways to make the most out of nutrients.* **2008**, *6* (8), 613–624.

180. Singh, K. D.; Schmalisch, M. H.; Stülke, J.; Görke, B. Carbon Catabolite Repression in *Bacillus subtilis*: Quantitative Analysis of Repression Exerted by Different Carbon Sources. *Journal of Bacteriology* **2008**, *190* (21), 7275–7284.

181. Tomás-Gallardo, L.; Santero, E.; Floriano, B. Involvement of a putative cyclic AMP receptor protein (CRP)-like binding sequence and a CRP-like protein in glucose-mediated catabolite repression of thn genes in Rhodococcus sp. strain TFB. *Applied and environmental microbiology* **2012**, *78* (15), 5460–5462.

182. Chi, B. K.; Busche, T.; Van Laer, K.; Bäsell, K.; Becher, D.; Clermont, L.; Seibold, G. M.; Persicke, M.; Kalinowski, J.; Messens, J.; Antelmann, H. Protein S-Mycothiolation Functions as Redox-Switch and Thiol Protection Mechanism in Corynebacterium glutamicum Under Hypochlorite Stress. *Antioxidants & Redox Signaling* **2013**, *20* (4), 589–605.
183. Hillion, M.; Imber, M.; Pedre, B.; Bernhardt, J.; Saleh, M.; Loi, V. V.; Maaß, S.; Becher, D.; Astolfi Rosado, L.; Adrian, L.; Weise, C.; Hell, R.; Wirtz, M.; Messens, J.; Antelmann, H. The glyceraldehyde-3-phosphate dehydrogenase GapDH of Corynebacterium diphtheriae is redox-controlled by protein S-mycothiolation under oxidative stress. *Sci Rep* **2017**, *7* (1), 5020, 1–14.
184. Schieber, M.; Chandel; Navdeep, S. ROS Function in Redox Signaling and Oxidative Stress. *Current Biology* **2014**, *24* (10), R453–R462.
185. Urbano, S. B.; Di Capua, C.; Cortez, N.; Farías, M. E.; Alvarez, H. M. Triacylglycerol accumulation and oxidative stress in Rhodococcus species: differential effects of pro-oxidants on lipid metabolism. *Extremophiles* **2014**, *18* (2), 375–384.
186. Zhang, L.; Liao, C.; Yang, Y.; Wang, Y.-Z.; Ding, K.; Huo, D.; Hou, C. Response of lipid biosynthesis in Chlorella pyrenoidosa to intracellular reactive oxygen species level under stress conditions. *Bioresource Technology* **2019**, *287*, 121414, 1–8.
187. Rajakumar, S.; Nachiappan, V. Lipid droplets alleviate cadmium induced cytotoxicity in Saccharomyces cerevisiae. *Toxicol Res (Camb)* **2016**, *6* (1), 30–41.
188. Yu, Y.; Li, T.; Wu, N.; Ren, L.; Jiang, L.; Ji, X.; Huang, H. Mechanism of Arachidonic Acid Accumulation during Aging in Mortierella alpina: A Large-Scale Label-Free Comparative Proteomics Study. *Journal of Agricultural and Food Chemistry* **2016**, *64* (47), 9124–9134.
189. Wang, T.; Ge, H.; Liu, T.; Tian, X.; Wang, Z.; Guo, M.; Chu, J.; Zhuang, Y. Salt stress induced lipid accumulation in heterotrophic culture cells of Chlorella protothecoides: Mechanisms based on the multi-level analysis of oxidative response, key enzyme activity and biochemical alteration. *Journal of Biotechnology* **2016**, *228*, 18–27.
190. Xin, L.; Hong-ying, H.; Yu-ping, Z. Growth and lipid accumulation properties of a freshwater microalga Scenedesmus sp. under different cultivation temperature. *Bioresource Technology* **2011**, *102* (3), 3098–3102.
191. Chokshi, K.; Pancha, I.; Ghosh, A.; Mishra, S. J. B. f. b. *Nitrogen starvation-induced cellular crosstalk of ROS-scavenging antioxidants and phytohormone enhanced the biofuel potential of green microalga Acutodesmus dimorphus.* **2017**, *10* (1), 60, 1–12.
192. Nguyen, T. T. M.; Iwaki, A.; Ohya, Y.; Izawa, S. Vanillin causes the activation of Yap1 and mitochondrial fragmentation in Saccharomyces cerevisiae. *Journal of Bioscience and Bioengineering* **2014**, *117* (1), 33–38.
193. Simon, O.; Klaiber, I.; Huber, A.; Pfannstiel, J. Comprehensive proteome analysis of the response of Pseudomonas putida KT2440 to the flavor compound vanillin. *Journal of Proteomics* **2014**, *109*, 212–227.
194. Ask, M.; Bettiga, M.; Mapelli, V.; Olsson, L. The influence of HMF and furfural on redox-balance and energy-state of xylose-utilizing Saccharomyces cerevisiae. *Biotechnol Biofuels* **2013**, *6* (1), 22, 1–13.
195. Riyadi, F. A.; Tahir, A. A.; Yusof, N.; Sabri, N. S. A.; Noor, M. J. M. M.; Akhir, F. N. M. D.; Othman, N. a.; Zakaria, Z.; Hara, H. Enzymatic and genetic characterization of lignin depolymerization by Streptomyces sp. S6 isolated from a tropical environment. *Sci Rep* **2020**, *10* (1), 7813, 1–9.

Chapter 5

Electrocatalytic and Photocatalytic Approaches to Lignin Conversion

Shuya Li,[1] Kayla Davis,[1] and Gyu Leem[1,2,*]

[1]Department of Chemistry, State University of New York College of Environmental Science and Forestry, Syracuse, New York 13210, United States

[2]The Michael M. Szwarc Polymer Research Institute, Syracuse, New York 13210, United States

[*]Email: gyleem@esf.edu.

Lignin is a heterogeneous aromatic macromolecule and represents a potential feedstock suitable to produce valuable chemicals. Molecular catalysts hold great potential for lignin fragmentation. Numerous catalytic systems have been developed for the cleavage of C–O and/or C–C interunit linkages. Especially, selective cleavage of C–O and/or C–C bond linkages is pursued as a useful chemical fragmentation method for lignin valorization to afford valuable products. Traditional lignin depolymerization methods require harsh conditions, resulting in poor product selectivity and unwanted products. Thus, novel methods to convert lignin to valuable chemicals under mild conditions are still appealing. In this book chapter, we highlight recent advances in the selective degradation of lignin via photo- and electro-catalysis to achieve predominant C–O/C–C bond cleavages and obtain highly desirable and valuable aromatic products. These depolymerization processes are promising and can serve potentially as a viable method for the conversion of lignin into value-added chemicals.

Introduction

Lignin is the largest non-carbohydrate component in lignocellulosic biomass, representing a promising resource for renewable biofuels and valuable chemicals (*1*). It is a three-dimensional amorphous and interlinked biopolymer comprised of three primary monomeric subunits: *p*-hydroxyphenyl (*H*), guaiacyl (*G*), and syringyl (*S*), moieties that are present in the β-1, β-5, β-β, α-O-4, and β-O-4 untis shown in Figure 1 (*2*). The most common recurring linkage is the β-O-4 up to around 50% of all linkages.

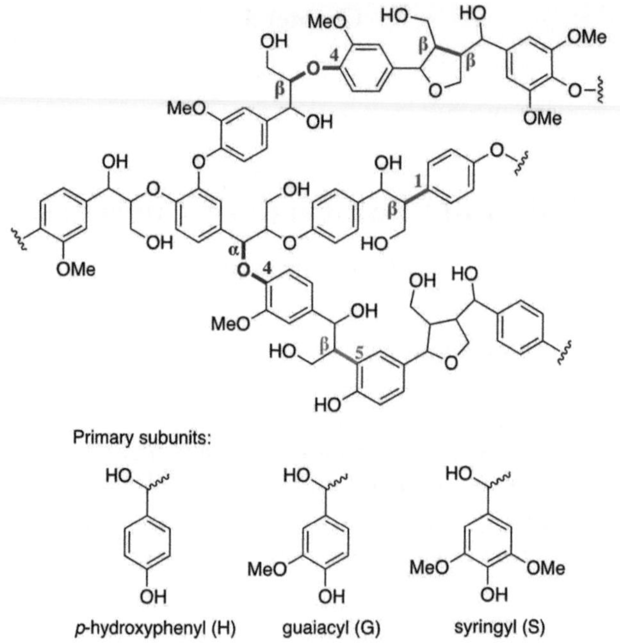

Figure 1. General lignin structure, containing various linkages including β-O-4, α-O-4, β-1, β-5, and β-β (top) and primary subunits (bottom).

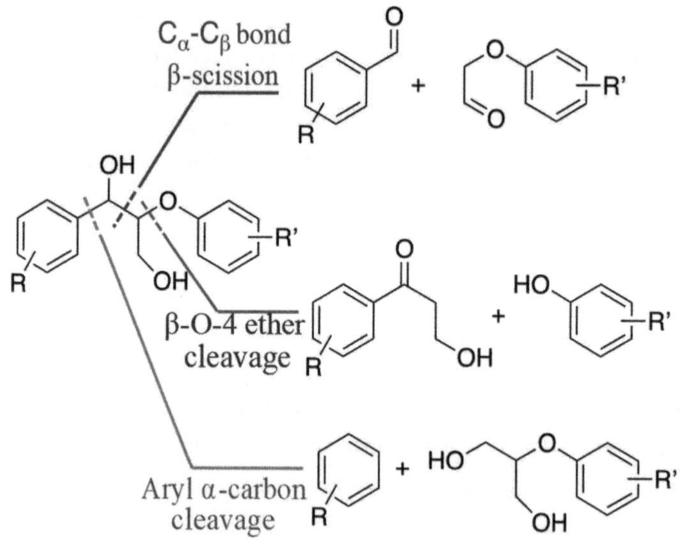

Figure 2. Lignin basic structure with C–C/C–O bond cleavage and the predicted possible products.

The selective cleavage of these relatively stable C–O and C–C linkages is crucial in facilitating the lignin depolymerization because it can provide for highly valuable aromatic compounds as renewable feedstocks (3). Traditional biorefinary processes (4) (e.g., acidic hydrolysis, alkaline-oxidative pretreatment) have been investigated to fragment lignin model compounds and native lignin. Generally, these processes require high temperature and pressure (e.g., >180 °C and 1.5 MPa) which leads to low functionalized aromatics or alcohols and unavoidably poor selectivity (5). Furthermore, under harsh conditions, radicals may evolve and result in repolymerization process resulting in decreased product yield. In other words, repolymerization reactions can be attributed to these harsh conditions (*6, 7*). For example, recondensation of formed fragments occurs at high temperatures (*6*). This recondensation is an undesired pathway resulting in diverse aromatic compounds with low selectivity. Hence, recent studies are focused on the development of novel processes and strategies for highly selective lignin depolymerization and utilization under mild conditions, leading to the production of well-defined value-added products (*8*).

The highly efficient catalytic cleavage of C−O and C−C linkages in lignin under mild conditions remains a central challenge. Recently, there are numerous promising catalytic cleavage methods including thermochemical (*9, 10*), electrocatalytic (*11, 12*), or photocatalytic depolymerization (*13, 14*). Most of these depolymerization methods target the β-aryl ether bonds, aryl α-carbon, or vicinal C_α-C_β bond cleavages shown in Figure 2. This book chapter gives an overview of photocatalytic and electrocatalytic depolymerization processes for C−O and/or C−C bond cleavages in lignin. First, the discussion of the recent advances in chemical catalysis using homogeneous and heterogeneous catalysts. Next, an overview of photocatalytic approaches of the selective, oxidative cleavage of aromatic C−O bonds under UV or visible light illumination. Then direct electrooxidation approaches will be introduced followed by electrocatalytic oxidation of lignin via hydrogen atom transfer agents as mediators. Lastly, a photocatalysis technique coupled with a dye-sensitized photoelectrochemical cell demonstrating chemoselective benzylic alcohol oxidation of lignin models and real lignin will be discussed.

Catalytic Bond Cleavage of Lignin

The β-O-4 linkage dominates in native lignin (40-50% in softwood and 60% in hardwood). Cleavage of the C−O/C−C bonds in the aryl ether β-O-4 linkages of lignin is facing a challenge due to the relatively stable C_α-O (Bond Dissociation Enthalpy, BDE: 70.5 kcal/mol) and vicinal C_α-C_β (BDE: 69.2 kcal/mol) bonds (*15*). Recently, numerous catalytic methods have been explored for the fragmentation of lignin model compounds under a homogeneous or heterogeneous catalysis system with metal catalysts such as oxovanadium (V) complexes (*16*), copper (Cu) (*17*), and the pentamethylcyclopentadienyl iridium(III) bipyridonate (Cp*Ir-bipyridonate) (*18*). These transition metal-based catalysts have high selectivity of lignin oxidation and reductive cleavage. Figure 3 shows selected examples of C−O/C−C bond cleavages (including C_α-C_β bond β-scission, β-O-4 ether cleavage, and aryl ethyl cleavage) in lignin for depolymerization. Different biomass conversion pathways can be controllable depending on the ligand structure and reaction temperature. This section will focus on homogeneous and heterogeneous catalysis for catalytic cleavage of aliphatic C−O/C−C bonds in aryl ethers of lignin.

Stein et.al reported ruthenium complexes with triphos-type phosphine ligands as being highly active catalysts for dehydrogenation-initiated retro-aldol C−C bond cleavage under relatively high temperature (e.g., 160 °C) shown in Figure 3a (*3*). Lancefield et. al. reported Ir-catalysed primary alcohol oxidation, a retro-aldol C_α-C_β bond cleavage reaction with excellent yields of cleaved aromatic compounds under basic conditions (*18*). Their system features relatively low catalyst loadings (less than 1 mol%), and short reaction times in the absence of stoichiometric oxidants.

Ruthenium (Ru)-catalyzed, redox-neutral C−O bond cleavage of various 2-aryloxy-1-arylethanols containing β-O-4 linkages in lignin models was reported (*19*). With a wide, bite-angel ligand—9,9-dimethyl-bis(diphenylphos-)-xanthene (Ph-xantphos)—the Ru-based catalyst yields up to 99% cleavage of β-O-4 ether bond for the period of 1.75 h at 135 °C. This excellent yield was due to the redox-neutral structural bond transformation by a tandem dehydrogenation/reductive ether cleavage. An Ir-based catalyst was performed for the highly chemoselective cleavage of C−O bond in β-O-4 lignin (*20*). In addition, homogeneous vanadium(V)-based complexes carried out non-oxidative C−O bond cleavage of β-O-4 linkages in dimeric lignin model compounds over benzylic C_α oxidation under aerobic conditions below 100 °C (*21*, *22*). Furthermore, the vanadium catalyst was tested with a trimeric lignin model compound for C−O bond cleavage, demonstrating the conversion of trimeric lignin models to three monomeric compounds in the presence of air. This result could open the possibility of application to real lignin.

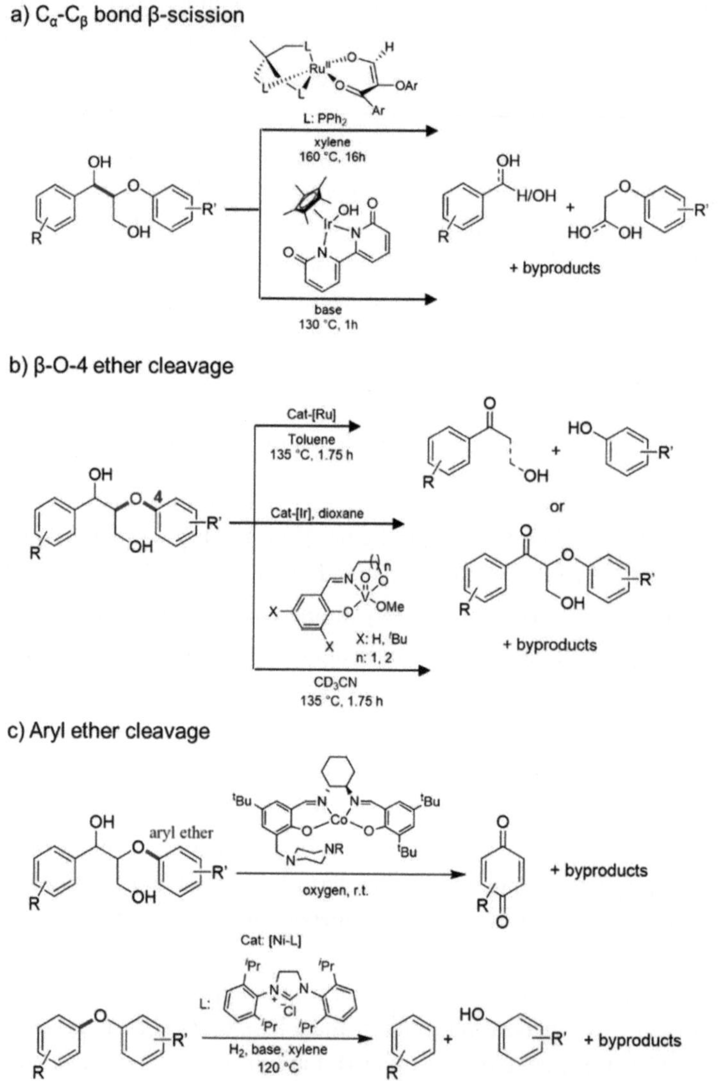

Figure 3. Selective examples of C−O/C−C bond cleavages. a) C_α-C_β bond β-scission (3, 18), b) β-O-4 ether cleavage (19–21), and c) aryl ethyl cleavage (23, 24) for lignin depolymerization.

As shown in Figure 3c, the cobalt-based catalyst displays a high catalytic reactivity in performing oxidative cleavage of lignin model dimers to benzoquinones under aerobic conditions (*23*). A nickel carbene complex as the heterogeneous catalyst showed selective hydrogenolysis of aliphatic C−O bonds in aryl ethers (*24*). At 1 bar of hydrogen and below 120 °C, a series of nickel catalysts containing 1,5-cyclooctadiene with the combination of an N-heterocyclic carbene and strong base (e.g., ^{t}BuONa) allowed for selective hydrogenolysis of the aromatic and benzylic carbon-oxygen (C−O) bonds of the 4-O-5 linkage and α-O-4 linkage in lignin. The strong base in hydrogenolysis can facilitate the formation of an anionic nickel complex, which is active for the cleavage of C−O bonds.

The catalytic processes above have high selectivity and excellent yield of cleavage products. However, these thermal catalytic reactions, in general, require high temperature (> 100 °C) with a specific atmosphere (e.g., H_2) or/and stoichiometric oxidants. As a result, radical coupling or condensation reactions occur and resulting in the production of unwanted products.

Photocatalytic Conversion

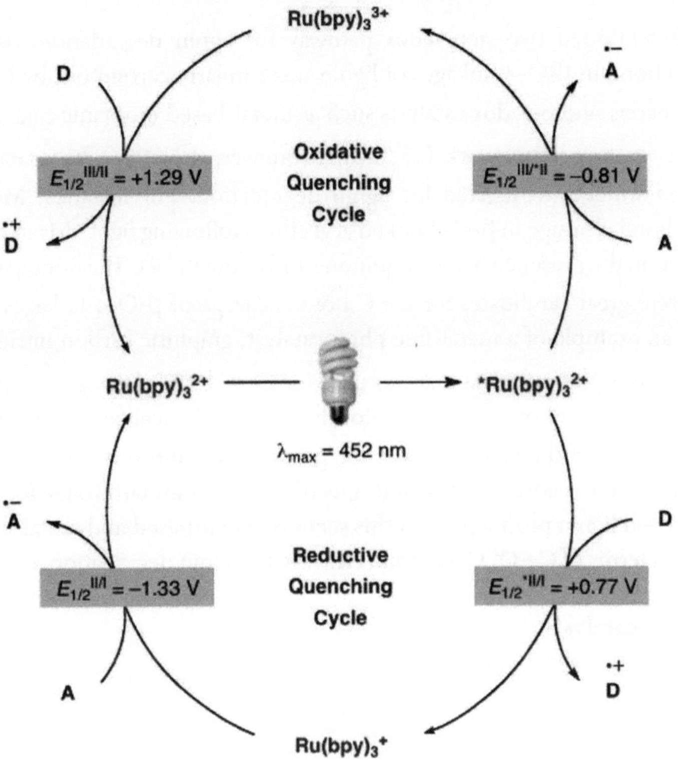

Figure 4. Oxidative and reductive quenching cycles for Ru(bpy)$_3$$^+$. Reprinted with the permission from reference (26). Copyright 2013 American Chemical Society.

Photocatalysis has been widely studied and is considered to be a promising approach for the depolymerization of lignin (25). Photocatalytic reactions are attractive, eco-friendly alternatives due to their mild reaction conditions (e.g. room temperature), the enhancement of chemoselectivity for the desired products, and avoidance of strong oxidants or reducing agents. A photocatalytic chemical transformation of lignin is initiated by a photoexcited metal complex following quenching processes such as oxidative quenching (electron transfer to quencher) or reductive quenching (acceptance of electron from quencher). For example, ruthenium(II) tris(2,2′-bipyridine) ($Ru(bpy)_3^{2+}$) is one of the most commonly utilized photocatalysts. Due to the unique property of the $Ru(bpy)_3^{2+}$ photoexcited state, the Ru photocatalyst acts as an oxidant or reductant due to a higher energy electron in the π* in ligands and a lower-energy hole in the Ru metal-centered t_{2g} orbital (26).

Two quenching pathways from photoexcited $^*Ru(bpy)_3^{2+}$ occur as shown in Figure 4. Upon the absorption of visible light, $Ru(bpy)_3^{2+}$ is photoexcited to form excited-state Ru(II)*, which is more oxidative or reductive than the ground-state species. The excited photocatalyst then performs an oxidative or a reductive quenching cycle. In the oxidative quenching cycle, Ru(II)* transfers the excited electron to an electron acceptor (A), which renders the oxidized form Ru(III) with the acceptor being reduced to a radical anion of A. The oxidized Ru(III) is then regenerated to the initial Ru(II) by accepting an electron from an electron donor (D) to form a radical cation of D in the meanwhile. The reductive quenching process of redox catalysis also relies on two consecutive intermolecular single electron transfer (SET) processes, in which Ru(II) is reduced by an electron donor to form Ru(I), followed by being oxidized back to the original form by oxidation with an electron acceptor.

In the aforementioned two-step redox pathway for lignin degradation, oxidative/reductive cleavage of C–O bond in β-O-4 linkages of lignin was similarly carried out by homogeneous (*21*, *22*) and heterogeneous photo-redox catalysts such as metal-based inorganic catalysis, $ZnIn_2S_4$, and carbazolic porous organic framework (*27*, *28*). Moreover, selective photocatalytic C–C bond cleavage has been widely investigated for lignin degradation. For instance, Moody et. al. have developed C–C bond cleavage in β-O-4 linked aryl ethers following light-driven oxidation of lignin model compounds in the presence of benzoquinone and copper (*29*). The homogeneous vanadium-based catalysts were great candidates for C–C bond cleavage of β-O-4 linkages with visible light illumination. As an example of a metal-free photocatalyst, graphitic carbon nitride ($g\text{-}C_3N_4$) is an efficient heterogeneous photocatalyst that was used for C–C bond cleavage in lignin via conjugated stacking interactions (*30*). Photocatalytic approaches allow for excellent control over the reaction, and thus the great yield of the resulting product/product distributions. Therefore, these methods can be highly considered as cost-effective and 'green' processes and utilizable for the industrialized processes of valorized lignin production. In this section, metal-based and metal-free photocatalysts will be discussed in terms of C–O/C–C bond cleavage for lignin degradation.

Metal-Based Photocatalysts

Titanium Dioxide

Titanium dioxide (TiO_2) is a common semiconductor for photocatalysis in many organic reactions due to its robust advantages such as stable catalytic performance and low cost (*31*). Upon irradiation of ultraviolet light ($\lambda_{ex} < 400$ nm), it generates photoexcited electron (e^-) and hole (h^+) pairs in the conduction band and valence band, respectively. The degradation of lignin occurs via a free radical reaction process with the high oxidative ability of the holes in TiO_2 (*32*). A photocatalytic pathway for β-O-4 cleavage in lignin has been developed using TiO_2 coupled with a nickel catalyst and pyridinium chlorochromate (PCC) as the oxidant under relatively mild aerobic conditions and at room temperature, under UV light illumination (30 W) (*33*). Loading of silver (Ag) onto TiO_2 can boost the photocatalytic activities of TiO_2 for lignin conversion under UV illumination (*34*). Mechanistic studies indicate that the photocatalytic process is initiated with the conversion of C_α-OH alcohol to C_α=O ketone in β-O-4 linkages, followed by photocatalytic cleavage of β-O-4 ether bond.

Cadmium Sulfide

Cadmium sulfide quantum dots (CdS QDs) were found to be one of the efficient photocatalysts for lignin degradation by catalyzing the cleavage of β-O-4 linkage in the lignin system (*35*). In this colloid solution, electron-hole couples were generated under visible light illumination. Then this redox process induced the formation of a C_α radical intermediate in β-O-4 linkages followed by cleavage of aryl ether C−O bonds in lignin via single electron transfer. Oxidized products and lower molecular weight aromatic compounds including benzaldehyde and guaiacol were directly produced in the one-pot reaction. The main advantage of this heterogeneous photocatalyst is its excellent recyclability for the lignin degradation reaction.

The surface structure of CdS QDs was modified with organic ligands, resulting in enhanced photocatalytic cleavage of β-O-4 linkage (*36*). The hydrophilicity or hydrophobicity of ligands (e.g., 6-mercaptohexanoic acid (MHA)) was tuned to obtain a proper QD colloidal solution for intimate contact between QDs and lignins. The anchoring groups and the length of ligands were manipulated in this study to determine the effects of the ligand in the electron-transfer process. According to the kinetic study, electron decay behavior indicates the electron-transfer process through ligands via an electron-tunneling pathway. Sun et. al. reported that Ni-decorated CdS QDs performed highly selective photocatalytic cleavage of β-O-4 linkages in a lignin model compound in the mixtures of CH_3CN and H_2O. As a result, the products of acetophenone and phenol were obtained (*37*). Interestingly, the use of acetonitrile (ACN) only allowed for selective oxidation of benzylic alcohols, yielding α-carbonyl products. However, a mixture of ACN and water as the solvent showed the α-carbonyl products and cleaved products. Moreover, with a basic solution containing 0.1 M KOH, direct photocleavage of lignin model compounds occurred without the production of α-carbonyl products. This work is a rare example of controlling the photocatalytic selectivity of lignin cleavage by tuning the reaction conditions with metal-decorated CdS QDs.

Ceria

Ceria is the metal oxide of cerium (Ce^{IV}) with a high oxidative ability and strong visible-light absorption. Wang et. al. reported a photo-induced strategy for C−C bond cleavage of the β-1 linkage in lignin under aerobic conditions (*38*). In this study, Ce^{IV} and CuO_x clusters are deposited onto TiO_2 nanotubes to induce the separation of electron-proton pairs upon photo-excitation of ceria. With the presence of CuO_x clusters, the enhancement of photocatalytic activity was demonstrated, purportedly due to the increase in the population of ceria surface defects (Ce^{3+} ions and oxygen vacancies).

Zinc–Indium–Sulfide

The energy band of metal-based photocatalysts can be tuned by controlling the atomic ratio of two metal atoms. The atomic ratio of Zn/In was adjusted to produce zinc–indium–sulfide, $Zn_4In_2S_7$ photocatalyst showing excellent photocatalytic performance for C−C/C−O bond cleavage of β-O-4 linkages in lignin under visible-light illumination due to its proper energy bandgap and suitable redox capability (*27, 39*). The thiol groups on the surface of $Zn_4In_2S_7$ have a strong effect on the cleavage of a β-O-4 bond, which can form the C_α-radical intermediate followed by C−O bond cleavage.

A Dual Pd/Ir System

Two-step oxidative and reductive cleavage approaches have been developed for the fragmentation of lignin models (*40–42*). Firstly, the C_α benzylic alcohol moiety in the β-O-4 linkages is selectively oxidized followed by the conversion to a ketone. The presence of the alpha-carbonyl group—the C_α ketone—activates the β-O-4 linkage for cleavage of the C−O/C−C bond. Stephenson's research group reported that the Pd-catalyzed oxidation reaction allowed for the activation of β-O-4 systems in lignin models to form an oxidized benzyl ketone (*40*). Then, the C−O bond reductive cleavage in β-O-4 systems occurred in the presence of Ir(II)-based photocatalyst and base-pair under mild conditions. This method produced ketone and phenol forms as fragmentation products. Moreover, chemoselective C−C bond cleavage has been developed via redox-neutral photocatalytic strategies using transition metal-based catalysts via proton-coupled electron transfer (PCET) (*8, 43*). Ir-based catalysts were used to form C_α-oxygen radical intermediate via this PCET process, followed by β-scission of C_α -C_β bond.

Vanadium

Despite many efforts to develop highly efficient photocatalysts, many of these systems do not take into account the abundance of the metal-catalysts that they utilize. The earth-abundance metal catalysts can be determined whether or not a particular reaction system can be considered to be industrially-relevant. Thus, much scientific exploration is still required regarding the photocatalytic efficiency of earth-abundant metal-based photocatalysts. Vanadium (V) is one of the most earth-abundant transition metals. V-based metal complexes were reported to exhibit a high photocatalytic reactivity and chemoselectivity for the cleavage of β-O-4 lignin with heat (e.g., 80 °C) (*44*). Soo and coworkers successfully developed a V^V oxo complex for the highly-chemoselective, aliphatic C−C bond cleavage of lignin models under relatively mild conditions (*45*). The product analysis is based on isotopically labeled lignin model compounds. This analysis confirms that the V^V oxo complex photocatalyst plays an important role to cleave the aliphatic C−C bond regioselectively. Besides, theoretical density functional calculations confirm that the V^V photocatalyst *via* photoexcitation exhibits transition state T_1 energy surface generation and great selectivity for C−C bond cleavage reactions. Interestingly, the combination of the T_1 excited state and carbonyl groups in the oxidized models activates the cleavage of the C−C bond. The homolytic C−C bond cleavage is then driven by the ligand metal charge transfer (LMCT), in which an efficient single electron transfer (SET) occurs due to the lower energy of the ligand's π-orbital. This study sheds light on an eco-friendly approach to harvest and convert solar energy while utilizing renewable biomass.

Metal-Free Photocatalysts

Quinones

Quinones are a group of organo-oxidants in natural redox processes with the combination of both plastoquinone and phylloquinone acting as electron acceptors in photosynthesis processes (*46*). 1,4-Benzoquinone (BQ) has been used in multiple research-based oxidation reactions due to its low redox potential, which can be reduced to catalytic hydroquinone under visible light illumination. Mitchell et. al explored a photochemical oxidation pathway for secondary benzylic alcohols using a BQ photocatalyst and copper nanoparticles as the electron transfer mediator (ETM) under an oxygen atmosphere (*29*). Photocatalytic oxidation of benzylic alcohols in lignin model compounds is an important step for bond cleavages. In order to activate benzylic alcohol oxidation, it requires the formation of active BQ. 1,4-Hydroquinone (BQH_2) can be oxidized to generate the active oxidant BQ with an ETM and molecular oxygen (Scheme 1a). More interestingly, this BQ-BQH_2 redox process allows for simultaneous alcohol oxidation and C−C bond cleavage in lignin models due to the operation of two competing reaction pathways. In Scheme 1b, the hydrogen atom on the benzylic α-C-H bond is initially abstracted by the photoexcited triplet quinone (^{3}BQ), resulting in the formation of a C_α radical intermediate. Sequential hydrogen abstraction from the O-H bond of C_α results in formation of the ketone product and hydroquinone BQH_2. In the case of C−C bond cleavage, hydrogen abstraction from the O-H bond initiates the formation of an alkoxy radical that favors C−C bond fragmentation (Scheme 1c).

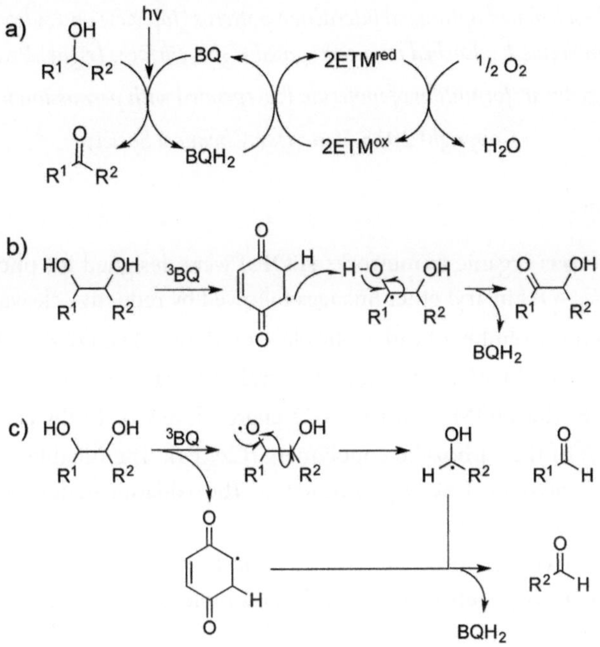

Scheme 1. a) Schematic BQ-BQH_2 redox process incorporated with an electron transfer mediator (ETM). Proposed mechanistic cleavage pathways for b) ketone product and c) C−C homolytic fragmentation. Reprinted with permission from reference (29). Copyright 2014 American Chemical Society.

Graphitic Carbon Nitride

Wang's research group explored a mild photocatalytic strategy for the C−C bond cleavage of β-O-4 and β-1 linkages in lignin model compounds by using a mesoporous graphitic carbon nitride (mpg-C_3N_4) photocatalyst (*30*). A strong $\pi - \pi$ stacking interaction between the flexible carbon nitride surface and lignin model molecule occurs according to DFT calculation (Figure 5). Also, a high specific surface area of mpg-C_3N_4 and low recombination of electrons and holes largely contribute to the enhancement of its photocatalytic activity. As shown in Figure 5, following photoexcitation of mpg-C_3N_4, photogenerated holes trigger the oxidation and C−C bond cleavage of lignin β-O-4 on the mpg-C_3N_4 surface.

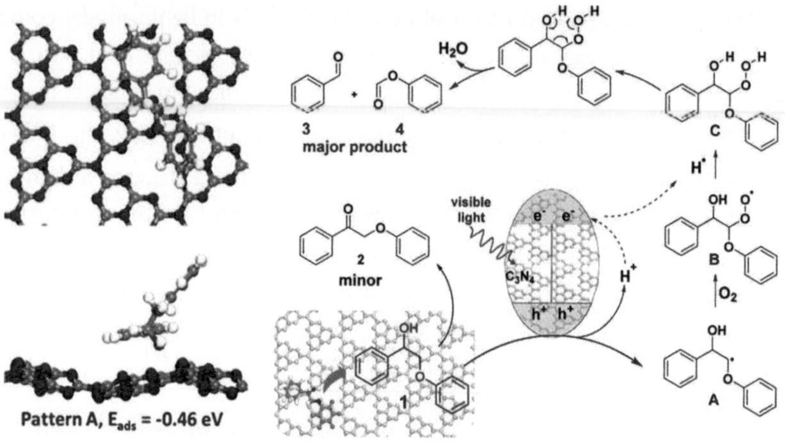

Figure 5. (left) DFT-calculated optimized interaction patterns (top view and side view) and adsorption energy of substrate molecule 1 adsorbed on a corrugated C_3N_4 surface; (right) Proposed mechanism of mpg-C_3N_4 catalyzed transformation of molecule 1. Reprinted with permission from reference (30). Copyright 2018, American Chemical Society.

Porous Organic Frameworks

π-Conjugated porous organic frameworks (POFs) were designed for photocatalytic oxidation of benzylic alcohols C_α-OH in aryl ether linkages followed by reductive cleavage of the C−O bond in β-O-4 ketones via the assembly of carbazolic electron donor (D), DCB (1,4-bis(9*H*-carbazol-9-yl)benzene), and acceptor (A), 4CzIPN (1,2,3,5-tertrakis(carbazol-9-yl)4,6-dicyanobenzene) (*28*). As shown in Figure 6, the HOMO and LUMO energy levels of POFs could be directly tuned by the ratio of D to A of the carbozylic copolymers (CzCPs). The tunable energy levels rendered adjustable redox potentials of the CzCPs photocatalyst. The oxidation-reduction processing was then monitored at different ratios of the D and A comonomers. In the presence of *N*-hydroxyphthalimide (NHPI) and oxygen, the conversion of benzylic alcohols to ketones in β-O-4 linkages occurs under visible light by the interaction between the NHPI and molecular oxygen. Especially, the resulting CzCPs with a D: A ratio of 66:33 possesses the highest photocatalytic efficiency for the reductive cleavage of β-O-4 ketones with formic acid. CzCPs can be recycled without loss of photocatalytic activity.

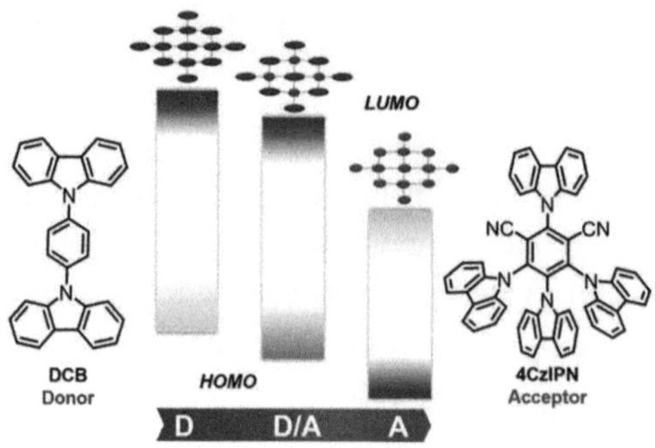

Figure 6. Schematic diagram of the carbazolic copolymer containing DCB and 4CzIPN. Reprinted with permission from reference (28). Copyright 2017 American Chemical Society.

Phenothiazine (PTH)

Modified lignin substrates containing oxime derivatives as internal nucleophiles, which were prepared via oximation and acetylation of lignin models, can provide a novel pathway for the cleavage of aryl ether lignin bonds (47). Aryl C−O bonds were cleaved to produce corresponding non-phenolic arylamines in the presence of methyl-substituted phenothiazines (PTH PC1) (Figure 7). The proposed mechanism proves that these pre-modified models can generate iminyl radicals from photocatalytic reduction, allowing for aryl migration from O to the N atom. This method enables the production of non-phenolic nitrogen-containing aromatics from lignin models.

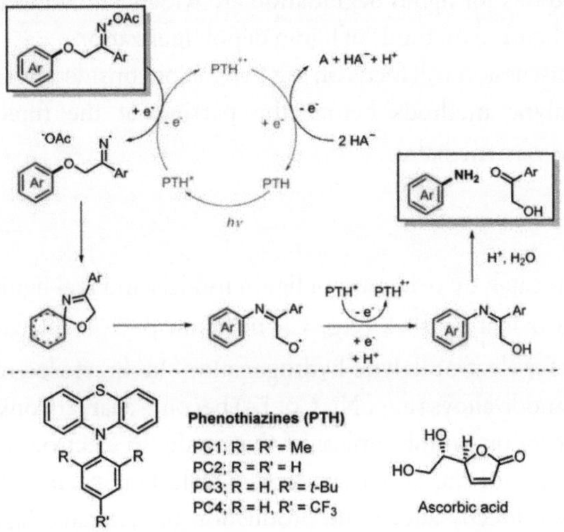

Figure 7. Proposed pathway for amination of aryl ether to primary arylamine using PTH with different substitutes as photocatalysts, E_{red} ($PTH^{\cdot+}/PTH^*$) ~ −2.1 V vs. SCE). Reprinted with permission from reference (47). Copyright 2019 American Chemical Society.

Ionic Liquids

Ionic liquids (ILs) exhibit great chemical stability and solubility of lignin substrates. ILs are not only an ideal solvent for lignin fractionation from biomass but also an efficient catalyst for lignin depolymerization (48). Kang et. al. reported a metal-free photochemical route for lignin degradation under UV light illumination through a free-radical-mediated reaction in the binary-IL system: 1-propenyl-3-methylimidazoliumbis [(trifluoromethyl)-sulfonyl] imide ([PMim][NTf_2]) and Brønsted acid 1-propylsulfonic-3-methylimidazolium trifluoromethane-sulfonate ([$PrSO_3HMim$][OTf]) (49). The free-radical-mediated process drives the generation of the alkyl, hydroxyl, peroxyl free radicals in an oxidized lignin model compound. The C_β-H bond cleavage is initiated by [PMim][NTf_2] and [$PrSO_3HMim$][OTf], which triggers the subsequent C−O bond cleavage via a photochemical induced free-radical process. This IL system provides a novel strategy for photochemical, sustainable biomass conversion through a self-initiated radical strategy.

Electrocatalytic Conversion of Lignin

Electrocatalytic oxidative and/or reductive cleavage of lignin can be carried out directly or indirectly, where direct electrocatalytic methods use metal-based catalysts as electrodes, and indirect electrocatalytic methods utilize photoelectrodes incorporated with solvent-soluble mediators as co-catalysts in the presence of electrodes. There exists a broad spectrum of methods to achieve the electrocatalytic oxidation of lignin, which differ based on the varying identity of the electrodes, electrolyte, and mediator. For example, electrocatalytic reductive cleavage of C−O/C−C bonds in lignin requires specific metal or metal oxide cathodes and anodes (*50*). Electrocatalytic methods allow for fine-tuning of reactions based on the control of the aforementioned method variables, and thus electrocatalytic methods for lignin degradation are widely known to be more energy-efficient than traditional thermochemical methods of lignin depolymerization.

In this section, the discussion will focus on the two major considerations for developing efficient and specific electrocatalytic methods before the parsing of the direct electrooxidation and electrocatalytic oxidation via a mediator.

Electrodes

Anodes in the electrocatalytic oxidation of lignin models and real lignin should be catalytically active for specific lignin oxidation sites (e.g. C_α in lignin β-O-4 linkage), while the cathode is generally to be available for electrocatalytic hydrogenation. In direct electrooxidation, morphology and surface structure of anode alloys (e.g., Ni, Co, Fe) become a larger consideration, as most lignin oxidation occurs by adsorption on the surface of the anode. In electrocatalytic oxidation coupling with mediators, the identity of the anodic material is selected specifically to regenerate catalytic mediators in solution and directly affects the production of generated lignin radicals such as C_α-hydroxy radical.

For direct electrooxidation of lignin, chemoselectivities are straightly dependent on the identity of the anode, while the identity of solution-based catalytic mediators (e.g., (2,2,6,6-tetramethylpiperidin-1-yl)oxyl (TEMPO)) determines the oxidative selectivity of benzylic alcohols in lignin for indirect electrocatalytic lignin depolymerization. The effect of anodic material identity has been observed in the comparison between metal oxide and mixed metal oxide Ti/PbO_2 and Ti/Sb-SnO_2, with the former being the more oxidatively selective of the pair due to adsorption of lignin on the electrode surface (*51*). Ideally, an electrode would have high selectivity when being utilized in lignin degradation, as this creates ease of product isolation and recovery, and further reputes a method for industrialization, the ultimate goal. Moreover, to remain efficient throughout the reaction, or progressive cycles of reaction batches, metal-catalyst embedded electrodes should be resistive to corrosion and electrolytically stable. They need to be specially modified to fulfill these missions, and thus the potential for modification (e.g., nanoparticle deposition) must also be considered. Electrodes that have been modified by deposition or integration are considered doped and have enhanced characteristics of catalysis and stability. FeCN-PbO_2 has been reported (*52*) (as PbO_2 electrode doped with redox ion $[Fe(CN)_6]^{3-}$) to have a higher surface area and can maintain an electrocatalytic oxidative reaction longer. Structural support modifications, such as the multi-walled carbon nanotubes loaded with PbO_2 reported by Bateni et al., can increase catalyst characteristics such as maximum reaction rate of lignin electrooxidation (*53*).

Various transition metals, metal oxides, and mixed metal oxides have been reported for use of anode composition in electrocatalytic C−O/C−C bond cleavages for lignin degradation (*54*, *55*).

Predominantly, nickel has been used as anodic material in electrocatalytic oxidative lignin conversion with high catalytic capability, stability, and high surface area as an electrode in general (*56, 57*). In addition, metal oxides (e.g, TiO_2, and IrO_2) are also catalytically efficient and particularly resistant to corrosion (*58–60*). IrO_2-based anode in the electrooxidation of lignin can achieve a high rate constant with a value of 9.9×10^{-3} min^{-1}, obtained by plotting $\ln(C/C_0)$ vs. reaction time, where C stands for the concentration of lignin substrate and C_0 is the initial substrate concentration (*59*). Combining metals and metal oxides as a redox couple allows for greater selectivity of lignin oxidation sites (*55*). In the electrochemical reduction process using Cu/Ni-Mo-Co cathode and Pb/PbO_2 anode, C–O bonds were hydrocracked selectively.

Direct Electrooxidation of Lignin

Direct electrooxidation of lignin in which both C–O and/or C–C bonds cleave using electrocatalytic methods has been widely reported (*49*). Notably, cleavage of C_α–C_β with the oxidation of β–O–4 linkage was also reported (*61*). Heterogeneous catalysts such as Ni, Co, Fe, and Ti have been reported in Kraft-lignin degradation (*62*). Ni has higher corrosion resistance compared to Co which has a higher yield; both select vanillin and acetovanillone as only monomers, indicating ideal narrow product distribution (*57*). Wang et. al. reported successful oxidative degradation of lignin by OH radical generation on Pb/PbO_2 electrode surface in alkali solution (*63*). Electrochemical conversion of corn stover lignin to biomass-based chemicals between Cu/Ni-Mo-Co cathode and Pb/PbO_2 anode in sodium hydroxide solution is an example of electrochemical oxidation at the anode and electrocatalytic hydrogenation at the cathode (*55*). Another example of electrochemical oxidation and electrocatalytic hydrogenation was reported with Cu and Pb/PbO_2 electrodes in alkali solution with yields of trans-ferulic acid, vanillin, 3-hydroxy-4-methoxyphenyl-ethanone, syringaldehyde, acetosyringone, and 4-methoxy-3-methyl-phenol (*54*).

This electrocatalytic oxidation of lignin in heterogeneous systems offers several advantages over its homogeneous counterpart, especially concerning the stability and recyclability of catalysts (*63*). However, side reactions often occur alongside lignin oxidation and degradation via electrocatalysis, such as condensation of C–C bond with an applied voltage of 1.0 V vs. SCE under 70 °C (*64*). The generation of byproducts can be ascribed to radical coupling reactions or intermolecular condensation reactions (*65*). Moreover, these direct electrochemical processes are limited by surface catalysts, and the final products are often varying in their distribution, and non-predictable.

Electrocatalytic Oxidation via Mediator

In comparison to direct (mediator-free) electrooxidation systems, the additional presence of a mediator achieves the same types of lignin conversion while significantly reducing the necessarily applied potential and thus offering milder oxidation conditions. Besides the generous homogeneous oxidants, e.g., polyoxometalates, ferric chloride, and laccase mediators (*66–68*), hydrogen atom transfer (HAT) agents are found to be effective as the oxidant and proton/electron reservoir for lignin oxidation in recent studies (*69*).

The mission of mediators in the indirect mediator-assisted electrocatalytic oxidation of lignin is to provide an easy route for proton/electron transfer from electrode/electrolytic solution to target lignin oxidation sites. During the redox cycle of the mediators, lignin is oxidized and the mediator is reduced. Figure 8 shows the mechanism of mediator-assisted electrocatalytic oxidation of lignin using NHPI via proton coupled electron transfer (PCET) under the applied bias (*69*). As shown in Figure 8, the electrooxidation process is initiated by anodic oxidation of NHPI to in situ-generated N-oxyl radical, phthalimide N-oxyl (PINO), which has a suitable O−H bond strength (BDE_{NHPI} = 86 kcal/mol) for lignin oxidation. While PINO abstracts the proton of benzylic C_{α} in lignin model compound (**1a**) to form C_{α} radical intermediate (**1′**), the PINO radical is subsequently reduced by **1a**, regenerating NHPI. Ideally, mediators would be regenerated in solution by the electrode without the need for further mediator addition, thereby avoiding waste generation and cost build-up. The identity of the mediator can be manipulated in order to selectively oxidize or cleave lignin linkages.

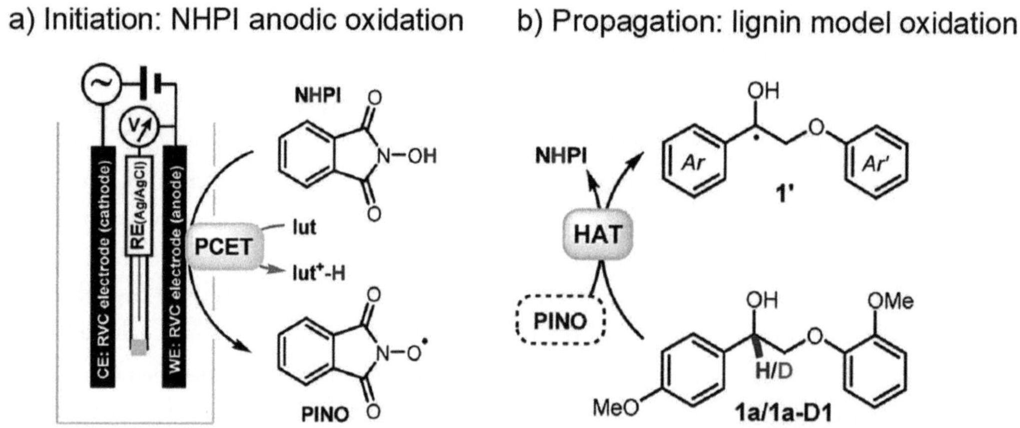

Figure 8. Schematic illustrations of NHPI/2,6-lutidine-mediated electrocatalytic oxidation of lignin systems. a) Initiation step of NHPI anodic oxidation via PCET. b) Propagation pathway proposed under aerobic and anaerobic conditions. Reprinted with permission from reference (69). Copyright 2017 American Chemical Society.

Traditional chemical oxidants, such as HAT agents (e.g., NHPI), and nitroxyls (e.g., TEMPO), can catalyze the oxidation of carbonyl groups in lignin, and different catalysts under various conditions can achieve target chemoselectivities for chemical transformation in lignin. These HATs can be used as mediators (or co-catalysts) in indirect electrocatalytic oxidation systems for lignin oxidation and subsequent C−O/C−C cleavage in lignin. In addition, nitroaromatics (*70*), aqueous polyoxometalates (POMs) (*71*), ferric redox couples (*68, 72*), and halides (*73*) have been reported in more traditional use, but laccase mediators such as NHPI and TEMPO are emerging as the most successful mediators in electrochemical methods (*69, 74–79*).

NHPI-mediated electrocatalytic oxidation of lignin yields high chemoselectivity of benzylic α-carbon oxidation in lignin model compounds and native lignin (*69*). Organic aminoxyls (also called nitroxyls), e.g., 2,2,6,6-tetramethylpiperidine N-oxyl (TEMPO) and TEMPO derivatives, can catalyze the oxidation of benzylic alcohols in different positions depending on the reaction condition, e.g., pH (*74*). This group of catalysts can selectively oxidize benzylic alcohol, which involves the oxidation of secondary alcohols under acidic conditions and the oxidation of primary alcohols with the presence of base (see Scheme 2) (*74, 6–78*). Stahl's group studied the performance of different TEMPO derivatives on the oxidation of alcohols. 4-Acetamido-TEMPO (ACT), one of the TEMPO derivatives (Scheme 2a), was found to have a better performance in lignin oxidation due to its electron-withdrawing substituent in the 4-position and higher nitroxyl/oxoammonium redox potential (*79*).

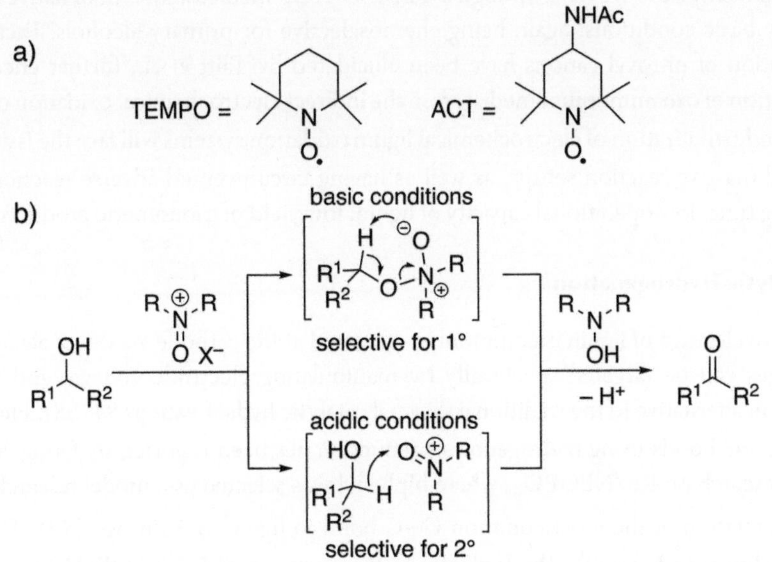

Scheme 2. a) Structure of TEMPO and ACT, (b) Mechanism of alcohol oxidation by oxoammonium under basic and acidic conditions. Reprinted with permission from reference (74). Copyright 2019 American Chemical Society.

The presence of α-carbonyl and γ-carbonyl groups in lignin is essential for bond cleavage and preventing condensation during lignin depolymerization. Density functional theory (DFT) calculations confirm that the energy of C—O bonds can be significantly weakened (by ~14 kcal/mol) with the adjacent carbon being oxidized (*80*). A two-step redox lignin depolymerization method has been studied to cleave the C—O bond in β-O-4 linkage recently. The first step involves the oxidation of benzylic alcohol groups in lignin to reduce the bond dissociation energy of certain bonds (e.g., ether bonds in β-O-4 linkage). Since the binding energy of the ester bond is reduced more with oxidized secondary alcohol, researchers chose secondarily oxidized products for the further cleavage of the β-O-4 linkage using a metal-based photosensitizer (*42*). Meanwhile, the left secondary carbon in primarily oxidized products favors nucleophilic attack. Stahl' s research group performed the retro-aldol reaction for primarily oxidized products, obtaining aromatic carboxylic acids and aromatic aldehydes from C—C bond cleavage (*75*). Accordingly, two main pathways for lignin degradation were developed: C—O bond cleavage of secondary alcohol oxidized lignin and C—C bond cleavage on primary alcohol oxidized lignin. Applying this method to lignin system, a two-step lignin degradation in solution included a chemoselecive benzylic oxidation with a recyclable oxidant ([4-AcNH-TEMPO]BF_4) and a catalytic reductive C—O bond cleavage via the photocatalyst [Ir[ppy]$_2$(dtbbpy)]PF_6] (*15*). Multiple β—O—4-type lignin model compounds with *H*, *G*, or *S* units in the dimers were tested for β-O-4 linkage cleavage. High selectivity and excellent yield of isolated compounds have been obtained. When working as the oxidant in solution, ACT favors the oxidation of primary alcohols in lignin under basic conditions and performs the oxidation of secondary alcohols under acidic conditions (*74*). The study of Sannami et al. about the oxidation of β—O—4-type lignin models indicates that the electrolytes also have a strong effect on the oxidation selectivity between α-carbonylation and γ-carboxylation (*81*). With the combination of TEMPO-type oxidants and heterogeneous electrooxidation, lignin-type aldehyde and acetic acid were produced (*75*). Rafiee et al. also reported the production of a novel polyelectrolyte material with high water solubility and nearly 30% wt yield through a TEMPO-ACT mediator system of native poplar lignin under mildly basic conditions, again being chemoselective for primary alcohols. Factors affecting the regeneration of nitroxyl cations have been elucidated by Taitt et al., further encouraging the experimentation of oxoammonium mediators in the indirect electrochemical oxidation of lignin (*82*).

Future industrialization of electrochemical lignin oxidation systems will face the issue of needing complex and massive reaction setups, as well as having circumvented adverse reaction conditions such as a long time, low operational capacity of lignin, low yield of monomeric products (*83*).

Electrocatalytic Hydrogenation

Reductive cleavage of lignin is commonly performed at the cathode via direct electroreduction. These linkages can be targeted specifically by manipulating electrode voltage, and this method functions as an alternative to the traditional thermal catalytic hydrotreating (*84–88*). Electrocatalytic cleavage of C–C bonds using hydrogenolysis reduction has been reported by Dong et al. through their novel research on $Ru/NbOPO_4$, where biphenyl was selected as a model research compound for its representation of the most common C–C bond in lignin, 5-5 linkage (*89*). Three possible routes were shown in Figure 9. The biphenyl in the presence of Ru/NbOPO4 first converted to phenylcyclohexene by partial hydrogenation. Biphyenyl is cracked into benzene, cyclohexane, and methylcyclopetane (Route 1). Biphenyl is hydrogenated to phenylcyclohexene followed by isomerization and hydrogenation to produce (1-methylcyclopentyl)benzene and then is cracked to the production of benzene and methyl cyclopentane (Route 2). Hydrogenation of biphenyl produces bicyclohexyl (Route 3). It is interesting to note that route 1 is the dominant pathway that shows the efficient electrocatalytic cleavage of C–C bonds, resulting into the produciton of monocyclic hydrocarbons. Moreover, almost complete removal of oxygen was achieved—for the sake of increasing the combustion capabilities of resulting product bio-fuel—and the density of the phosphatic, Bronstead acid sites and the presence of Ru nanoparticles were crucial for optimizing monomer yields. Hydrogenation typically occurs alongside electroreductive hydrogenolysis, as the latter is able to add chemisorbed hydrogen to aromatic rings after the former cleaves ether linkages within lignin polymer structure.

Hydrogenation cells are used to extract excess oxygen from lignin-derived bio-oil for fuel refinery. Phenolic monomers of lignin present in the oil have been traditionally converted through pyrolysis, but electrochemical hydrogenation cells offer a mild alternative method via lignin reduction of adsorbed hydrogen produced by various carbon-supported catalysts and electrolytic solutions (*90*).

A summary table for photocatalytic and electrocatalytic approaches to lignin conversion is shown in Table 1.

Table 1. Summary of Photocatalytic and Electrocatalytic Approaches to Lignin Conversion

	Method	Reaction conditions: Temp.	Light	Potential	Solvent	Atom.	Time	Conversion (%)	Major Product(s)	Ref.
Electro-catalytic method	Pb/PbO_2 anode and Cu cathode	40 °C	-	20 mA cm^{-2}	alkali solution	air	2h	~90%	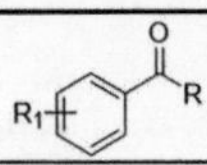 R = H or CH_3 or OH	[54]
	Ni anode	25 - 95 °C	-	38 mA cm^{-2}	alkali solution	air	-	1.5 wt%	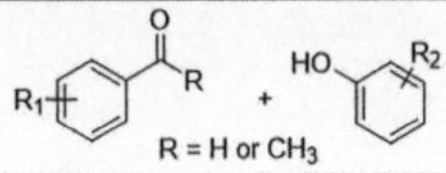 R = H or CH_3	[57]
	Electrolysis with TEMPO or ACT mediator: 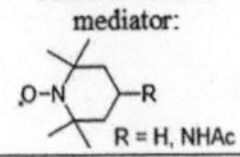 R = H, NHAc	RT	-	0.7 V vs. Ag/AgCl	acetonitrile /H_2O with base	air	5h	98%	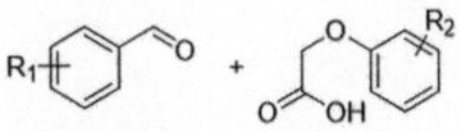	[74]
Photo-catalytic method	$Zn_mIn_2S_{m+3}$	RT	Xe lamp (λ = 400-780 nm)	-	methanol	N_2	4h	99%	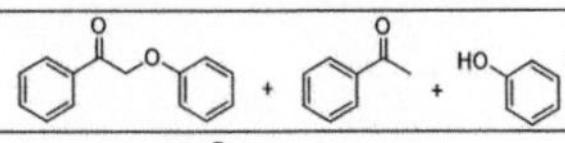	[27]
	Pd/$ZnIn_2S_4$, TiO_2	RT	LED light (λ = 455/365 nm)	-	ethanol	O_2	18h	>99%	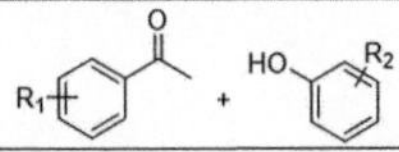	[39]
	Porous Organic Frame 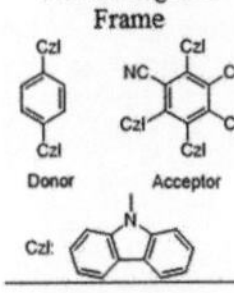	RT	Blue LED	-	acetone	O_2	24h	>99%	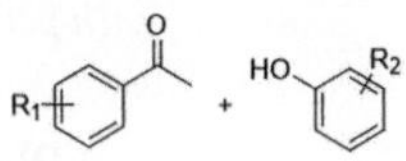	[28]
	Benzoquinone/Cu: 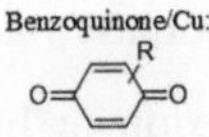	RT	HQI-T[a] metal halide lamp	-	acetone	O_2	16h	84%	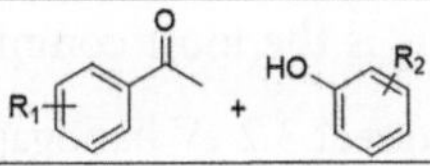	[29]
	Mesoporous graphitic carbon nitride (mpg -C_3N_4)	RT	LED light (λ = 455 nm)	-	acetonitrile	O_2	12h	>98%	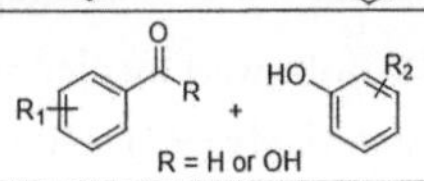 R = H or OH	[30]
	Cadmium sulfide quantum dots (CdS QDs)	RT	LED light (λ = 395 nm)	-	acetonitrile	N_2	20h	>99%	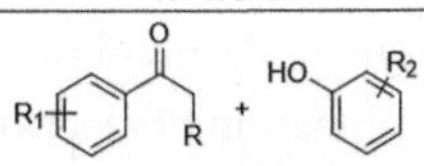 R = H or CH_2OH	[35]
	Ceria and CuO_x clusters	RT	LED light (λ = 455 nm)	-	acetonitrile	O_2	24h	>99%	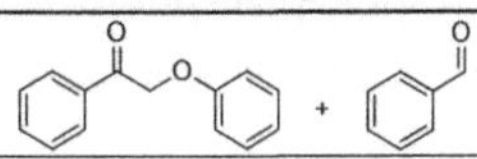	[38]
	Ir-based catalyst 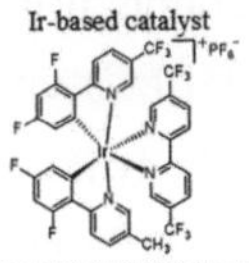	RT	Blue LED	-	acetonitrile	air	12h	>97%	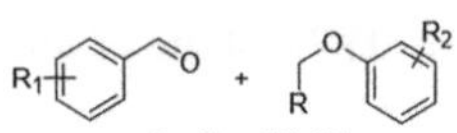 R = H or CH_2OH	[43]
	Vanadium-based catalyst: 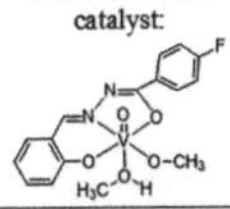	80 °C	AM1.5G (1 sun)	-	acetonitrile	air	24h	100%	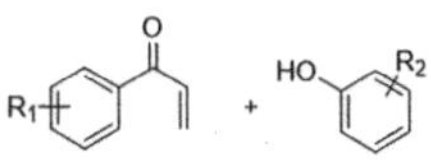	[45]
Redox catalysis	Electrochemical oxidation with reticulated vitreous carbon (RVC) electrode	RT	-	0.64 V vs. Fc^+/Fc	acetonitrile	O_2	4h	87%	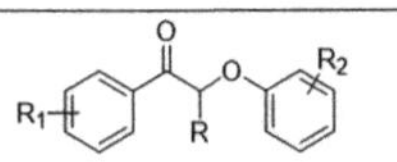 R = H or CH_2OH	[69]
	Ir-based catalyst 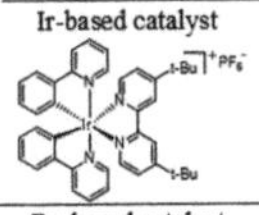	RT	Blue LED	-	acetonitrile	air	-	~90%	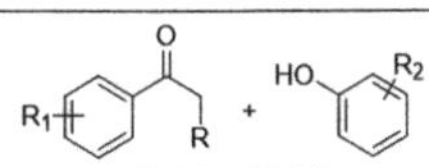 R = H or CH_2OH	
Photo-electro-catalysis	Ru-based catalyst 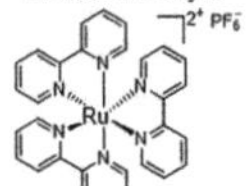	RT	AM1.5 solar light	0.75 V vs. SHE	ACN	air	20h	>91%	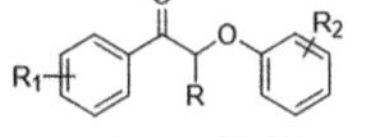 R = H or CH_2OH	[94]

[a]hydrargyrum medium-arc iodide metal halide lamp

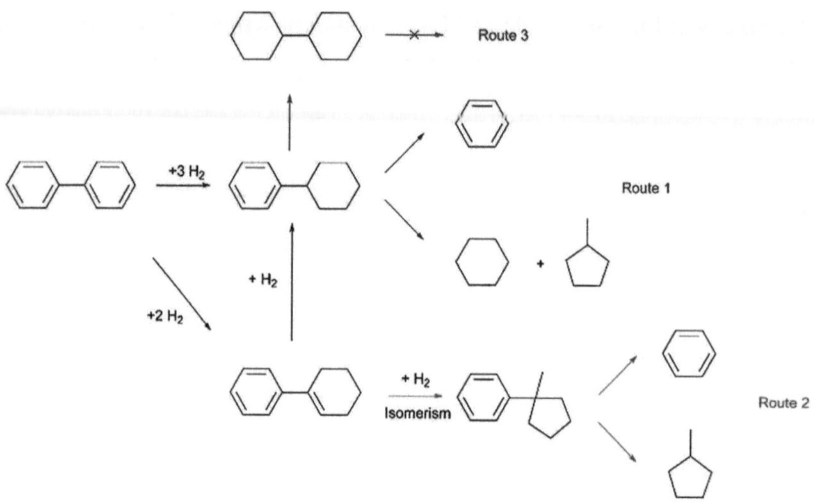

Figure 9. Proposed reaction network of the conversion of biphenyl over the Ru/NbOPO$_4$ catalyst. Reprinted with permission from reference (89). Copyright 2019 Elsevier.

Photoelectrocatalytic Oxidation of Lignin

TiO_2 is the most common semiconductor for the use of photocatalytic biomass valorization (*91*). A direct 3.2 eV bandgap excitation of anatase TiO_2 under light generates oxidizing holes (h^+) under UV, followed by electron-hole separation to the conduction band (CB) and the valence band (VB), respectively (*92*). Due to the wide bandgap of TiO_2, it has been widely reported to be utilized as the photoanode in a photoelectrochemical (PEC) cell for biofuel applications. However, due to the fast recombination of electron-hole pairs prior to the desired photocatalytic oxidative reaction and its wide bandgap, its photocatalytic efficiency is limited (*93*). For enhancing these limitations, a dye-sensitized TiO_2 semiconductor system has been developed. We recently explored the chemoselective oxidation of the 2° benzyl alcohol derivatives in the presence of primary alcohols using a dye-sensitized TiO_2 photoanode in a dye-sensitized photoelectrochemical cell (DSPEC) under mild conditions (*94*). DSPECs include a polypyridyl Ru photocatalyst containing dicarboxylic acid (RuC, ((bis-2,2′-bipyridine)(2,2′-bipyridine-4,4′-dicarboxylic acid) ruthenium(II)), NHPI co-catalyst, and mesoporous TiO_2 electrode onto a fluorine-doped tin oxide glass (TiO_2/FTO) (Figure 10). This photoelectrochemical heterogeneous catalytic process provides a unique foundation to perform the selective C–O bond cleavage essential to the development of lignin conversion technologies.

The possible reaction mechanism for the oxidation of secondary alcohols of a lignin model compound was proposed (Figure 10). The oxidation process is performed via five steps. Under AM1.5G solar light illumination, visible light absorption by the photocatalyst RuC on the TiO_2/FTO surface results in the formation of the excited state (RuC*). The RuC* is sufficiently reduced to sensitize a mesoporous TiO_2 interface. Then, upon the photoexcited electron injection to TiO_2, the RuC^{3+} was formed. It regenerates the ground state following hole transfer to the hydrogen-atom-transfer NHPI/base co-catalyst. The formation of oxidized NHPI affords the active form of the NHPI radical, which is able to selectively abstract a hydrogen atom from the C_α of the β-O-4 linkage. The resulting α-hydroxybenzylic radical species were produced after sequential 1-electron/1-proton oxidation. As a result, the formation of the ketone intermediate is shown. The RuC photocatalyst and

NHPI are regenerated in this process and can react with additional lignin equivalents in a DSPEC system. This combination of photocatalysis and electrocatalytic oxidation enables lignin oxidation with low applied bias at room temperature.

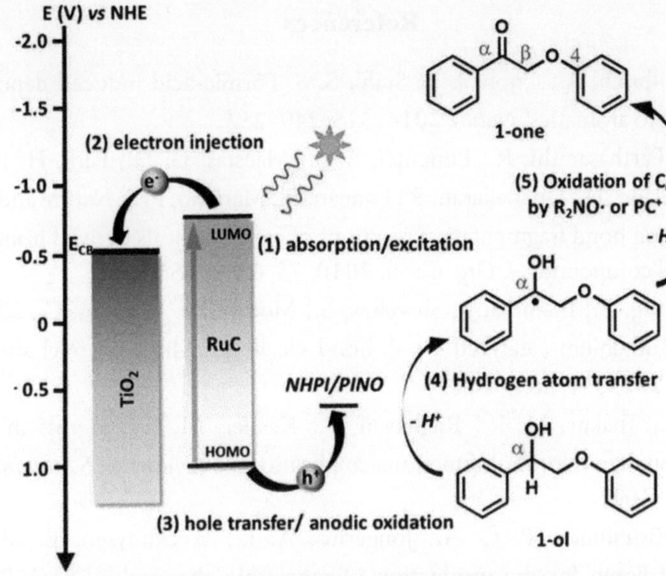

*Figure 10. Proposed mechanism for the photocatalytic oxidative conversion of the secondary alcohol in the model compound, 2-phenoxy-1-phenyl ethanol (**1-ol**), to a ketone (**1-one**). Reprinted with permission from reference (94). Copyright 2020 American Chemical Society.*

Summary

Lignin is a by-product in biorefinery industries and underutilized, wasting precious carbon as a heat source. Lignin valorization has been largely attractive due to the generation of important aromatic chemicals and valuable products as renewable materials. Conventional thermally activated homogeneous or heterogeneous catalysis has been developed for lignin valorization at relatively high temperatures and pressures. In terms of the selectivity of the products, the electro- and photocatalytic processes are advantageous to chemoselective cleavage of the C_β-O and C_α-C_β bonds of lignin model compounds and real lignin for lignin degradation under milder conditions such as room temperature and atmospheric pressure. Especially, the concept of utilizing visible light energy at room temperature is an innovative approach to selectively convert lignin to small molecular weight aromatic compounds. Photocatalytic or photoelectrocatalytic processes can minimize unfavorable side reactions and enhance catalytic efficiency of lignin depolymerization under mild conditions, as opposed to electrochemically activated catalysis with high applied bias and the limited options for lignin bond-selection. The challenge still lies in understanding the mechanistic processes of the complex systems containing semiconductors, electrolytes, mediators, photocatalysis, and light sources.

Acknowledgments

Financial support from State University of New York, College of Environmental Science and Forestry (SUNY ESF) is gratefully acknowledged. Oak Ridge National Laboratory is managed by UT-Battelle, LLC, under contract DE-AC05-00OR22725 with the U.S. Department of Energy

(DOE). This work was supported and performed as part of the Center for Bioenergy Innovation (CBI). The CBI is a U.S Department of Energy Bioenergy Research Center supported by the Office of Biological and Environmental Research in the DOE Office of Science.

References

1. Rahimi, A.; Ulbrich, A.; Coon, J. J.; Stahl, S. S. Formic-acid-induced depolymerization of oxidized lignin to aromatics. *Nature* **2014**, *515*, 249–252.
2. Cho, D. W.; Parthasarathi, R.; Pimentel, A. S.; Maestas, G. D.; Park, H. J.; Yoon, U. C.; Dunaway-Mariano, D.; Gnanakaran, S.; Langan, P.; Mariano, P. S. Nature and kinetic analysis of carbon-carbon bond fragmentation reactions of cation radicals derived from SET-oxidation of lignin model compounds. *J. Org. Chem.* **2010**, *75*, 6549–6562.
3. Stein, T.; Hartog, T.; Buendia, J.; Stoychev, S.; Mottweiler, J.; Bolm, C.; Klankermayer, J.; Leitner, W. Ruthenium-catalyzed C−C bond cleavage in lignin model substrates. *Angew. Chem. Int. Ed.* **2015**, *54*, 5859–5863.
4. Thakur, V. K.; Thakur, M. K.; Raghavan, P.; Kessler, M. R. Progress in green polymer composites from lignin for multifunctional applications: a review. *ACS Sustainable Chem. Eng.* **2014**, *2*, 1072–1092.
5. Zakzeski, J.; Bruijnincx, P. C. A.; Jongerius, A. L.; Weckhuysen, B. M. The catalytic valorization of lignin for the production of renewable chemicals. *Chem. Rev.* **2010**, *110*, 3552–3599.
6. Deuss, P. J.; Scott, M.; Tran, F.; Westwood, N. J.; Vries, J. G.; Barta, K. Aromatic monomers by in situ conversion of reactive intermediates in the acid-catalyzed depolymerization of lignin. *J. Am. Chem. Soc.* **2015**, *137*, 7456–7467.
7. Jastrzebski, R.; Constant, S.; Lancefield, C. S.; Westwood, N. J.; Weckhuysen, B. M.; Bruijnincx, P. C. A. Tandem catalytic depolymerization of lignin by water-tolerant lewis acids and rhodium complexes. *ChemSusChem* **2016**, *9*, 2074–2079.
8. Nguyen, S. T.; Murray, P. R. D.; Knowles, R. R. Light-driven depolymerization of native lignin enabled by proton-coupled electron transfer. *ACS Catal.* **2020**, *10*, 800–805.
9. Liu, W.-J.; Jiang, H.; Yu, H.-Q. Thermochemical conversion of lignin to functional materials: a review and future directions. *Green Chem.* **2015**, *17*, 4888–4907.
10. Pandey, M. P.; Kim, C. S. Lignin depolymerization and conversion: a review of thermochemical methods. *Chem. Eng. Technol.* **2011**, *34*, 29–41.
11. Ma, R.; Xu, Y.; Zhang, X. Catalytic oxidation of biorefinery lignin to value-added chemicals to support sustainable biofuel production. *ChemSusChem* **2015**, *8*, 24–51.
12. Behling, R.; Valange, S.; Chatel, G. Heterogeneous catalytic oxidation for lignin valorization into valuable chemicals: what results? what limitations? what trends? *Green Chem.* **2016**, *18*, 1839–1854.
13. Lanzalunga, O.; Bietti, M. Photo- and radiation chemical induced degradation of lignin model compounds. *J. Photochem. Photobiol. B* **2000**, *56*, 85–108.
14. Granone, L. I.; Sieland, F.; Zheng, N.; Dillert, R.; Bahnemann, D. W. Photocatalytic conversion of biomass into valuable products: a meaningful approach? *Green Chem.* **2018**, *20*, 1169–1192.

15. Nguyen, J.; Matsuura, B. S.; Stephenson, C. R. J. A photochemical strategy for lignin degradation at room temperature. *J. Am. Chem. Soc.* **2014,** *136*, 1218–1221.
16. Hanson, S. K.; Baker, R. T. Knocking on wood: base metal complexes as catalysts for selective oxidation of lignin models and extracts. *Acc. Chem. Res.* **2015,** *48*, 2037–2048.
17. Mottweiler, J.; Puche, M.; Räuber, C.; Schmidt, T.; Concepción, P.; Corma, A.; Bolm, C. Copper- and vanadium-catalyzed oxidative cleavage of lignin using dioxygen. *ChemSusChem* **2015,** *8*, 2106–2113.
18. Lancefield, C. S.; Teunissen, L. W.; Weckhuysen, B. M.; Bruijnincx, P. C. A. Iridium-catalysed primary alcohol oxidation and hydrogen shuttling for the depolymerization of lignin. *Green Chem.* **2018,** *20*, 3214–3221.
19. Nichols, J. M.; Bishop, L. M.; Bergman, R. G.; Ellman, J. A. Catalytic C–O bond cleavage of 2-aryloxy-1-arylethanols and its application to the depolymerization of lignin-related polymers. *J. Am. Chem. Soc.* **2010,** *132*, 12554–12555.
20. Zhu, R.; Wang, B.; Cui, M.; Deng, J.; Li, X.; Ma, Y.; Fu, Y. Chemoselective oxidant-free dehydrogenation of alcohols in lignin using Cp*Ir catalysts. *Green Chem.* **2016,** *18*, 2029–2036.
21. Hanson, S. K.; Baker, R. T.; Gordon, J. C.; Scott, B. L.; Thorn, D. L. Aerobic oxidation of lignin models using a base metal vanadium catalyst. *Inorg. Chem.* **2010,** *49*, 5611–5618.
22. Son, S.; Toste, F. D. Non-oxidative vanadium-catalyzed C–O bond cleavage: application to degradation of lignin model compounds. *Angew. Chem. Int. Ed.* **2010,** *49*, 3791–3794.
23. Biannic, B.; Bozell, J. J. Efficient cobalt-catalyzed oxidative conversion of lignin models to benzoquinones. *Org. Lett.* **2013,** *15*, 2730–2733.
24. Sergeev, A. G.; Hartwig, J. F. Selective, nickel-catalyzed hydrogenolysis of aryl ethers. *Science* **2011,** *332*, 439–443.
25. Liu, X.; Duan, X.; Wei, W.; Wang, S.; Ni, B.-J. Photocatalytic conversion of lignocellulosic biomass to valuable products. *Green Chem.* **2019,** *21*, 4266–4289.
26. Prier, C. K.; Rankic, D. A.; MacMillan, D. W. C. Visible light photoredox catalysis with transition metal complexes: applications in organic synthesis. *Chem. Rev.* **2013,** *113*, 5322–5363.
27. Lin, J.; Wu, X.; Xie, S.; Chen, L.; Zhang, Q.; Deng, W.; Wang, Y. Visible-light-driven cleavage of C–O linkages for lignin valorization to functionalized aromatics. *ChemSusChem* **2019,** *12*, 5023–5031.
28. Luo, J.; Zhang, X.; Lu, J.; Zhang, J. Fine tuning the redox potentials of carbazolic porous organic frameworks for visible-light photoredox catalytic degradation of lignin β-O-4 models. *ACS Catal.* **2017,** *7*, 5062–5070.
29. Mitchell, L. J.; Moody, C. J. Solar photochemical oxidation of alcohols using catalytic hydroquinone and copper nanoparticles under oxygen: oxidative cleavage of lignin models. *J. Org. Chem.* **2014,** *79*, 11091–11100.
30. Liu, H.; Li, H.; Lu, J.; Zeng, S.; Wang, M.; Luo, N.; Xu, S.; Wang, F. Photocatalytic cleavage of C–C bond in lignin models under visible light on mesoporous graphitic carbon nitride through $\pi-\pi$ stacking interaction. *ACS Catal.* **2018,** *8*, 4761–4771.
31. Guo, Q.; Ma, Z.; Zhou, C.; Ren, Z.; Yang, X. Single molecule photocatalysis on TiO_2 surfaces. *Chem. Rev.* **2019,** *119*, 11020–11041.

32. Li, S.-H.; Liu, S.; Colmenares, J. C.; Xu, Y.-J. A sustainable approach for lignin valorization by heterogeneous photocatalysis. *Green Chem.* **2016**, *18*, 594–607.

33. Chen, C.; Liu, P.; Xia, H.; Zhou, M.; Zhao, J.; Sharma, B. K.; Jiang, J. Photocatalytic cleavage of β-O-4 ether bonds in lignin over Ni/TiO_2. *Molecules* **2020**, *25*, 2109.

34. Hongdilokkul, P.; Chuangchote, S.; Laosiripojana, N.; Sagawa, T. Conversion of lignin via photocatalysis using synthesized Ag-TiO_2 photocatalysts sintered under different atmospheres. *J. Sustainable. Energy Environ.* **2017**, *8*, 101–105.

35. Wu, X.; Fan, X.; Xie, S.; Lin, J.; Cheng, J.; Zhang, Q.; Chen, L.; Wang, Y. Solar energy-driven lignin-first approach to full utilization of lignocellulosic biomass under mild conditions. *Nat. Catal.* **2018**, *1*, 772–780.

36. Wu, X.; Xie, S.; Liu, C.; Zhou, C.; Lin, J.; Kang, J.; Zhang, Q.; Wang, Z.; Wang, Y. Ligand-controlled photocatalysis of CdS quantum dots for lignin valorization under visible light. *ACS Catal.* **2019**, *9*, 8443–8451.

37. Han, G.; Yan, T.; Zhang, W.; Zhang, Y. C.; Lee, D. Y.; Cao, Z.; Sun, Y. Highly selective photocatalytic valorization of lignin model compounds using ultrathin metal/CdS. *ACS Catal.* **2019**, *9*, 11341–11349.

38. Hou, T.; Luo, N.; Li, H.; Heggen, M.; Lu, J.; Wang, Y.; Wang, F. Yin and yang dual characters of CuO_x clusters for C–C bond oxidation driven by visible light. *ACS Catal.* **2017**, *7*, 3850–3859.

39. Luo, N.; Wang, M.; Li, H.; Zhang, J.; Liu, H.; Wang, F. Photocatalytic oxidation-hydrogenolysis of lignin β-O-4 models via a dual light wavelength switching strategy. *ACS Catal.* **2016**, *6*, 7716–7721.

40. Magallanes, G.; Kärkäs, M. D.; Bosque, I.; Lee, S.; Maldonado, S.; Stephenson, C. R. J. Selective C–O bond cleavage of lignin systems and polymers enabled by sequential palladium-catalyzed aerobic oxidation and visible-light photoredox catalysis. *ACS Catal.* **2019**, *9*, 2252–2260.

41. Figueiédo, M. B.; Venderbosch, R. H.; Deuss, P. J.; Heeres, H. J. A two-step approach for the conversion of technical lignins to biofuels. *Adv. Sustainable Syst.* **2020**, *1900147* (1-13)

42. Liu, Y.; Li, C.; Miao, W.; Tang, W.; Xue, D.; Li, C.; Zhang, B.; Xiao, J.; Wang, A.; Zhang, T.; Wang, C. Mild redox-neutral depolymerization of lignin with a binuclear Rh complex in water. *ACS Catal.* **2019**, *9*, 4441–4447.

43. Wang, Y.; Liu, Y.; He, J.; Zhang, Y. Redox-neutral photocatalytic strategy for selective C–C bond cleavage of lignin and lignin models via PCET process. *Sci. Bull.* **2019**, *64*, 1658–1666.

44. Chan, J. M. W.; Bauer, S.; Sorek, H.; Sreekumar, S.; Wang, K.; Toste, F. D. Studies on the vanadium-catalyzed nonoxidative depolymerization of *Miscanthus giganteus*-drived lignin. *ACS Catal.* **2013**, 3, 1369–1377.

45. Gazi, S.; Ng, W. K. H.; Ganguly, R.; Moeljade, A. M. P.; Hirao, H.; Soo, H. S. Selective photocatalytic C–C bond cleavage under ambient conditions with earth abundant vanadium complexes. *Chem. Sci.* **2015**, *6*, 7130–7142.

46. Maruyama, K.; Kubo, Y. In *CRC Handbook of organic photochemistry and photobiology*; Horspool, W. M., Song, P.-S., Eds.; CRC Press: Boca Raton, FL, 1995, 748–756.

47. Li, H.; Bunrit, A.; Lu, J.; Gao, Z.; Luo, N.; Liu, H.; Wang, F. Photocatalytic cleavage of aryl ether in modified lignin to non-phenolic aromatics. *ACS Catal.* **2019**, *9*, 8843–8851.

48. Liu, F.; Liu, Q.; Wang, A.; Zhang, T. Direct catalytic hydrogenolysis of kraft lignin to phenols in choline-derived ionic liquids. *ACS Sustainable Chem. Eng.* **2016**, *4*, 3850–3856.
49. Kang, Y.; Lu, X.; Zhang, G.; Yao, X.; Xin, J.; Yang, S.; Yang, Y.; Xu, J.; Feng, M.; Zhang, S. Metal-free photochemical degradation of lignin-derived aryl ethers and lignin by autologous radicals through ionic liquid induction. *ChemSusChem* **2019**, *12*, 4005–4013.
50. Stiefel, S.; Marks, C.; Schmidt, T.; Hanisch, S.; Spalding, G.; Wessling, M. Overcoming lignin heterogeneity: reliably characterizing the cleavage of technical lignin. *Green Chem.* **2016**, *18*, 531–540.
51. Comninellis, C. Electrocatalysis in the electrochemical conversion/combustion of organic pollutants for waste water treatment. *Electrochim. Acta* **1994**, *39*, 1857–1862.
52. Hao, X.; Quansheng, Y.; Dan, S.; Honghui, Y.; Jidong, L.; Jiangtao, F.; Wei, Y. Fabrication and characterization of PbO_2 electrode modified with $[Fe(CN)_6]^{3-}$ and its application on electrochemical degradation of alkali lignin. *J. Hazard. Mater.* **2015**, *286*, 509–516.
53. NaderiNasrabadi, M.; Bateni, F.; Chen, Z.; Harrington, P. B.; Staser, J. A. Biomass-depolarized electrolysis. *J. Electrochem. Soc.* **2019**, *166*, F1037–F1046.
54. Liu, M.; Wen, Y.; Qi, J.; Zhang, S.; Li, G. Fine chemicals prepared by bamboo lignin degradation through electrocatalytic redox between Cu cathode and Pb/PbO_2 anode in alkali solution. *ChemistrySelect* **2017**, *2*, 4956–4962.
55. Cai, P.; Fan, H.; Cao, S.; Qi, J.; Zhang, S.; Li, G. Electrochemical conversion of corn stover lignin to biomass-based chemicals between Cu/Ni-Mo-Co cathode and Pb/PbO_2 anode in alkali solution. *Electrochim. Acta* **2018**, *264*, 128–139.
56. Oshckepkov, A. G.; Braesch, G.; Ould-Amara, S.; Rostamikia, G.; Maranzana, G.; Bonnefont, A.; Papaefthimiou, V.; Janik, M. J.; Chatenet, M.; Savinova, E. R. Nickel metal nanoparticles as anode electrocatalysts for highly efficient direct borohydride fuel cells. *ACS Catal.* **2019**, *9*, 8520–8528.
57. Zirbes, M.; Schmitt, D.; Beiser, N.; Pitton, D.; Hoffmann, T.; Waldvogel, S. R. Anodic degradation of lignin at active transition metal-based alloys and performance-enhanced anodes. *ChemElectroChem* **2019**, *6*, 155–161.
58. Zhang, Z.; Liu, J.; Gu, J.; Su, L.; Cheng, L. An overview of metal oxide materials as electrocatalysts and supports for polymer electrolyte fuel cells. *Energy. Environ. Sci.* **2014**, *7*, 2535–2558.
59. Tolba, R.; Tian, M.; Wen, J.; Jiang, Z.-H.; Chen, A. Electrochemical oxidation of lignin at IrO_2-based oxide electrodes. *J. Electroanal. Chem.* **2010**, *649*, 9–15.
60. Rauber, D.; Dier, T. K. F.; Volmer, D. A.; Hempelmann, R. Electrochemical lignin degradation in ionic liquids on ternary mixed metal electrodes. *Z. Phys. Chem.* **2018**, *232*, 189–208.
61. Shiraishi, T.; Takano, T.; Kamitakahara, H.; Nakatsubo, F. Studies on electrooxidation of lignin and lignin model compounds. Part 1: direct electrooxidation of non-phenolic lignin model compounds. *Holzforschung* **2012**, *66*, 303–309.
62. Du, X.; Zhang, H.; Sullivan, K. P.; Gogoi, P.; Deng, Y. Electrochemical lignin conversion. *ChemSusChem* **2020**, *13*, 1–27.
63. Chen, J.; Yang, H.; Fu, H.; He, H.; Zheng, Q.; Li, X. Electrochemical oxidation mechanisms for selective products due to C–O and C–C cleavages of β-O-4 linkages in lignin model compounds. *Phys. Chem. Chem. Phys.* **2020**, *22*, 11508–11518.

64. Lalvani, S. B.; Rajagopal, P. Hydrogen production from lignin-water solution by electrolysis. *Holzforschung* **1993**, *47*, 283–286.
65. Barbier, J.; Charon, N.; Dupassieux, N.; Loppinet-Serani, A.; Mahé, L.; Ponthus, J.; Courtiade, M.; Ducrozet, A.; Quoineaud, A.-A.; Cansell, F. Hydrothermal conversion of lignin compounds. A detailed study of fragmentation and condensation reaction pathways. *Biomass Bioenerg.* **2012**, *46*, 479–491.
66. Liu, W.; Cui, Y.; Du, X.; Zhang, Z.; Chao, Z.; Deng, Y. High efficiency hydrogen evolution from native biomass electrolysis. *Energy Environ. Sci.* **2016**, *9*, 467–472.
67. Du, X.; Liu, W.; Zhang, Z.; Mulyadi, A.; Brittain, A.; Gong, J.; Deng, Y. Low-energy catalytic electrolysis for simultaneous hydrogen evolution and lignin depolymerization. *ChemSusChem* **2017**, *10*, 847–854.
68. Yang, L.; Liu, W.; Zhang, Z.; Du, X.; Gong, J.; Dong, L.; Deng, D. Hydrogen evolution from native biomass with Fe^{3+}/Fe^{2+} redox couple catalyzed electrolysis. *Electrochimica Acta* **2017**, *246*, 1163–1173.
69. Bosque, I.; Magallanes, G.; Rigoulet, M.; Kärkäs, M. D.; Stephenson, C. R. J. Redox catalysis facilitates lignin depolymerization. *ACS Cent. Sci.* **2017**, *3*, 621–628.
70. Smith, C.; Utley, J. H. P.; Petrescu, M.; Viertler, H. Biomass electrochemistry: anodic oxidation of an organo-solv lignin in the presence of nitroaromatics. *J. Appl. Electrochem.* **1989**, *19*, 535–539.
71. Liu, W.; Cui, Y.; Du, X.; Zhang, Z.; Chao, Z.; Deng, Y. High efficiency hydrogen evolution from native biomass electrolysis. *Energy Environ. Sci.* **2016**, *9*, 467–472.
72. Mao, L.; Zhang, L.; Gao, N.; Li, A. $FeCl_3$ and acetic acid co-catalyzed hydrolysis of corncob for improving furfural production and lignin removal from residue. *Bioresour. Technol.* **2012**, *123*, 324–331.
73. Liang, S.; Zeng, C.-C.; Tian, H.-Y.; Sun, B.-G.; Luo, X.-G.; Ren, F.-Z. Electrochemically oxidative α-C-H functionalization of ketones: a cascade synthesis of α-amino ketones mediated by NH_4I. *J. Org. Chem.* **2016**, *81*, 11565–11573.
74. Rafiee, M.; Alherech, M.; Karlen, S. D.; Stahl, S. S. Electrochemical aminoxyl-mediated oxidation of primary alcohols in lignin to carboxylic acids: polymer modification and depolymerization. *J. Am. Chem. Soc.* **2019**, *141*, 15266–15276.
75. Rahimi, A.; Azarpira, A.; Kim, H.; Ralph, J.; Stahl, S. S. Chemoselective metal-free aerobic alcohol oxidation in lignin. *J. Am. Chem. Soc.* **2013**, *135*, 6415–6418.
76. Das, A.; Rahimi, A.; Ulbrich, A.; Alherech, M.; Motagamwala, A. H.; Bhalla, A.; Costa Sousa, L.; Balan, V.; Dumesic, J. A.; Hegg, E. L. Lignin conversion to low-molecular-weight aromatics via an aerobic oxidation-hydrolysis sequence: comparison of different lignin sources. *ACS Sustainable Chem. Eng.* **2018**, *6*, 3367–3374.
77. Wang, M.; Lu, J.; Zhang, X.; Li, L.; Li, H.; Luo, N.; Wang, F. Two-step, catalytic C–C bond oxidative cleavage process converts lignin models and extracts to aromatic acids. *ACS Catal.* **2016**, *6*, 6086–6090.
78. Dabral, S.; Hernández, J. G.; Kamer, P. C. J.; Bolm, C. Organocatalytic Chemoselective Primary Alcohol Oxidation and Subsequent Cleavage of Lignin Model Compounds and Lignin. *ChemSusChem* **2017**, *10*, 2707–2713.

79. Rafiee, M.; Miles, K. C.; Stahl, S. S. Electrocatalytic alcohol oxidation with TEMPO and bicyclic nitroxyl derivatives: driving force trumps steric effects. *J. Am. Chem. Soc.* **2015**, *137*, 14751–14757.
80. Kim, S.; Chmely, S. C.; Nimlos, M. R.; Bomble, Y. J.; Foust, T. D.; Paton, R. S.; Beckham, G. T. Computational study of bond dissociation enthalpies for a large range of native and modified lignins. *J. Phys. Chem. Lett.* **2011**, 2, 2846–2852.
81. Sannami, Y.; Kamitakahara, H.; Takano, T. TEMPO-mediated electro-oxidation reactions of non-phenolic β-O-4-type lignin model compounds. *Holzforschung* **2017**, *71*, 109–117.
82. Taitt, B. J.; Bender, M. T.; Choi, K.-S. Impacts of the regeneration pathways of the oxoammonium cation on electrochemical nitroxyl radical-mediated alcohol oxidation. *ACS Catal.* **2020**, *10*, 265–275.
83. Ha, J.-M.; Hwang, K.-R.; Kim, Y.-M.; Jae, J.; Kim, K. H.; Lee, H. W.; Kim, J.-Y.; Park, Y.-K. Recent progress in the thermal and catalytic conversion of lignin. *Renew. Sust. Energy Reviews* **2019**, *111*, 422–441.
84. Davisson, V. J.; Woodside, A. B.; Neal, T. R.; Stremler, K. E.; Muehlbacher, M.; Poulter, C. D. Phosphorylation of isoprenoid alcohols. *J. Org. Chem.* **1986**, *51*, 4768–4779.
85. Cyr, A.; Chiltz, F.; Jeanson, P.; Martel, A.; Brossard, L.; Lessard, J.; Ménard, H. Electrocatalytic hydrogenation of lignin models at Raney nickel and palladium-based electrodes. *Can. J. Chem.* **2000**, *78*, 307–315.
86. Mahdavi, B.; Lafrance, A.; Martel, A.; Lessard, J.; Ménard, H.; Brossard, L. Electrocatalytic hydrogenolysis of lignin model dimers at Raney nickel electrodes. *J. Appl. Electrochem.* **1997**, *27*, 605–611.
87. Wu, W.-B.; Huang, J.-M. Electrochemical cleavage of aryl ethers promoted by sodium borohydride. *J. Org. Chem.* **2014**, *79*, 10189–10195.
88. Dabo, P.; Cyr, A.; Lessard, J.; Brossard, L.; Ménard, H. Electrocatalytic hydrogenation of 4-phenoxyphenol on active powders highly dispersed in a reticulated vitreous carbon electrode. *Can. J. Chem.* **1999**, *77*, 1225–1229.
89. Dong, L.; Lin, L.; Han, X.; Si, X.; Liu, X.; Guo, Y.; Lu, F.; Rudic, S.; Parker, S. F.; Yang, S.; Wang, Y. Breaking the limit of lignin monomer production via cleavage of interunit carbon-carbon linkages. *Chem.* **2019**, *5*, 1521–1536.
90. Carneiro, J.; Nikolla, E. Electrochemical conversion of biomass-based oxygenated compounds. *Annu. Rev. Chem.* **2019**, *10*, 85–104.
91. Butburee, T.; Chakthranont, P.; Phawa, C.; Faungnawakij, K. Beyond artificial photosynthesis: prospects on photobiorefinery. *ChemCatChem* **2020**, *12*, 1873–1890.
92. Frischmann, P. D.; Mahata, K.; Würthner, F. Powering the future of molecular artificial photosynthesis with light-harvesting metallosupramolecular dye assemblies. *Chem. Soc. Rev.* **2013**, *42*, 1847–1870.
93. Colmenares, J. C.; Varma, R. S.; Nair, V. Selective photocatalysis of lignin-inspired chemicals by integrating hybrid nanocatalysis in microfluidic reactors. *Chem. Soc. Rev.* **2017**, *46*, 6675–6686.
94. Li, S.; Li, Z-J.; Yu, H.; Sytu, M. R.; Wang, Y.; Beeri, D.; Zheng, W.; Sherman, B. D.; Yoo, C. G.; Leem, G. Solar-driven lignin oxidation via hydrogen atom transfer with a dye-sensitized TiO_2 photoanode. *ACS Energy Lett.* **2020**, *5*, 777–784.

Chapter 6

Recent Advances in Lignin-Based Carbon Materials Preparation and Their Application in Catalysis

Huan Wang and Dong-Jie Yang*

South China University of Technology, Guangzhou 510640, China

***Email: cedjyang@scut.edu.cn.**

Lignin is a renewable resource with abundant natural reserves, high carbon content, three-dimensional network structure, and a large number of conjugated structures. Carbon nanomaterials have great application value in the fields of catalysts, especially in the photocatalysis, thermal catalysis and electrocatalysis. Using industrial lignin from the biorefining and pulping and paper industries as raw materials, preparing a lignin-based carbon catalyst with high catalytic activity through precise control of its microstructure characteristics is one of the most effective ways to realize the ultra-high added value of lignin. The research of lignin carbon-based catalytic materials is a cross-cutting subject involving chemistry, chemical engineering, physics, engineering and other disciplines. How to control the preparation of lignin-based carbon catalysts with excellent performance and good stability is still a challenging subject. This review article aims to summarize recent advances in lignin-based carbon materials preparation and their application as catalyst carrier and modifier in photocatalysis, thermocatalysis and electrocatalysis which are used for organic wastewater treatment and new energy. In future research, in order to prepare lignin-based carbon catalysts with excellent properties, the molecular structure, microaggregate structure and regulatory mechanism of lignin should be investigated in depth, so as to make better use of its three-dimensional network structure and a large number of aromatic structures. The research focuses on the structure-activity relationship between the structural properties of lignin-based carbon materials and their application performance, and establishes the control theory of the preparation process of lignin-based carbon catalysts, which lays a theoretical foundation for the realization of the application of lignin in the field of new catalytic materials with high added value.

Introduction

Biomass resources have the advantages of being cheap, easy to obtain, environmentally friendly and renewable. Its high value-added development and utilization has important practical significance for alleviating the increasingly severe resource crisis and environmental pollution problems. Lignin, as the second-largest component in natural plants, more than 100 million tons of industrial lignin is produced by pulping, papermaking and biorefining in the world every year (*1–6*). Only ~10% of them are effectively used, and most of them are directly discharged as industrial waste or burned as low-value fuels, which not only causes waste of resources but also creates new environmental pollution problems (*7–15*). Therefore, the high-value utilization of industrial lignin not only has important economic value, but also alleviates environmental pollution problems and resource crises. Lignin has broad application prospects in the fields of nano-functional materials, chemical additives, sunscreen agents, and polymer material dopants due to its characteristics of a three-dimensional network molecular structure, a large number of conjugated structures, and rich oxygen-containing functional groups (*16–23*). The high value utilization of lignin in the field of nano-functional materials mainly includes the following aspects: first, utilizing the amphiphilicity to prepare lignin nanocolloid spheres. Secondly, the preparation of lignin/oxide nanocomposites utilizing the large amount of functional groups and three-dimensional network molecular structure. Third, utilizing the high carbon content and large amount of benzene ring structure to prepare of lignin-based carbon nanomaterials. Among them, the preparation of lignin-based carbon nanomaterials and their high-value applications have gradually become a current research focus (*24–35*). Lignin is a three-dimensional network of natural polyaromatic macromolecular compounds connected by C-C bonds and ether bonds with phenylpropane units as the skeleton, which is an ideal carbon material precursor due to its high carbon content and phenylpropane units.

Lignin-based carbon material is one of the most potential catalyst substrate materials because of its adjustable micro-pore structure, stable chemical properties, good electrical conductivity and high specific surface area (*26, 27, 32, 33*). In recent years, researchers have reported many novel preparation technologies and high value-added applications of lignin carbon-based catalytic materials. The preparation of lignin carbon-based materials and their application in energy storage materials and new catalytic materials are mainly as follows: First, the preparation of heteroatom-doped lignin porous carbon has broad application prospects in the field of energy conversion and storage. The second is to load metal catalysts on lignin carbon materials to prepare lignin carbon/metal nanocomposites, and apply them in the field of electrocatalytic energy conversion. The third is to prepare lignin carbon/metal oxide nanocomposite materials and explore their high-value utilization in the field of photocatalytic wastewater treatment. This chapter reviews the international research progress and development trends in the preparation of lignin carbon-based materials and the application of catalysts. In combination with the research progress of the author's research team in recent years, mainly to introduce and summary of the preparation of lignin porous carbon, lignin-based carbon/metal nanocomposite, and lignin-based carbon/metal oxide nanocomposite and their applications in photocatalysis, thermal catalysis and electrocatalysis.

Preparation of Lignin-Derived Carbon Materials

Controllable preparation of lignin-based carbon materials has been widely reported and become a research hotspot in recent years, such as the preparation of lignin-based porous carbon materials due to its high carbon content and three-dimensional network structure to prepare, the utilization of rich benzene rings and conjugate structures contained to prepare lignin-based carbon fibers, use of its hydrophilic and lipophilic amphiphilic structural characteristics to control the preparation of lignin-based carbon nanosheets through self-assembly process, utilizing the abundant active sites to prepare lignin-based carbon/metal nanocomposites and lignin-based carbon/metal oxide nanocomposites. This section will give a detailed overview of the international research progress in the preparation of lignin carbon-based materials.

Lignin-Based Porous Carbon

The commonly used preparation methods of lignin-derived carbon materials mainly include template method, activation method, hydrothermal method, electrostatic spinning method, molten salt method, etc. The lignin-derived carbon obtained by different preparation methods has different microstructure characteristics, there has more macroporous and mesoporous by template method, there has more microporous obtained by the activation method, there has a special spherical or two-dimensional lamellar structure by hydrothermal method. The lignin-derived carbon has a regular fiber structure obtained by the electrospinning method. Furthermore, many researchers also developed some preparation methods that have both template and activation effects in order to control the microstructure characteristics of lignin-derived carbon materials.

Porous carbon has a wide range of applications in the field of catalysis and adsorption attributed to the high specific surface area, abundant pore structure and adjustable active sites. The preparation of lignin-derived porous carbon with hierarchical porous structure and high specific surface area and heteroatom-doped porous carbon has gradually become a research focus in recent years. *Xie, Dai, Chen, Liu, Ge, Ma, Zhang, Zhou, Tian, Li, and Yan* (36) prepared lignin-based porous carbon using sodium lignosulfonate as a carbon source and a large amount of NaCl as a hard template carbonized at 500 °C. The obtained sample had a specific surface area of 438 $m^2 \cdot g^{-1}$, and its pore structure is mainly mesoporous with a small amout of micropores. *Zhang, Lin H, Lin Z, Yin, Lu, Liu, and Zhao* (37) reported a lignin-derived porous carbon using with alkali lignin as raw material and potassium hydroxide as activator. The prepared lignin-derived porous carbon had hierarchical three-dimensional pore structure characteristics, including macropore, mesopore and micropore, the specific surface area is as high as 907 $m^2 \cdot g^{-1}$. *Guo, Li, Sun, Wang, and Yang* (38) prepared a three-dimensional hierarchical lignin-derived porous carbon by hydrothermal-activation two-step method using enzymatic lignin as raw material, the synthesis process and microstructure morphology characteristics are shown in Figure 1. Enzymatic lignin first undergoes hydrothermal carbonization at a lower temperature to obtain hydrothermal carbon rich in macroporous. Then, potassium hydroxide is used as an activator for secondary carbonization to make pores to obtain a three-dimensional hierarchical lignin-derived porous carbon containing macropores, mesopores, and small pores. And its specific surface area is as high as 1660 $m^2 \cdot g^{-1}$, and it has excellent electrical conductivity (4-5.4 $S \cdot cm^{-1}$). Additionally, industrial lignin contains a large number of active sites and functional groups due to chemical treatment during its production process, which can be modified with heteroatoms to prepare heteroatom-doped lignin-based porous carbon. *Shen, Li, Yang, Zhang, Liang, and Peng* (39) prepared a 3D lignin-derived porous carbon co-doped with N, S and Cl using alkali lignin as raw material by $NaCl/ZnCl_2$ molten salt activation method, with high specific surface area (289 $m^2 \cdot g^{-1}$), large pore volume (2.80 $cm^3 \cdot g^{-1}$) and stable structure with good connections. These results show that the microscopic pore structure characteristics and specific surface area of lignin-derived porous carbon are adjustable.

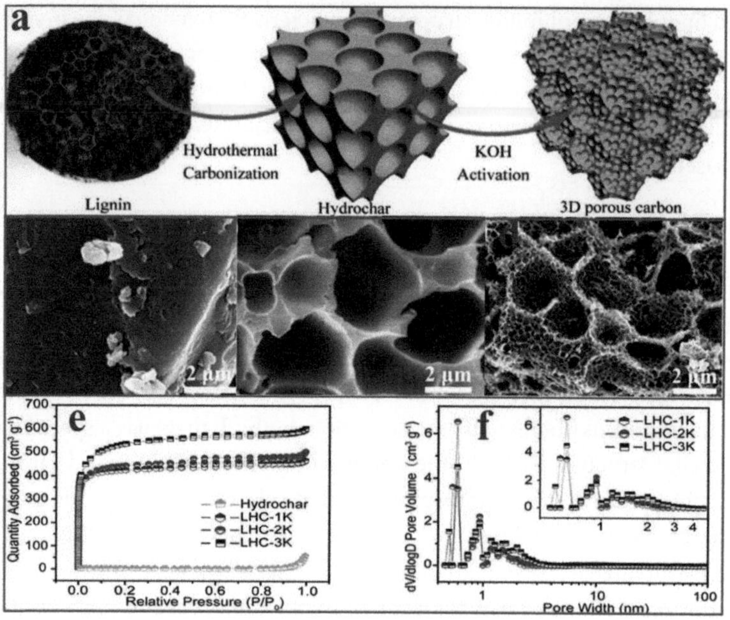

Figure 1. Schematic diagram of the preparation process of three-dimensional hierarchical lignin-derived porous carbon and its micro-morphology SEM image and pore structure characteristics. Reproduced with permission from reference (38). Copyright 2017 Royal Society of Chemistry.

Lignin-Based Carbon Fiber

Lignin-based carbon nanofibers is considered to be a catalyst carrier material with excellent performance due to its one-dimensional shape, which can provide extremely high specific surface area and enhanced electrical conductivity. Its main preparation methods include electrospinning and melt blowing law. *Shi, Jin, Wang, and Zhang* (*40*) used polyacrylonitrile (PAN) with an average molecular weight of 160,000, Pluronic P123 ($EO_{20}PO_{70}EO_{20}$) as soft template agent and a lignin with a molecular weight of 10,000 at a ratio of 1:1 to prepare lignin-based nanofibers. Then it was carbonized at a temperature of 1200 °C to give the lignin-based carbon nanofibers, and its synthesis process flow chart and micro-morphology were shown in Figure 2. The prepared lignin-derived carbon nanofiber has a large aspect ratio, with a diameter of about 200 nanometers, and is very uniform. The specific surface area and total pore volume are 68.8 $m^2 \cdot g^{-1}$ and 0.081 $cm^3 \cdot g^{-1}$, respectively. The pore size distribution is mainly concentrated in the 1.8 nm, 2.8 nm and 10 nm regions. The 1.8 nm microporous area is caused by the removal of soft template Pluronic P123 ($EO_{20}PO_{70}EO_{20}$) and the precipitation of volatiles during carbonization. *Wang, Yang, Stubbs, Li, and He* (*41*) used organic solvent lignin and polyethylene oxide soft template to prepare lignin-based carbon nanofibers with specific surface area and microporosity of 473 $m^2 \cdot g^{-1}$ and 87 % respectively through electrospinning. As the negative electrode of a lithium ion battery, showing a specific capacity of 445 $mA \cdot h \cdot g^{-1}$ after 50 cycles at a current density of 30 $mA \cdot g^{-1}$. Furthermore, urea was used as a nitrogen source to prepare lignin-based carbon nanofibers with a specific surface area of 381 $m^2 \cdot g^{-1}$. Its specific capacity is higher than that of the graphite electrode material under the same test conditions. In addition, *Tenhaeffet, Rios, More, and McGuire* (*42*) prepared lignin-based nanofibers by melt-blowing method using solvent extraction of lignin as raw materials. Then it is heated to 1000 °C, 1500 °C and 2000 °C at a heating rate of 10 °C/min to carbonize to obtain lignin-based carbon nanofibers. The controllable preparation of lignin-based carbon nanofibers has received more and more research and reports. However, its controllable preparation requires high purity of lignin, which may affect its wide application.

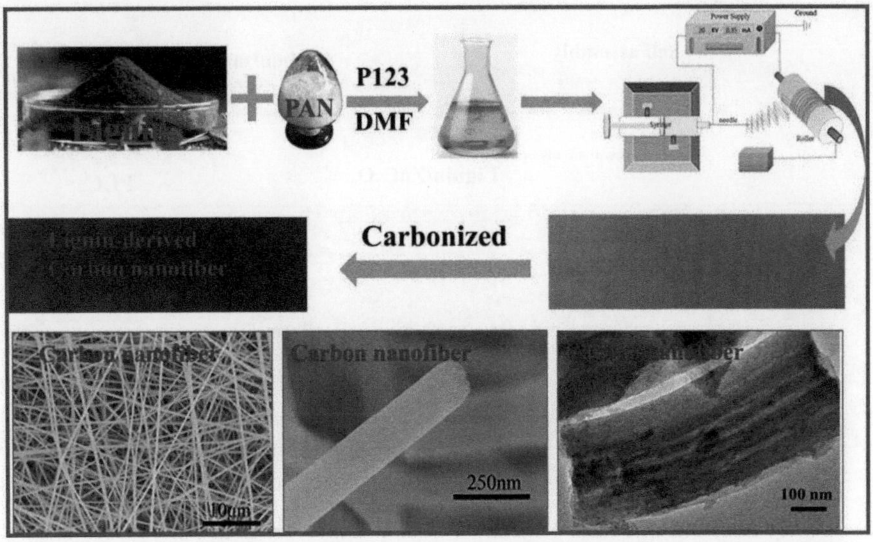

Figure 2. Synthesis process flow chart and micro-morphology SEM and TEM images of lignin-based carbon nanofibers. Reproduced with permission from reference (40). Copyright 2017 Elsevier.

Lignin-Based Carbon Nanosheets

In recent years, two dimensional porous carbons are emerging as greatly potential eletrochemical energy storage materials owing to their unique microstructures, reasonable pore structures, and stable physicochemical properties. Lignin with abundant aromatic rings has a high carbon content of 60%, which is viewed as ideal precursor for carbon material. However, modulating lignin molecular structure characteristic in a controllable manner to obtain high-performance two-dimensional lignin carbon nanosheet with regular structure remains challenging. *Fu, Yang, Zhang, Wang, and Qiu* (25) found that lignosulfonate with good solubility and dispersity was favorable for achieving the uniform dispersion of lignin and zinc oxalate. Following by a solvent induced effect in ethanol/water system, the hydrophobic phenylpropane skeleton and hydrophilic sulfonate of lignosulfonate self-assembled with zinc oxalate layer by layer to form a laminated composite structure. In the in-situ pyrolysis process, the thinning and drilling of the lignin skeleton was realized by means of the gas-phase stripping and the in-situ template effect of zinc oxalate, which effectively solved the problem that the skeleton is easy to collapse and agglomerate, successfully preparing lignin porous carbon with quasi nanosheet structure. The preparation flow chart and micro-morphology structure characteristics are shown in Figure 3. The prepared lignin-based carbon nanosheet has high specific surface area, interconnected micropore-mesopore-macropore network, and high electrical conductivity, providing sufficient active sites for charge accumulation. The electrolyte ions can not only be transported horizontally on the surface of the sheet, but also can be transported longitudinally through the pores between the layers, which shortens the diffusion distance of electrolyte ions, and significantly improves the transfer kinetic of electrolyte ions, thus leading to the great application potential in electrode materials. *Zhang B, Yang, Qiu, Qian, Wang, Yi, and Zhang D* (26) used enzymatic lignin (EHL) as precursor and nano-MgO as template to prepare a flower-like carbon (LFC) with a high degree of graphitization through a simple evaporation-induced self-assembly

(EISA) and carbonization process. The lignin flower-like carbon was composed of ultrathin porous nanosheets and a loose core, which not only possesses the advantages of two-dimensional carbon-based materials, but also avoids the problems of easy stacking and agglomeration.

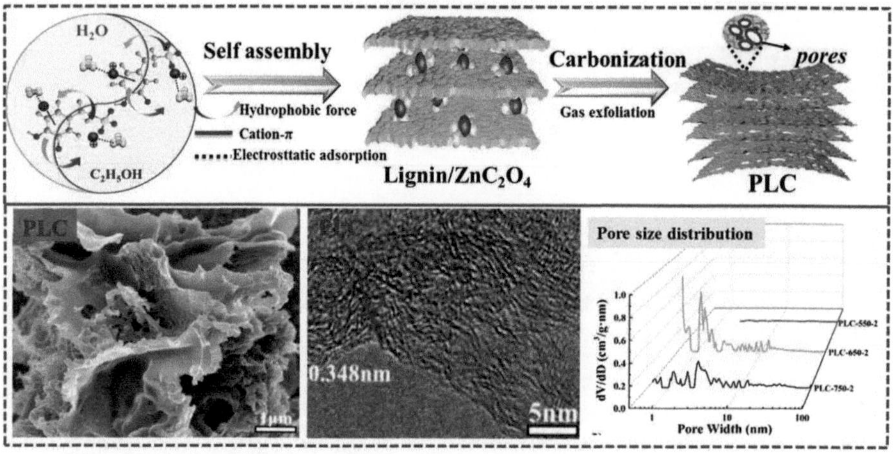

Figure 3. Preparation flow chart and microstructure characteristics of lignin porous carbon nanosheets (PLC). Reproduced with permission from reference (25). Copyright 2020 Elsevier.

Lignin-Based Carbon/Nano Metal Particle Composites

Lignin-based carbon not only has adjustable microscopic morphology and pore structure, but also stable chemical properties, wide range of pH adaptability, good electrical conductivity, and high specific surface area for active catalyst components, making it one of the most potential catalyst substrate materials. There have been many reports in recent years on the research of lignin carbon-supported precious metal catalysts. *Lai, Kolla, Zhao, Fong and Smirnova* (43) first prepared a kind of lignin carbon nanofibers (ECNFs) by electrospinning. The resulting ECNFs have a uniform diameter of 100 nm, and a high BET specific surface area of 583 $cm^2 \cdot g^{-1}$ with abundant pores. Subsequently, they loaded Ag nanoparticles by CO_2 supercritical method. As shown in Figure 4, the spherical AgNPs with the size between 2-10 nm are uniformly distributed on the surface of ECNFs, and the particle size can be effectively adjusted by changing the experimental parameters. More importantly, the prepared lignin-carbon/Ag composite can be used as an ORR catalyst for alkaline fuel cells, showing extremely high economic benefits and high-efficiency electrocatalytic activity, even exceeding the mass activity of the HiSPEC 4100TM commercial Pt/C. *Qin, Jian, Bai, Tang, Zhou Y, Zhu, Zhao, Ni, Wang, Liu, Zhou Q, and Li* (44) used lignin with different molecular weights as carbon sources to synthesize a series of ordered mesoporous carbons and investigated their catalytic performance as catalyst carriers for Fischer-Tropsch synthesis (FTS). The mesoporous carbon incorporating iron nanoparticles prepared by impregnation method showed ordered mesoporous characteristic and well dispersity. The Fe/CSL sample showed excellent CO conversion rate and high ratio of olefin to paraffin in FTS reaction. *Bedia, Rosas, Rodríguez-Mirasol, and Cordero* (45) prepared kraft lignin-derived mesoporous activated carbon by H_3PO_4 activation and applied it as a carrier for palladium catalyst. The results showed that the surface phosphorus and surface palladium accounted for 0.5 wt% in the composite catalyst. The highly dispersed palladium nanoparticles displayed a uniform size of 5-12 nm. Metal palladium and Pd^{2+} are the main chemical states of palladium on the surface of the catalyst prepared by heat treatment before and after palladium loading, respectively, marked as LT-Pd and L-PdT. L-PdT showed a higher conversion rate in the catalytic oxidation of toluene. Tremendous findings have demonstrated that the microstructures of lignin-based carbon, including specific surface area, pore size distribution, and surface charge, can be tailored for desired application, especially holding great promise in the field of metal catalyst supports. The performance of some lignin carbon/metal catalysts can even be comparable to commercial carbon/metal catalysts in laboratory research, but the controllable synthesis and stability of lignin carbon/metal composite catalyst are necessary to be further improved.

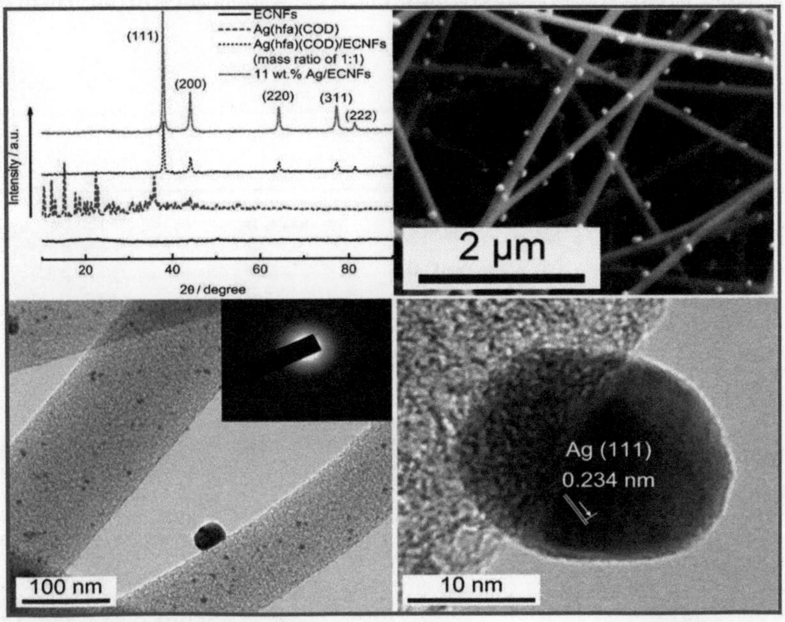

Figure 4. Micro-morphological and structural characteristics of lignin-based carbon/Ag nanocomposites. Reproduced with permission from reference (43). Copyright 2014 Elsevier.

Lignin-Based Carbon/Metal Oxide Nanocomposite

Nowadays, metal oxide/carbon composite materials show good application prospects in energy conversion and storage. Lignin and metal oxide precursors can efficiently combine through intermolecular interactions to obtain lignin/metal oxide precursors. By carbonizing the precursors at high temperature under inert atmosphere, lignin-based carbon/metal oxide composie materials, such as lignin carbon/nickel oxide, lignin-based carbon/zinc oxide, and lignin carbon/ titanium oxide, can be produced.

Wang, Qiu, Liu, and Yang (32) reported a lignin-carbon/zinc oxide hybrid composite by an electrostatic self-assembly co-precipitation method. The preparation process is shown in Figure 5. Alkali lignin derived from papermaking black liquor was used as the raw material. Its poor water solubility and the existence of a large number of negatively charge functional groups, for instance, carboxyl and hydroxyl, introduces an electrostatic repulsion with zinc oxalate so that they cannot effectively composite. Therefore, alkali lignin was first quaternized and modified to cationic lignin, and the electrostatic force between quaternary ammonium and zinc oxalate brought about self-assembly and co-precipitation to acquire lignin/zinc oxalate composite. After carbonization at the temperature of 550 °C under nitrogen, the composite was evolved into a lignin carbon/zinc oxide composite with typical frame structure, where ZnO nanoparticles are very uniformly embedded in the lignin carbon nanosheet. Similar to the structural characteristics of graphene/zinc oxide composite materials, it has good potential in the field of photocatalysis and energy storage. Zhang B, Yang, Qiu, Qian, Wang, Yi, and Zhang D (*26*) first used cheap and readily available enzymatic lignin (EHL) as a precursor to prepare a lignin-based "flower-like" carbon (LFC) with a high degree of graphitization. Further through in-situ hydrothermal reaction, ZnO particles with a particle size of about 10 nm could be uniformly loaded on the LFC to obtain a ZnO/LFC composite photocatalyst, which prevents the agglomeration of ZnO nanoparticles and greatly increases the active sites for photocatalytic reaction. The preparation process and microstructure characteristics are shown in Figure 6. The lignin-based carbon/ZnO composites has great application in the field in the treatment of organic waste water.

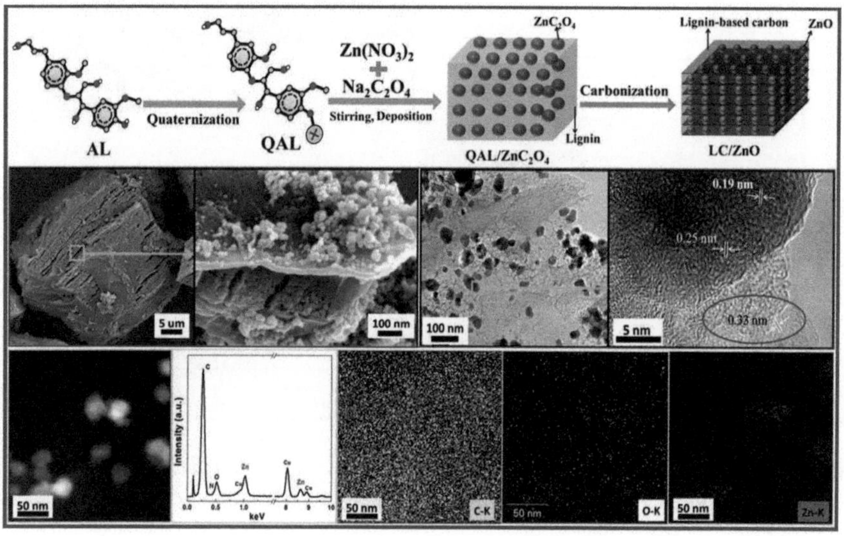

Figure 5. Preparation process and micro morphology of lignin-based carbon/ZnO composites. Reproduced with permission from reference (32). Copyright 2017 Elsevier.

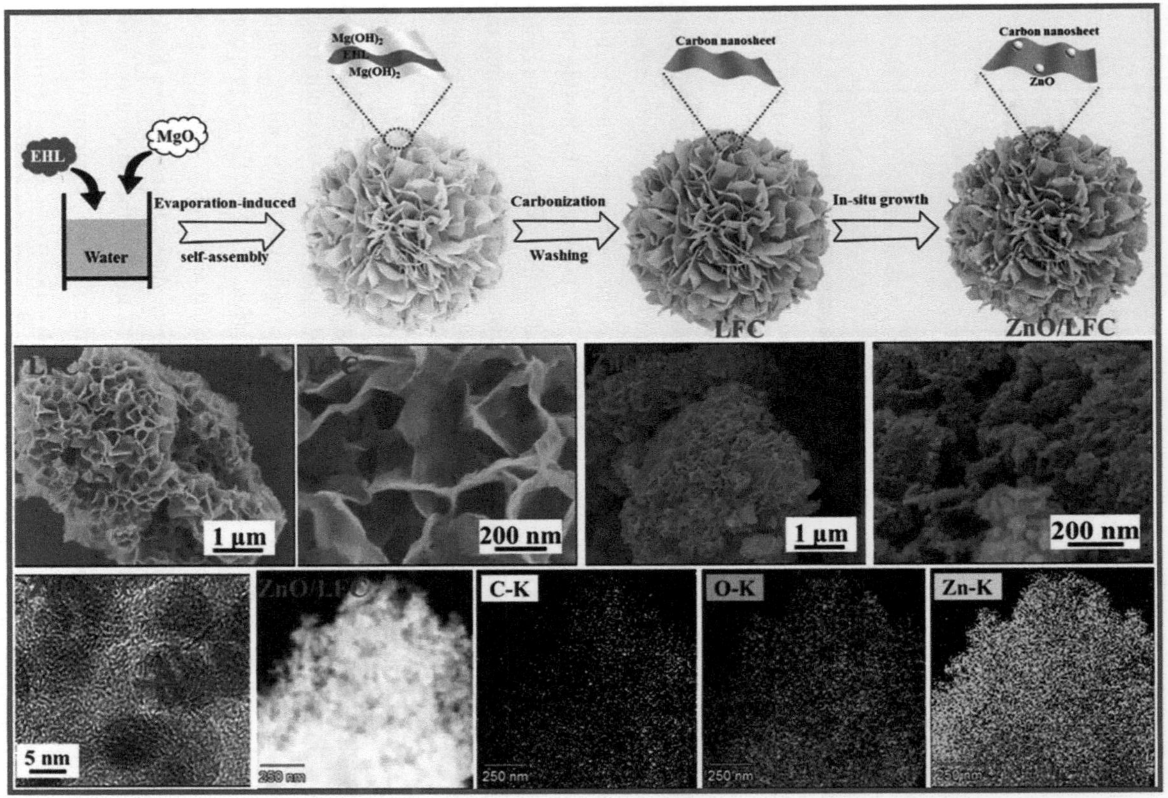

Figure 6. Preparation flow chart and micro morphology of ZnO/LFC composite. Reproduced with permission from reference (26). Copyright 2020 Elsevier.

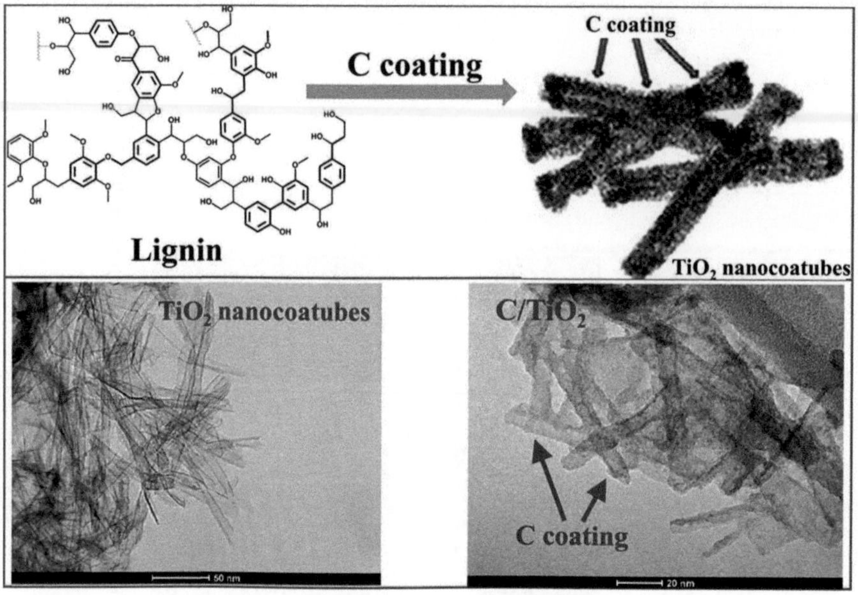

Figure 7. The microscopic morphology of the lignin carbon/TiO_2 nanotube composite. Reproduced with permission from reference (49). Copyright 2019 Elsevier.

The metal oxide composited with lignin-based carbon was widely investigated in recent years. *Chen, Zhou, Yao, Fan, Yang, Fei, and Zhong* (46) proposed a method for preparing lignin carbon/NiO composites using sodium lignosulfonate and nickel nitrate as the precursors. First, sodium lignosulfonate and nickel nitrate hexahydrate were dissloved in potassium hydroxide and formaldehyde aqueous solution at a temperature of 80 °C for 8 h to allow the two to undergo a cross-linking reaction, and then lignin carbon/NiO composite material was obtained by calcinating the reaction products at 600 °C under nitrogen. The composite material has a high specific surface area (503-802 $m^2 \cdot g^{-1}$), and the loading and dispersity of nickel oxide is controllable in the lignin carbon structure. However, this method requires a large amount of organic solvents (methanol, ethanol) and chemical cross-linking agents, which is not conducive to the green synthesis of lignin-carbon composite materials. Additionally, *Wang, Fan, and Ai* (47) prepared a porous carbon/cerium oxide composite using sodium lignosulfonate as dispersant and template at different temperatures, which has a regular structure and good dispersibility. *Chen, Kuo D, Lu, Hou, and Kuo Y* (48) prepared a kind of mesoporous TiO_2 nanocomposite particles with alkali lignin as template and precursor, showing high specific surface area and abundant pore structure. *Gómez-Avilés, Peñas-Garzón, Bedia, Rodriguez, and Belver* (49) used lignin as the carbon precursor to modify the surface of TiO_2 nanotubes, and prepared several lignin-carbon-modified TiO_2 materials. The preparation process and microscopic morphology are shown in Figure 7. It showed that the lignin-derived carbon is uniformly loaded on the surface of TiO_2. When the heat treatment was performed in an inert atmosphere, a material with a much higher carbon content is produced, thereby maintaining the anatase phase of TiO_2 and smaller crystal size, but the color is very dark, causing a large amount of light absorption and reducing its photocatalytic activity. On the contrary, the samples of C-Ti-400air and C-Ti-500air calcined in the air at 400 and 500 °C show the highest photolysis activity of acetaminophen. Although lignin-carbon/metal oxide composite materials have gained extensive research interest due to their good application prospects, and many innovated preparation methods have been reported, there are still problems such as uncontrollable microstructure, poor repeatability, and instability. Controllable preparation mechanism need to be developed in future research and more attention should be paid to the compound mechanism at molecular level.

Application of Lignin-Based Carbon in the Field of Catalysis

Carbon materials have broad application prospects in the field of catalysts and their carriers due to their excellent electronic conductivity, adjustable microscopic morphology, pore structure characteristics, and good acid-base stability. Lignin-based carbon materials exhibit great application potential in catalysts, especially as catalyst carriers attributed to the advantages of simple preparation, low cost, adjustable morphology and surface properties. Lignin is rich in oxygen-containing functional groups, which can be pre-combined with catalyst active component precursors through electrostatic adsorption and hydrogen bonding, and then carbonized to prepare lignin-based carbon catalysts through one-step carbonization, which greatly simplifies the conventional catalyst preparation process. In recent years, the application of lignin-based carbon materials in the fields of photocatalysis, thermal catalysis, and electrocatalysis has been widely reported.

Photocatalyst

Due to their wide applicability and high efficiency, photocatalysts provide an effective way to degrade toxic and harmful pollutants in the environment. Semiconductor metal oxides have the advantages of low cost, non-toxicity, and large band gap. For example, TiO_2, Bi_2O_3 and ZnO have been used as photocatalysts in the field of photocatalytic degradation of organic wastewater (*50–52*). However, many semiconductor photocatalysts suffer poor absorption performance for visible light, and photo-generated electron-hole pairs are easily recombined, which greatly reduces their photocatalytic performance (*53*). The utilization rate of visible light and the photocatalytic efficiency can be effectively improved by incorporating precious metals, graphene or plasma doping.

In recent years, researchers have found that loading lignin carbon can increase the yield of photogenerated electrons in semiconductors and also prevent the recombination of photogenerated electron-hole pairs, thereby improving photocatalytic performance. *Wang, Fan, and Ai* (*47*) used sodium lignosulfonate as the template and carbon source, and cerium nitrate as the cerium source. After co-calcination, a porous carbon/CeO_2 composite was prepared. The composite showed excellent photocatalytic desulfurization performance, which was about 1.6 times than that of pure CeO_2. *Srisasiwimon, Chuangchote, Laosiripojana, and Sagawa* (*54*) prepared a lignin/TiO_2 composite by sol-gel method and prepared a lignin carbon/TiO_2 composite photocatalyst by high-temperature calcination. The composite has the ability in photocatalytic conversion of lignin to prepare vanillin, which attributed to the presence of lignin carbon, preventing the recombination of TiO_2 photogenerated electron-hole pairs. *Wang, Qiu, Liu, and Yang* (*32*) prepared a lignin-carbon/zinc oxide (LC/ZnO) hybrid composite material using alkali lignin derived from papermaking black liquor as raw materials. ZnO nanoparticles were evenly embedded in the lignin carbon nanosheet, forming a cubic frame structure. It was then applied to photocatalytic degradation of organic dye wastewater, and the experimental results are shown in Figure 8. Compared with pure ZnO, the rate of LC/ZnO photocatalytic degradation of methyl orange was increased by 5.1 times, and could be completely degraded, and the photodegradation rate of rhodamine B was increased by 2.7 times. The excellent photocatalytic degradation of organic dye wastewater is even better than many of the GR/ZnO composite photocatalysts reported in the literature. The results of this study show that the introduction of lignin carbon can effectively improve the separation efficiency of ZnO photogenerated electron-hole pairs and can quickly transfer the photogenerated electrons of ZnO. This research provides a novel direction for the application of lignin in the field of carbon-based catalyst supports. *Chen, Kuo D, Lu, Hou, and Kuo Y* (*48*) prepared a mesoporous TiO_2 using alkali lignin purified from papermaking black liquor as a template. The lignin-carbon hybridized TiO_2 showed excellent photocatalytic performance and could degrade 97.9% of phenol within 120 min. *Gómez-Avilés, Peñas-Garzón, Bedia, Rodriguez, and Belver* (*49*) used lignin as the carbon precursor to modify the surface of TiO_2 nanotubes and calcined the photocatalysts (C-Ti-400air and C-Ti-500air) at 400 and 500°C in air. The photocatalysts showed high photolytic activity and acetaminophen could be completely degraded in 1 hour under sunlight. The main reactive oxygen species involved in photodegradation of acetaminophen were OH^- and O^{2-} radicals, and the contribution of electrons was relatively weak. The above research results show that the loading of lignin carbon can significantly improve the photocatalytic performance of metal oxide semiconductor catalysts, and it has broad application prospects in the field of non-toxic and harmless treatment of pollutants. However, the weak binding force between lignin and semiconductor catalyst makes it difficult for the composite to realize the uniform dispersion of the semiconductor on the lignin carbon, resulting in the loss of the active surface area, and the photocatalytic performance is not maximized.

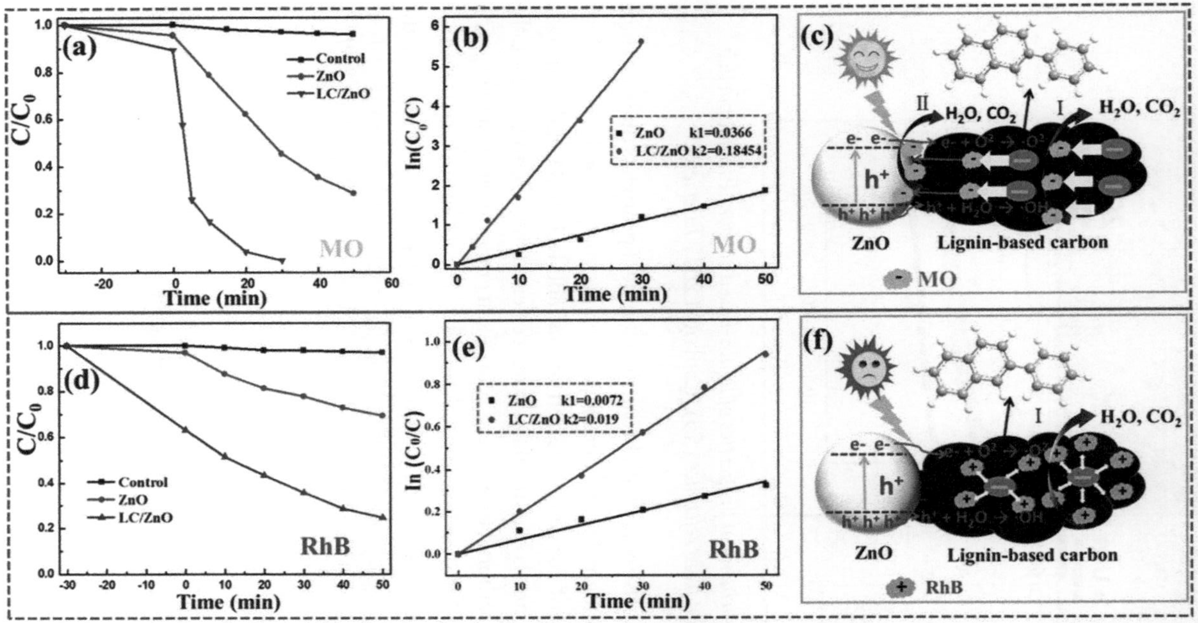

Figure 8. Photocatalytic degradation of organic dyes methyl orange (MO) and Rhodamine B (RhB) by lignin carbon /ZnO composite. Reproduced with permission from reference (32). Copyright 2017 Elsevier.

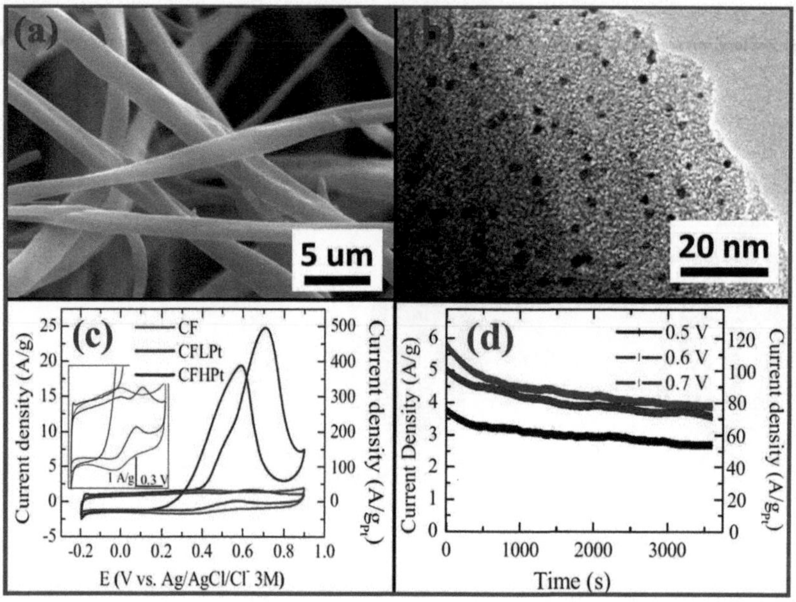

Figure 9. Micro-morphology and electrochemical properties of lignin-based carbon/Pt nanoparticle. Reproduced with permission from reference (56). Copyright 2017 Elsevier.

Hydrogen energy is a new type of green energy with broad development prospects due to its high energy density, non-toxicity and environmental protection. Hydrogen production by electrocatalytic hydrolysis includes two half reactions of oxygen evolution at the anode (OER) and hydrogen evolution at the cathode (HER). And it needs electrocatalysts to accelerate the reaction because the kinetics is too slow. The lignin-derived carbon has excellent acid/alkali environmental stability, and abundant functional groups can be used for structure control and precise coordination of transition metal catalysts, which can increase the charge distribution on the catalyst surface and improve the catalytic performance. The application of lignin-based carbon materials in the field of electrocatalysis has recived more and more attention and investigation. *Peng, Zhang, Chen, Zhong, Zhao D, Chi, Zhao, Li, Lu, Leng, Liu C, Liu W, Tang, and Loh* (55) prepared a metal-free mesoporous N-doped 3D carbon catalyst material using primary lignocellulosic biomass as a raw material, which shows high mechanical strength, high conductivity, cross-linked network and natural ion transport channels. It shows that the specific capacity is 801 mAh·g^{-1}, the energy density is 955 Wh·kg^{-1}, and the long-term stability is up to 110 h when it is used as the porous carbon plate material for ORR/OER electrodes in Zn-air battery. *Shen, Li, Yang, Zhang, Liang, and Peng* (39) prepared a 3D lignin-derived porous carbon co-doped with N, S, and Cl using alkali lignin as raw material by molten salt activation method. It is an excellent electrocatalyst for oxygen reduction reaction in proton exchange membrane fuel cells, showing high activity and stability in acidic media, and is superior to Pt/C catalysts and most non-noble metal catalysts. And it was used as a cathode catalyst to assemble the H_2-O_2 single fuel cell for the first time, showing a maximum power density of 779 mW·cm^{-2}. *García-Mateos, Cordero-Lanzac, Berenguer, Morallón, Cazorla-Amorós, Rodríguez-Mirasol, and Cordero* (56) reported a lignin-based carbon fiber supported Pt metal catalyst, and its microstructure and electrochemical performance are shown in Figure 9. Firstly, lignin is grafted and modified, and phosphoric acid groups are inserted into the lignin molecular structure, which is conducive to the uniform dispersion of Pt metal nanoparticles (~2 nm). The resulting Pt-containing carbon fiber can be directly used as an electrode without a binder or conductivity promoter. It shows obvious electrocatalytic response in the electrooxidation reaction of methanol and ethanol, and both show lower overpotential, higher specific oxidation current and stable catalytic response. *Bedia, Rosas, Rodríguez-Mirasol, and Cordero* (45) loaded the noble metal Pd on the lignin-derived porous carbon and showed excellent catalytic oxidation performance of toluene in VOCs. The application of lignin carbon-based materials in the field of electrocatalysis is of great significance to the development of high-value utilization of biomass.

Thermal Catalyst

Lignin-based carbon materials also have potential application value in the field of thermal catalysis. *Qin, Kang, Wang, Liu, Ni, Huang, Li Y, and Li X* (57) explored the application of lignin-based carbon catalysts in Fischer-Tropsch synthesis. First, a core-shell structure Co@C catalyst was prepared by low-temperature carbonization. As a Fischer-Tropsch synthesis catalyst, it exhibits a high C_{5+} selectivity of 56.8%. In addition, highly graphitized and iron nanoparticles coated lignin-based carbon was prepared through pyrolysis using alkali lignin and ferric nitrate as precursors (58). It shows a high C_2-C_4 olefin selectivity (52.4%) when used as a Fischer-Tropsch synthesis catalyst. On the other hand, ordered mesoporous lignin-derived carbon was prepared by soft template method using different molecular weight lignin extracted from rice straw, and mesoporous lignin-based carbon/iron nanoparticle composite catalyst was prepared by impregnating with ferric nitrate and carbonizing at 350 °C (44). When the molecular weight of lignin is the smallest, the selectivity of C_2-C_4 olefins is the highest, reaching 43.1%. *Martin-Martinez, Barreiro, Silva, Figueiredo, Faria, and Gomes* (59) researched the performance of lignin-based carbon as a metal-free catalyst for the oxidative degradation of 4-nitrophenol. It showed that with the assistance of hydrogen peroxide, the lignin-based carbon obtained from high temperature carbonization can effectively degrade 70% of 4-nitrophenol after 24 h. The good oxidative degradation effect is mainly due to the fact that lignin-based carbon promotes the formation of hydroxyl radicals during the decomposition of hydrogen peroxide and reduces the recombination of active free radicals.

Conclusion

In summary, the controllable preparation of lignin-based carbon materials and their high value-added applications in photocatalysis, thermal catalysis and electrocatalysis have been extensively studied. Lignin carbon as a catalyst carrier still faces the following problems: First, the molecular structure of lignin is complex, and the pyrolysis process is difficult to control, resulting in uneven morphology and serious agglomeration of the formed carbon material, which is not conducive to the loading of active components. Second, although there is an interaction force between the lignin and the active component precursor, the binding force between the two is weak, which leads to the easy loss of the active component of the prepared composite catalyst and poor cycle stability. The third is that the purity of lignin raw materials is low, containing some inorganic salts, and some impurities are difficult to remove after carbonization, which has a negative impact on subsequent structural characterization and catalytic testing. Therefore, obtaining high-purity lignin raw materials, regulating and enhancing the interaction between lignin and catalyst precursors to achieve controllable loading of catalyst active components has become the key to the preparation and application of lignin carbon-based catalysts. The research of lignin carbon-based functional materials is a frontier subject involving multiple disciplines. Although many researchers have done a lot of work, how to efficiently and controllably prepare lignin-based carbon materials with regular and controllable microstructures and excellent performance is still a challenging topic. It is of great social, economic and environmental significance to prepare ultra-high value-added carbon materials using industrial lignin from pulp waste liquid and biomass refining.It is of great significance for the large-scale application of lignin to realize the sustainable green development of papermaking and biomass refining.

Acknowledgments

This work was supported by the National Natural Science Foundation of China (NSFC) (No. 21878114, 21908071), China Postdoctoral Science Foundation (2019M652901), Basic scientific research operating expenses of central universities of South China University of Technology (2191960), Science and Technology Program of Guangzhou (201704030126) and Natural Science Foundation of Guangdong, (2018B030311052, 2020A1515011319), and State Key Laboratory of Pulp and Paper Engineering (202005).

References

1. Zhou, Y.; Qian, Y.; Wang, J.; Qiu, X.; Zeng, H. Bioinspired Lignin-Polydopamine Nanocapsules with Strong Bioadhesion for Long-Acting and High-Performance Natural Sunscreens. *Biomacromolecules* **2020**, *21* (8), 3231–3241.
2. Xiong, W.; Yang, D.; Zhong, R.; Li, Y.; Zhou, H.; Qiu, X. Preparation of lignin-based silica composite submicron particles from alkali lignin and sodium silicate in aqueous solution using a direct precipitation method. *Industrial Crops and Products* **2015**, *74*, 285–292.
3. Xiong, W.; Qiu, X.; Yang, D.; Zhong, R.; Qian, Y.; Li, Y.; Wang, H. A simple one-pot method to prepare UV-absorbent lignin/silica hybrids based on alkali lignin from pulping black liquor and sodium metasilicate. *Chemical Engineering Journal* **2017**, *326*, 803–810.
4. Chen, K.; Qiu, X.; Yang, D.; Qian, Y. Amino acid-functionalized polyampholytes as natural broad-spectrum antimicrobial agents for high-efficient personal protection. *Green Chemistry* **2020**, *22* (19), 6357–6371.
5. Wang, J.; Qian, Y.; Li, L.; Qiu, X. Atomic Force Microscopy and Molecular Dynamics Simulations for Study of Lignin Solution Self-Assembly Mechanisms in Organic-Aqueous Solvent Mixtures. *ChemSusChem* **2020**, *13*, 4420–4427.
6. Yang, D.; Huang, W.; Qiu, X.; Lou, H.; Qian, Y. Modifying sulfomethylated alkali lignin by horseradish peroxidase to improve the dispersibility and conductivity of polyaniline. *Applied Surface Science* **2017**, *426*, 287–293.
7. Zhang, B.; Yang, D.; Wang, H.; Qian, Y.; Huang, J.; Yu, L.; Qiu, X. Activation of Enzymatic Hydrolysis Lignin by NaOH/Urea Aqueous Solution for Enhancing Its Sulfomethylation Reactivity. *ACS Sustainable Chemistry & Engineering* **2018**, *7* (1), 1120–1128.
8. Xi, Y.; Yang, D.; Wang, Y.; Huang, J.; Yan, M.; Yi, C.; Qian, Y.; Qiu, X. Effect of structure of technical lignin on the electrochemical performance of lignin-derived porous carbon from K_2CO_3 activation. *Holzforschung* **2020**, *74* (3), 293–302.
9. Wang, H.; Qiu, X.; Liu, W.; Fu, F.; Yang, D. A Novel Lignin/ZnO Hybrid Nanocomposite with Excellent UV-Absorption Ability and Its Application in Transparent Polyurethane Coating. *Industrial & Engineering Chemistry Research* **2017**, *56* (39), 11133–11141.

10. Wang, H.; Lin, W.; Qiu, X.; Fu, F.; Zhong, R.; Liu, W.; Yang, D. In Situ Synthesis of Flowerlike Lignin/ZnO Composite with Excellent UV-Absorption Properties and Its Application in Polyurethane. *ACS Sustainable Chemistry & Engineering* **2018**, *6* (3), 3696–3705.
11. Zhang, B.; Yang, D.; Qiu, X.; Qian, Y.; Yan, M.; Li, Q. Influences of aggregation behavior of lignin on the microstructure and adsorptive properties of lignin-derived porous carbons by potassium compound activation. *Journal of Industrial and Engineering Chemistry* **2020**, *82*, 220–227.
12. Lu, S.; Yang, D.; Wang, M.; Yan, M.; Qian, Y.; Zheng, D.; Qiu, X. Pickering emulsions synergistic-stabilized by amphoteric lignin and SiO_2 nanoparticles: Stability and pH-responsive mechanism. *Colloids and Surfaces A: Physicochemical and Engineering Aspects* **2020**, *585*, 124158.
13. Zhang, D.; Qiu, S.; Huang, W.; Yang, D.; Wang, H.; Fang, Z. Mechanically strong and electrically stable polypyrrole paper using high molecular weight sulfonated alkaline lignin as a dispersant and dopant. *J Colloid Interface Sci* **2019**, *556*, 47–53.
14. Wu, Y.; Qian, Y.; Lou, H.; Yang, D.; Qiu, X. Enhancing the Broad-Spectrum Adsorption of Lignin through Methoxyl Activation, Grafting Modification, and Reverse Self-Assembly. *ACS Sustainable Chemistry & Engineering* **2019**, *7* (19), 15966–15973.
15. Qian, Y.; Zhou, Y.; Li, L.; Liu, W.; Yang, D.; Qiu, X. Facile preparation of active lignin capsules for developing self-healing and UV-blocking polyurea coatings. *Progress in Organic Coatings* **2020**, *138*, 105354.
16. Li, Y.; Qiu, X.; Qian, Y.; Xiong, W.; Yang, D. pH-responsive lignin-based complex micelles: Preparation, characterization and application in oral drug delivery. *Chemical Engineering Journal* **2017**, *327*, 1176–1183.
17. Ding, Z.; Qiu, X.; Fang, Z.; Yang, D. Effect of Molecular Weight on the Reactivity and Dispersibility of Sulfomethylated Alkali Lignin Modified by Horseradish Peroxidase. *ACS Sustainable Chemistry & Engineering* **2018**, *6* (11), 14197–14202.
18. Li, Y.; Yang, D.; Lu, S.; Lao, S.; Qiu, X. Modified Lignin with Anionic Surfactant and Its Application in Controlled Release of Avermectin. *J Agric Food Chem* **2018**, *66* (13), 3457–3464.
19. Li, Y.; Yang, D.; Lu, S.; Qiu, X.; Qian, Y.; Li, P. Encapsulating TiO_2 in Lignin-Based Colloidal Spheres for High Sunscreen Performance and Weak Photocatalytic Activity. *ACS Sustainable Chemistry & Engineering* **2019**, *7* (6), 6234–6242.
20. Wang, H.; Wang, Y.; Fu, F.; Qian, Y.; Xiao, Y.; Yang, D.; Qiu, X. Controlled preparation of lignin/titanium dioxide hybrid composite particles with excellent UV aging resistance and its high value application. *Int J Biol Macromol* **2020**, *150*, 371–379.
21. Xiong, W.; Yang, D.; Hoang, T. K. A.; Ahmed, M.; Zhi, J.; Qiu, X.; Chen, P. Controlling the sustainability and shape change of the zinc anode in rechargeable aqueous $Zn/LiMn_2O_4$ battery. *Energy Storage Materials* **2018**, *15*, 131–138.
22. Xiong, W.; Yang, D.; Zhi, J.; Hoang, T. K. A.; Qiu, X.; Chen, P. Improved performance of the rechargeable hybrid aqueous battery at near full state-of-charge. *Electrochimica Acta* **2018**, *271*, 481–489.

23. Yu, J.; Li, L.; Qian, Y.; Lou, H.; Yang, D.; Qiu, X. Facile and Green Preparation of High UV-Blocking Lignin/Titanium Dioxide Nanocomposites for Developing Natural Sunscreens. *Industrial & Engineering Chemistry Research* **2018**, *57* (46), 15740–15748.
24. Xi, Y.; Huang, S.; Yang, D.; Qiu, X.; Su, H.; Yi, C.; Li, Q. Hierarchical porous carbon derived from the gas-exfoliation activation of lignin for high-energy lithium-ion batteries. *Green Chemistry* **2020**, *22* (13), 4321–4330.
25. Fu, F.; Yang, D.; Zhang, W.; Wang, H.; Qiu, X. Green self-assembly synthesis of porous lignin-derived carbon quasi-nanosheets for high-performance supercapacitors. *Chemical Engineering Journal* **2020**, *392*, 123721.
26. Zhang, B.; Yang, D.; Qiu, X.; Qian, Y.; Wang, H.; Yi, C.; Zhang, D. Fabricating ZnO/lignin-derived flower-like carbon composite with excellent photocatalytic activity and recyclability. *Carbon* **2020**, *162*, 256–266.
27. Zhang, B.; Yang, D.; Qian, Y.; Pang, Y.; Li, Q.; Qiu, X. Engineering a lignin-based hollow carbon with opening structure for highly improving the photocatalytic activity and recyclability of ZnO. *Industrial Crops and Products* **2020**, *155*, 112773.
28. Xi, Y.; Wang, Y.; Yang, D.; Zhang, Z.; Liu, W.; Li, Q.; Qiu, X. K_2CO_3 activation enhancing the graphitization of porous lignin carbon derived from enzymatic hydrolysis lignin for high performance lithium-ion storage. *Journal of Alloys and Compounds* **2019**, *785*, 706–714.
29. Xi, Y.; Yang, D.; Liu, W.; Qin, Y.; Qiu, X. Preparation of porous lignin-derived carbon/carbon nanotube composites by hydrophobic self-assembly and carbonization to enhance lithium storage capacity. *Electrochimica Acta* **2019**, *303*, 1–8.
30. Xi, Y.; Yang, D.; Qiu, X.; Wang, H.; Huang, J.; Li, Q. Renewable lignin-based carbon with a remarkable electrochemical performance from potassium compound activation. *Industrial Crops and Products* **2018**, *124*, 747–754.
31. Fu, F.; Yang, D.; Wang, H.; Qian, Y.; Yuan, F.; Zhong, J.; Qiu, X. Three-dimensional Porous Framework Lignin-Derived Carbon/ZnO Composite Fabricated by a Facile Electrostatic Self-Assembly Showing Good Stability for High-Performance Supercapacitors. *ACS Sustainable Chemistry & Engineering* **2019**, *7* (19), 16419–16427.
32. Wang, H.; Qiu, X.; Liu, W.; Yang, D. Facile preparation of well-combined lignin-based carbon/ZnO hybrid composite with excellent photocatalytic activity. *Applied Surface Science* **2017**, *426*, 206–216.
33. Wang, H.; Qiu, X.; Zhong, R.; Fu, F.; Qian, Y.; Yang, D. One-pot in-situ preparation of a lignin-based carbon/ZnO nanocomposite with excellent photocatalytic performance. *Materials Chemistry and Physics* **2017**, *199*, 193–202.
34. Puziy, A M.; Poddubnaya, O. I.; Sevastyanova, O. Carbon materials from technical lignins: Recent advances. *Topics in Current Chemistry* **2018**, *376*, 33.
35. Nar, M.; Rizvi, H. R.; Dixon, R. A.; Chen, F.; Kovalcik, A.; D'Souza, N. Superior plant based carbon fibers from electrospun poly-(caffeyl alcohol) lignin. *Carbon* **2016**, *103*, 372–383.
36. Xie, A.; Dai, J.; Chen, Y.; Liu, N.; Ge, W.; Ma, P.; Zhang, R.; Zhou, Z.; Tian, S.; Li, C.; Yan, Y. NaCl-template assisted preparation of porous carbon nanosheets started from lignin for efficient removal of tetracycline. *Advanced Powder Technology* **2019**, *30* (1), 170–179.

37. Zhang, W.; Lin, H.; Lin, Z.; Yin, J.; Lu, H.; Liu, D.; Zhao, M. 3 D Hierarchical Porous Carbon for Supercapacitors Prepared from Lignin through a Facile Template-Free Method. *ChemSusChem* **2015**, *8* (12), 2114–22.
38. Guo, N.; Li, M.; Sun, X.; Wang, F.; Yang, R. Enzymatic hydrolysis lignin derived hierarchical porous carbon for supercapacitors in ionic liquids with high power and energy densities. *Green Chemistry* **2017**, *19* (11), 2595–2602.
39. Shen, Y.; Li, Y.; Yang, G.; Zhang, Q.; Liang, H.; Peng, F. Lignin derived multi-doped (N, S, Cl) carbon materials as excellent electrocatalyst for oxygen reduction reaction in proton exchange membrane fuel cells. *Journal of Energy Chemistry* **2020**, *44*, 106–114.
40. Shi, Z.; Jin, G.; Wang, J.; Zhang, J. Free-standing, welded mesoporous carbon nanofibers as anode for high-rate performance Li-ion batteries. *Journal of Electroanalytical Chemistry* **2017**, 795, 26–31.
41. Wang, S. X.; Yang, L.; Stubbs, L. P.; Li, X.; He, C. Lignin-derived fused electrospun carbon fibrous mats as high performance anode materials for lithium ion batteries. *ACS Appl Mater Interfaces* **2013**, *5* (23), 12275–82.
42. Tenhaeff, W. E.; Rios, O.; More, K.; McGuire, M. A. Highly Robust Lithium Ion Battery Anodes from Lignin: An Abundant, Renewable, and Low-Cost Material. *Advanced Functional Materials* **2014**, *24* (1), 86–94.
43. ai, C.; Kolla, P.; Zhao, Y.; Fong, H.; Smirnova, A. L. Lignin-derived electrospun carbon nanofiber mats with supercritically deposited Ag nanoparticles for oxygen reduction reaction in alkaline fuel cells. *Electrochimica Acta* **2014**, *130*, 431–438.
44. Qin, H.; Jian, R.; Bai, J.; Tang, J.; Zhou, Y.; Zhu, B.; Zhao, D.; Ni, Z.; Wang, L.; Liu, W.; Zhou, Q.; Li, X. Influence of Molecular Weight on Structure and Catalytic Characteristics of Ordered Mesoporous Carbon Derived from Lignin. *ACS Omega* **2018**, 3 (1), 1350–1356.
45. Bedia, J.; Rosas, J. M.; Rodríguez-Mirasol, J.; Cordero, T. Pd supported on mesoporous activated carbons with high oxidation resistance as catalysts for toluene oxidation. *Applied Catalysis B: Environmental* **2010**, *94* (1-2), 8–18.
46. Chen, F.; Zhou, W.; Yao, H.; Fan, P.; Yang, J.; Fei, Z.; Zhong, M. Self-assembly of NiO nanoparticles in lignin-derived mesoporous carbons for supercapacitor applications. *Green Chemistry* **2013**, *15* (11), 3057.
47. Wang, N.; Fan, H.; Ai, S. Lignin templated synthesis of porous carbon–CeO_2 composites and their application for the photocatalytic desulphuration. *Chemical Engineering Journal* **2015**, *260*, 785–790.
48. Chen, X.; Kuo, D.-H.; Lu, D.; Hou, Y.; Kuo, Y.-R. Synthesis and photocatalytic activity of mesoporous TiO_2 nanoparticle using biological renewable resource of un-modified lignin as a template. *Microporous and Mesoporous Materials* **2016**, *223*, 145–151.
49. Gómez-Avilés, A.; Peñas-Garzón, M.; Bedia, J.; Rodriguez, J. J.; Belver, C. C-modified TiO_2 using lignin as carbon precursor for the solar photocatalytic degradation of acetaminophen. *Chemical Engineering Journal* **2019**, *358*, 1574–1582.
50. Liu, B.; Khare, A.; Aydil, E. S. TiO_2-B/anatase core-shell heterojunction nanowires for photocatalysis. *ACS Appl Mater Interfaces* **2011**, 3 (11), 4444–50.

51. Han, C.; Chen, Z.; Zhang, N.; Colmenares, J. C.; Xu, Y.-J. Hierarchically CdS Decorated 1D ZnO Nanorods-2D Graphene Hybrids: Low Temperature Synthesis and Enhanced Photocatalytic Performance. *Advanced Functional Materials* **2015**, *25* (2), 221–229.
52. Su, Y.; Yang, Y.; Zhang, H.; Xie, Y.; Wu, Z.; Jiang, Y.; Fukata, N.; Bando, Y.; Wang, Z. L. Enhanced photodegradation of methyl orange with TiO_2 nanoparticles using a triboelectric nanogenerator. *Nanotechnology* **2013**, *24* (29), 295401.
53. Zhu, Q.; Sun, Y.; Xu, S.; Li, Y.; Lin, X.; Qin, Y. Rational design of 3D/2D In_2O_3 nanocube/ $ZnIn_2S_4$ nanosheet heterojunction photocatalyst with large-area "high-speed channels" for photocatalytic oxidation of 2, 4-dichlorophenol under visible light[J]. *Journal of hazardous materials* **2020**, *382*, 121098.
54. Srisasiwimon, N.; Chuangchote, S.; Laosiripojana, N.; Sagawa, T. TiO_2/Lignin-Based Carbon Composited Photocatalysts for Enhanced Photocatalytic Conversion of Lignin to High Value Chemicals. *ACS Sustainable Chemistry & Engineering* **2018**, *6* (11), 13968–13976.
55. Peng, X.; Zhang, L.; Chen, Z.; Zhong, L.; Zhao, D.; Chi, X.; Zhao, X.; Li, L.; Lu, X.; Leng, K.; Liu, C.; Liu, W.; Tang, W.; Loh, K. P. Hierarchically Porous Carbon Plates Derived from Wood as Bifunctional ORR/OER Electrodes. *Advanced Materials* **2019**, *31* (16), e1900341.
56. García-Mateos, F. J.; Cordero-Lanzac, T.; Berenguer, R.; Morallón, E.; Cazorla-Amorós, D.; Rodríguez-Mirasol, J.; Cordero, T. Lignin-derived Pt supported carbon (submicron)fiber electrocatalysts for alcohol electro-oxidation. *Applied Catalysis B: Environmental* **2017**, *211*, 18–30.
57. Qin, H.; Kang, S.; Wang, Y.; Liu, H.; Ni, Z.; Huang, Y.; Li, Y.; Li, X. Lignin-Based Fabrication of Co@C Core–Shell Nanoparticles as Efficient Catalyst for Selective Fischer–Tropsch Synthesis of C5+ Compounds. *ACS Sustainable Chemistry & Engineering* **2016**, *4* (3), 1240–1247.
58. Qin, H.; Zhou, Y.; Bai, J.; Zhu, B.; Ni, Z.; Wang, L.; Liu, W.; Zhou, Q.; Li, X. Lignin-Derived Thin-Walled Graphitic Carbon-Encapsulated Iron Nanoparticles: Growth, Characterization, and Applications. *ACS Sustainable Chemistry & Engineering* **2017**, *5* (2), 1917–1923.
59. Martin-Martinez, M.; Barreiro, M. F. F.; Silva, A. M. T.; Figueiredo, J. L.; Faria, J. L.; Gomes, H. T. Lignin-based activated carbons as metal-free catalysts for the oxidative degradation of 4-nitrophenol in aqueous solution. *Applied Catalysis B: Environmental* **2017**, *219*, 372–378.

Chapter 7

Recent Advances in Lignin Modification and Its Application in Wastewater Treatment

Bin Wang,[1] Dan Sun,[1] Tong-Qi Yuan,[1] Guoyong Song,[*,1] and Run-Cang Sun[*,2]

[1]Beijing Key Laboratory of Lignocellulosic Chemistry, Beijing Forestry University, Beijing 100083, P. R. China
[2]Center for Lignocellulose Chemistry and Materials, Dalian Polytechnic University, Dalian 116034, P. R. China
[*]Email: rcsun3@dlpu.edu.cn.

In recent years, lignin and its derivatives as wastewater treatment agents have received extensive attention due to their many advantages, including their renewable raw materials, superior performance, biodegradability, and no secondary pollution. The chapter systematically reviews the recent progress of lignin and its derivatives as adsorbents, flocculants and sterilants, and other wastewater treatment agents. It is emphatically introduced that chemical modifications can effectively control and adjust the structure of lignin to improve its wastewater treatment capacity. Moreover, the purification mechanisms and their influencing factors are discussed in detail. In addition, the chapter also critically points out the deficiencies of lignin-based wastewater treatment agents at the present stage and proposes corresponding improvement approaches. Furthermore, the future research emphasis and development direction of lignin and its derivatives in wastewater treatment are discussed.

Introduction

Water shortage is one of the most pervasive problems plaguing the world, and it would continue to worsen in the foreseeable future. The reuse of wastewater is one of the most effective ways to solve the shortage of water resources (*1*, *2*). However, the discharge of wastewater from both industry and living is increasing significantly, and the polluting components contained in wastewater are more complicated, which makes it arduous to recycle water resources (*3*, *4*). Fortunately, numerous technologies such as adsorption (*5*), coagulation/flocculation (*6*), sterilization (*7*), ion exchange (*8*), electrochemistry (*9*) and membrane filtration (*10*) have been widely applied to remove contaminations from wastewater. However, the current commercial wastewater treatment agents are basically petrochemical products, and their use usually will generate secondary pollution, causing a severe burden on the environment (*11–13*). Moreover, the non-renewable and high prices of raw materials are unavoidable difficulties. Natural polymers such as lignin (*14*), chitosan (*15*), starch (*16*) and cellulose (*17*) and their derivatives are considered as ideal candidates for wastewater treatment agents due to their biodegradability, renewable, abundant source, low cost and other advantages (*18–20*).

Lignin is composed of phenylpropane units via enzymatic polymerization of three aromatic alcohol precursors: coniferyl, sinapyl, and *p*-coumaryl alcohols, and it is the most abundant renewable aromatic polymer and the second richest renewable terrestrial polymer after cellulose (*21*). However, industrial lignin, as a by-product from the pulping and bio-refinery industries, is mainly burned for obtaining thermal energy or directly discharged as waste (*22, 23*). The industrial lignin used to produce value-added products only accounts for 1%-2% of the annual output (*24*). In the context of the irreversible depletion of petrochemical products, it is of great significance to increase the use value of lignin resources in order to replace petrochemical products (*25*). In particular, the application of lignin in the field of wastewater purification not only alleviates the problem of resource shortage, but also promotes the recycling of water resources, which has a double significance for environmental protection, thereby attracting widespread attention.

Industrial lignins contain a large number of functional groups such as hydroxyl, methoxy, carboxyl, etc. and chemical bonds such as ether bonds, carbon-carbon bonds, etc., which give it a degree of wastewater treatment capacity (*26*). Among them, lignosulfonate with sulfonated groups shows better water solubility and higher charge density, which has been emphatically studied and applied to purify wastewater. It can act as a flocculant to remove colloids from suspension (*24*). Moreover, Sylvie et al. used lignosulfonate to flocculate sulfur slurry from copper heap leaching and successfully increased the density of sulfur slurry from 40% to 67% (*27*). In addition, the adsorption capacity of some industrial lignins for metal ions has also been studied. In one study, kraft lignin exhibited precise selective adsorption to Au(III) in hydrochloric acid (*28*), in which Au(III) existed in the form of $AuCl_4^-$ in hydrochloric acid solution. Simultaneously, the ether functional group on kraft lignin was protonated and formed an oxonium salt, which made the lignin surface exhibit positive electricity. The anion exchange of chloride ion with chloroaurate anion, and electrostatic interaction of Au(III) with oxonium centre on kraft lignin caused a large amount of Au(III) to be adsorbed on the surface of kraft lignin, achieving the separation and recovery of gold. Moreover, the results showed that kraft lignin could not adsorb Pb(II), Zn(II), Cu(II), Fe(III) and other metal ions in hydrochloric acid, and exhibited outstanding adsorption selectivity for Au(III).

Although unmodified industrial lignins can treat some specific types of wastewater, they often show inefficiency for actual wastewater with complex components. In order to improve the wastewater treatment capacity of lignin and realize the commercial application of lignin-based water treatment agents, chemical modification of industrial lignin is essential (*29*). The present chapter summarizes a series of recent chemical modification methods to improve lignin wastewater treatment capacity, and evaluates the prospect of lignin derivatives as adsorbents, flocculants and sterilants, and other wastewater treatment agents. Moreover, we also discuss the effect of the lignin derivative structure on its wastewater treatment performance.

Lignin Modification Methods

Graft Copolymerization

Graft copolymerization is the most commonly used method for preparing lignin-based wastewater treatment agent, which belongs to a free radical polymerization reaction (*30*). In the process of graft polymerization, the primary free radicals are usually first generated via various initiation ways, and then the hydroxyl groups with high activity on the lignin will be attacked by these primary free radicals, forming lignin free radicals (*31*). Figure 1 shows where the lignin free radicals can occur. Graft copolymerization of lignin generally can be divided into one-step and two-step methods. The one-step method is that lignin is first dissolved in the reaction medium, and then initiator, reducing agent, and unsaturated monomer are added together to co-react, which has advantages of a simple process, fast reaction rate, and high production efficiency (*32*). However, part of unsaturated monomers will polymerize to form homopolymers due to the high reactivity ratio. As a result, only a small number of monomers are grafted onto lignin skeleton, making it difficult to obtain products with high grafting degrees. Moreover, lignin derivatives prepared by one-step method usually have larger viscosity, which increases the difficulty of purification (*33*). The two-step method is that the lignin and reducing agent are simultaneously added to the reaction medium. After heating, the unsaturated monomer and initiator are added slowly, so that the unsaturated monomer has enough time to contact with lignin molecules. The two-step method has the advantages of easy control, products with low viscosity and high solid content, high conversion rate (*34*). However, long reaction time, complicated steps and low molecular weight products make it difficult to be applied in the preparation of lignin-based wastewater treatment agents on a large scale (*35*).

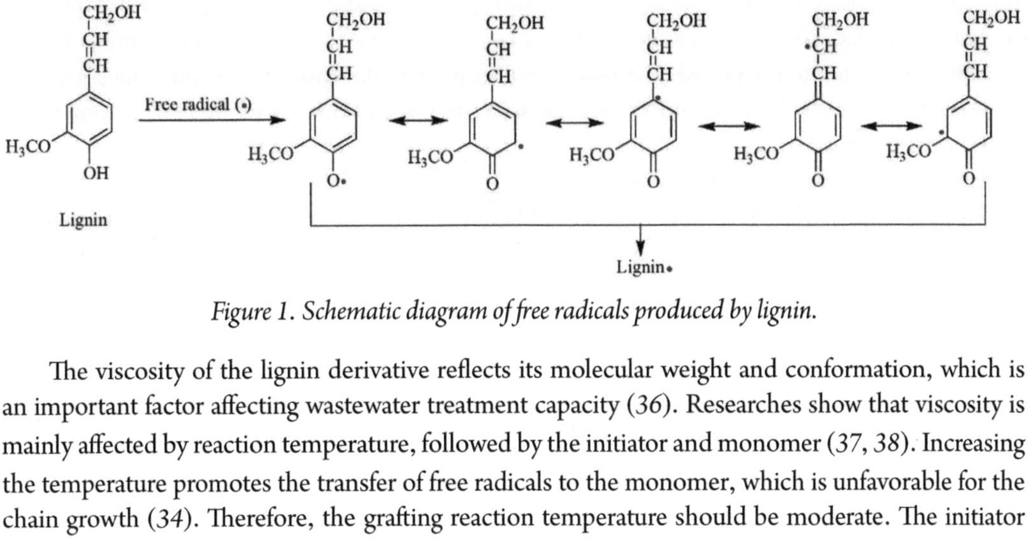

Figure 1. Schematic diagram of free radicals produced by lignin.

The viscosity of the lignin derivative reflects its molecular weight and conformation, which is an important factor affecting wastewater treatment capacity (*36*). Researches show that viscosity is mainly affected by reaction temperature, followed by the initiator and monomer (*37, 38*). Increasing the temperature promotes the transfer of free radicals to the monomer, which is unfavorable for the chain growth (*34*). Therefore, the grafting reaction temperature should be moderate. The initiator system affects the viscosity of lignin derivative from two aspects: the dose and the type of initiator. Increasing the initiator concentration can generate more free radicals and trigger more lignin, resulting in increased yield of the graft copolymer. However, when the initiator concentration exceeds the optimum value, the monomer radicals react with each other, resulting in a large consumption of free radicals and the production of homopolymer (*39*). Moreover, the graft polymerization of lignin is mostly initiated by initiators. There are many types of initiators, and their initiation conditions and mechanisms are different, which greatly affect the viscosity of the grafted polymers. Wang et al. used several commonly radical polymerization initiators, such as Fe^{2+}-H_2O_2, high cerium salt, $K_2S_2O_8$/ $Na_2S_2O_3$, and thiosulfate, to initiate the graft polymerization of lignin and acrylamide (AM) (*40*). The results showed that the grafted polymer induced by $K_2S_2O_8$/ $Na_2S_2O_3$ had the highest viscosity, higher reaction yield, and lower surface tension, which had better graft efficiency and flocculation performance.

The effects of grafted monomers on the wastewater treatment capacity of lignin derivatives are multidirectional. There are a variety of monomers can be grafted onto lignin, and the vinyl monomers are the most common, which are summarized in Table 1. The reaction medium of graft copolymerization needs to be selected according to the grafted monomer. For example, a monomer with positive charge has a stronger grafting effect using methanol as the medium as opposed to water (*41*). Conversely, a negatively charged monomer is more suitable for water as the medium (*42*). Although different types of monomers can give lignin diverse properties, they usually have some similar characteristics. Generally, the molecular weight is significantly increased and the water solubility is changed by introducing monomers, which determine the dynamic diameter and bridging effect of lignin-based wastewater treatment agents (*43*). Moreover, when the monomer grafted to lignin carries charges, the charge properties or charge density of the lignin derivatives will change accordingly, which is conducive to the charge attraction of lignin-based wastewater treatment agents (*44*).

Acrylamide (AM) is the most frequently used monomer in the preparation of lignin-based flocculants. AM has quite high reactivity ratio due to the presence of carbon-carbon double bonds and amide groups, and it is easily grafted onto the lignin skeleton by free radical polymerization (*45*). However, the AM monomer is toxic and expensive, and it can react with the guanine reaction on DNA, leading to genetic material damage and gene mutations. Moreover, excessive amounts of AM can damage the nervous system (*46*). Therefore, the application of polyacrylamide (PAM) in wastewater is limited. Grafting AM onto lignin to prepare flocculant can reduce the amount of acrylamide, reducing the cost and the risk of secondary pollution caused by AM. Although the AM is non-ionic and cannot provide charge, the hydrophilicity and molecular weight of the lignin have been greatly improved after grafting AM, which usually exhibit excellent flocculation efficiency for suspension (*47*). In other study, a series of lignin-based water treatment agents with different molecular weight and charge density were prepared by grafting polymerization of lignin and acrylamide (*48*). The experimental results showed that the lignin derivative with the highest molecular weight and charge density had the highest flocculation effect for kaolin suspension, and the molecular weight was the dominant factor affecting the flocculation capacity. However, lignin derivatives prepared by graft copolymerization of lignin and AM are anionic, which are generally inefficient for anionic wastewater with small diameter pollution particles. In this case, cationization of these anionic lignin derivatives or the addition of cationic coagulants can significantly improve the flocculation capacity of lignin-AM. For example, Li et al. synthesized a cationic lignin-based flocculant via graft polymerization of kraft lignin, AM, and dimethyl diallyl ammonium chloride (DADMAC) (*49*). The lignin-based flocculant combined use with polyaluminium chloride (PAC) had excellent decolorization efficiency for the disperse and reactive dyes wastewater.

Unsaturated monomers with ionizing groups can give lignin stronger wastewater treatment capacity as compared with AM. Acrylic acid (AA) is the most commonly used anionic unsaturated monomer, which belongs to unsaturated carboxylic acid (*50*). The introduction of carboxylic acid groups derived from AA can increase hydrophilicity and chelating ability of the lignin (*51*). Kong et al. produced a soluble lignin derivative with high molecular weight by polymerizing kraft lignin and AA in an acidic aqueous suspension system (*52*). The lignin derivative as flocculant had stronger flocculation efficiency and pH independent for aluminum oxide suspensions as compared with polyacrylic acid (PAA). Research showed that acidic conditions could facilitate the free radical polymerization of AA and lignin.

Table 1. Examples of Monomers Used in Graft Polymerization Reactions for the Preparation of Lignin-Based Wastewater Treatment Agents

Types	*Vinyl Monomer*	*Structure*	*Ref.*
Non-ionic	Acrylamide (AM)	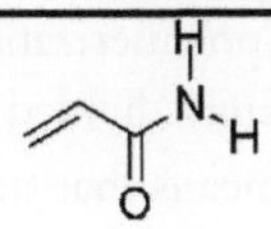	(53-55)
	N, N-dimethylacrylamide (DMAA)	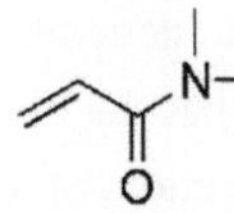	(56)
Anionic	Acrylic acid (AA)	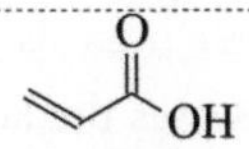	(57-59)
	Methacrylic acid	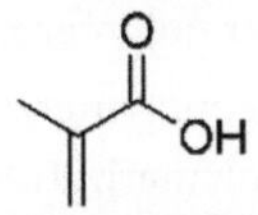	(60, 61)
Cationic	Dimethyl diallyl ammonium chloride (DADMAC)	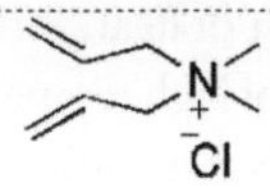	(62)
	(2-methacryloyloxyethyl) Trimethyl ammonium chloride (DMC)	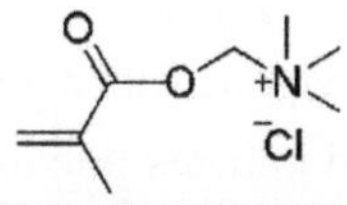	(63, 64)

Some cationic unsaturated monomers, such as dimethyl diallyl ammonium chloride (DADMAC) and (2-methacryloyloxyethyl) trimethyl ammonium chloride (DMC), have been extensively studied in recent years due to their high positive charge density and reactivity, as shown in Table 1. The introduction of cationic unsaturated monomers into lignin can increase its molecular mass and enhance the positive charge density, thus enhancing the effects of charge attraction and bridging (*65*). The cationic unsaturated monomers are mainly quaternary ammonium salts, and lignin grafted unsaturated quaternary ammonium salts can achieve the ideal flocculation effect without adding coagulants when treating most anionic wastewater (*66*). Moreover, since the quaternary ammonium salt group has the bactericidal capacity, the lignin derivatives with quaternary ammonium salts can not only effectively flocculate and precipitate bacteria form wastewater, but also kill them, which greatly reduces the number of subsequent disinfectants (*67*).

With the deepening of the research on the preparation of lignin-based wastewater treatment agent by graft copolymerization, the grafting of two or more monomers onto lignin, or the grafting of two natural polymers with single monomer have attracted wide attention, which can simultaneously give lignin multiple properties, thereby significantly improving the wastewater treatment capacity and efficiency (*62*). However, it is the most challenging problem to prepare the multiple copolymers that the reactivity ratios among the reactants are different. The reactivity ratio of the reactant determines its reactivity, which affects the final structure of the copolymer (*44*). The reactant with a high reactivity ratio has stronger reaction capacity and often requires a larger feed ratio in order to achieve the best polymerization efficiency. It is difficult for monomer with low reactivity ratio to perform effective polymerization reaction when used together with monomer having a high reactivity ratio (*68*). Therefore, monomers with different reactivity ratios are generally grafted onto lignin in sequence instead of one-step method.

In addition to the common initiator-initiated polymerization, there are photoinitiated polymerization (*69*), electrochemical initiated polymerization (*70*), and radiation-induced polymerization (*71*). Moreover, several initiation methods can be used simultaneously to improve grafting efficiency.

The photoinitiated polymerizations of lignin are dominated by ultraviolet (UV) initiation, and the UV initiation is mainly divided into direct photoinitiation and indirect photoinitiation (*72*). Direct photoinitiation means that the monomer is first converted to an excited state by absorbing the photon of UV light, generating free radicals and initiating polymerization (*73*). Indirect photoinitiation refers to that the photoinitiator first undergoes molecular cleavage to form primary free radicals under the action of UV light. Afterwards, these primary free radicals initiate the formation of monomer free radicals and eventually initiate the polymerization (*69*). Photoinitiated graft polymerization possesses the advantages of shorter polymerization time, less initiator, higher reaction rate, and a larger polymer molecular weight than other initiation systems (*74*). At present, the photoinitiated graft copolymerization of lignin is mainly initiated by long-wavelength UV light. However, since the copolymerization reaction induced by the common long-wave UV light usually produces a large amount of heat, it is necessary to install condensation equipment. In addition, the preparation conditions of high temperature and high pressure increase the production of by-products caused by explosive polymerization (*69*). The short-wavelength UV light has stronger energy, and the polymerization reaction induced by short-wavelength UV that generates little heat, can be carried out under normal pressure, which greatly reduces the cost (*75*).

Electrochemical initiated polymerization can be carried out in an aqueous or organic solution (*76*). The polymerization of lignin, initiated by electrochemical action, takes place at the electrode. In the electrochemical initiated polymerization reaction, lignin and monomer are first adsorbed on the surface of the electrode, and their functional groups undergo electrolytic oxidation reactions (or reduction reaction), after which the polymerization occurs between the activated functional groups (*70*). Zhang et al. (1998) grafted an acrylic monomer onto lignin by electrolysis in a diaphragm-free electrolytic cell using lead dioxide as the anode and stainless steel as the cathode, preparing a lignin derivative with good surface activity and dispersibility (*77*). Electrochemical initiated polymerization has the following advantages as compared with traditional initiator-initiated polymerization: low cost, mild reaction conditions, high selectivity, and environmentally friendly (*78*).

Radiation-induced polymerization is a reaction in which high-energy ionizing radiation (α-ray, β-ray, γ-ray, x-ray, and electron beam) radiates the monomer or solution to generate radicals or ions, and the primary active particles further initiate the polymerization of lignin and monomer (*79*). The radiation-induced polymerization reaction is characterized by fast reaction rate and high purity products, and it does not require the addition of an initiator, eliminating the risk of metal residues (*80*). Moreover, it can be carried out at room temperature or lower, which is advantageous for generating active radical polymerization of thermally unstable monomers. Some radiation-induced polymerization systems can generate free radicals, anions, and cations at the same time. Due to the cage effect in the cloud cluster, the particles can recombine rapidly, and only a small part escapes from the cloud to form a steady state substance. Since free radicals are much more likely to escape from clouds than ions, most of the systems without special drying are dominated by free radical polymerization (*71*). There have been studies using radiation-induced polymerization to prepare chitosan-based flocculants. However, the radiation-induced polymerization has hardly been mentioned in the preparation of lignin-based wastewater treatment agents, which deserves further exploration.

Mannich Reaction

Lignin macromolecule has a large number of active hydrogen atoms, which are located at the ortho and para positions of the phenolic hydroxyl group and the α position of the hydroxyl group on the side chain. Lignin with active hydrogen atoms can undergo a condensation reaction with amines and aldehydes to form lignin amine, which is known as the Mannich reaction (Figure 2) (*81*). The hydrophobicity and cationicity of lignin amine synthesized by the Mannich reaction are significantly improved due to the introduction of long-chain alkyl groups and cationic amino groups. When sufficient cationic groups are introduced into the lignin, the lignin begins to exhibit positively charged characteristics and becomes a typical cationic surfactant (*82*). There are many types of fatty amines that can undergo Mannich reactions, as shown in Table 2, where dimethylamine is the most widely used fatty amine. Aldehydes act as crosslinkers between fatty amines and lignin during the Mannich reaction. In terms of cost and production, formaldehyde is the most commonly used aldehyde (*83*, *84*). In addition, the acetaldehyde, propionic aldehyde, benzaldehyde, furfural, dialdehyde (such as glutaraldehyde, glyoxal), and unsaturated aldehydes (such as acrolein, butenal) are also used (*85*, *86*).

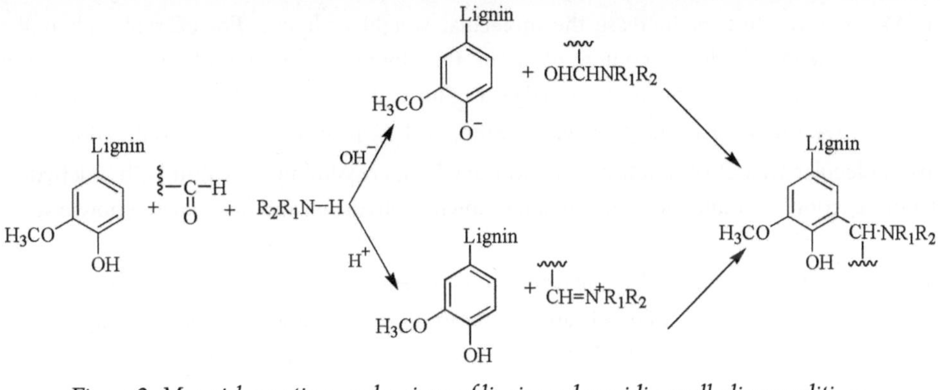

Figure 2. Mannich reaction mechanisms of lignin under acidic or alkaline conditions.

The mechanism of the Mannich reaction is complicated. The amine is first reacted with an aldehyde to form an intermediate N-hydroxymethylamine, and then condensed with lignin containing an active hydrogen atom (*87*). The reaction can be carried out in an aqueous phase (usually an aqueous alkaline solution), an organic phase (alcohols, dioxanes, etc.), and a mixture of water and organic solvent. In general, the ratio of the active hydrogen atom, the amine, and the formaldehyde is 1.0: (1.05-1.10): (1.5-2.0). Concentrated hydrochloric acid is typically used as the catalyst. The reaction temperature is usually maintained at around the boiling point of the solution. Excessive reaction temperature will lead to the production of more byproducts. The Mannich reaction also depends on the pH value. When the pH value is closer to 7, the products have a greater degree of substitution and higher nitrogen content (*88*). Mannich reaction of lignin skeleton can only occur on carbon with a high electron density, that is, the ortho and para positions of the lignin phenolic hydroxyl group. Therefore, lignin with more phenolic hydroxyl groups is beneficial to the Mannich reaction. In one study, lignin having more phenolic hydroxyl groups and fewer aryl ether bonds was separated from the corn stover via the cosolvent enhanced lignocellulosic fractionation (CELF) method (*89*). Whereafter, a lignin-based adsorbent was prepared by Mannich modification of lignin using diethylenetriamine (DETA) and formaldehyde, which could effectively decolor the methylene blue dye wastewater and direct blue 1 dye wastewater.

Table 2. Types of Fatty Amines in the Mannich Reaction

Classification of Fatty Amines	*Example*	*Ref*
Linear, branched alkylamine	Octadecylamine, acrylamide, animal fat amine	(*90–92*)
Naphthenic amino	Cyclohexyl amine	(*93*)
Cyclic amine	Pyrrolidine, piperazine, morpholine, imidazoline	(*94–96*)
Polyamine	Hexylenediamine, ethylenediamine, triethylenetetramine	(*97, 98*)
Hydroxyl alkyl amine	Hydroxyethyl primary amine, dihydroxyethyl amine, aminoethyl ethanolamine	(*99–101*)
Amino acids	Creatine, asparagine, aspartic acid, tyrosine	(*102–104*)

However, lignin derivatives prepared by the Mannich reaction still has disadvantages of low reactivity and molecular weight, and their ability to treat wastewater is inferior to commonly commercial wastewater treatment agents. Therefore, based on application, lignin can be appropriately modified before subjecting it to Mannich reaction in order to increase the active point of the Mannich reaction or increase the molecular weight of lignin. For example, phenolization can increase phenolic hydroxyl content of lignin, thereby increasing the nitrogen content of lignin derivatives from the Mannich reaction (*105*). Lignin can remove part of the methoxy group by strong alkali treatment or co-heating with amine, and its reactivity is increased (*106*, *107*). The relative molecular weight of lignin can be improved via crosslinking reaction with aldehydes, and alkylation reaction with dihalogenated alkanes, epichlorohydrins, and halogenated polyoxyethylene (*108, 109*). In one study, alkaline lignin was first phenolized to strengthen reaction activity, and then it was sequentially modified by Mannich reaction and sulfomethylation (*110*). Finally, an efficient lignin-based adsorbent was obtained, and its removal efficiency and adsorption capacity for Pb (II) were as high as 100% and 130.2 mg/g, respectively. Moreover, it presented excellent cycling capability, which was considered as a promising bio-adsorbent for removing heavy metal ions from wastewater. In other study, the Mannich reaction was first performed with acetone, formaldehyde and dimethylamine, and then the product was grafted onto hydroxymethylated lignin to synthesize a lignin-based flocculant (*111*). The dye removal rates of the lignin-based flocculant for three simulated dye wastewaters (Acid Black, Reactive Red and Direct Red) could all exceed 95%, and over 85% of the chemical oxygen demand (COD) was removed, which is comparable to commercial flocculants. In short, lignin modified by a single Mannich cannot meet the demand for commercial wastewater treatment agents. Fortunately, the lignin-based wastewater treatment agents with excellent wastewater treatment capacity can be obtained by modifying lignin with Mannich reaction in cooperation with other modification methods.

Etherification

The active hydroxyl group on lignin can be etherified by the etherification agents with the lower density electron cloud, which belongs to the nucleophilic substitution reaction. Etherified lignin can be classified into non-ionic and ionic types (*112*). Hydroxyethyl lignin and hydroxypropyl lignin are common non-ionic etherified polymers (*113, 114*). The hydrogen bonding force of lignin is weakened due to the introduction of hydroxyethyl or hydroxypropyl, so its hydrophilicity and pH independence are improved, and the ability to adsorb metal ions is enhanced. However, non-ionic etherified lignin is generally less efficient as wastewater treatment agent because it lacks sufficient charge attraction.

Table 3. Synergistic Absorption of Carboxymethyl Lignin and Metal Ion for the Dye Wastewater

			Adsorption Dye			
Lignin Type	*Metal Ion*	*Dye Type*	*Adsorption Capacity*	*Removal Rate*	*Regeneration Rate*	*Ref*
Carboxymethyl Lignin	Fe^{3+}	Brilliant Red 2BE	73.6 mg g^{-1}	-	99.5% (One cycle)	(*115*)
	Al^{3+}	Reactive Blue-4	73.52 mg g^{-1}	93.97%	98.33% (Four cycles)	(*116*)
	MnMn^{2+}	Reactive Blue-4	55.16 mg g^{-1}	75.91%	98.08% (Four cycles)	

At present, many researches focus on the ionic etherified lignin in the field of wastewater treatment. Carboxymethyl lignin is a typical ionic etherate, which mainly exists in the form of sodium salt (*117*). The introduction of carboxymethyl can improve the hydrophilicity of lignin, and its molecular surface activity and negative charge density are increased, which enhances the wastewater treatment capacity for cationic wastewater. In one study, a carboxymethyl lignin was synthesized via the carboxymethylation of irradiated lignin and monochloroacetic acid, and it had effective adsorption effect on graphite (*118*). In addition, the combination of carboxylated lignin and metal ions can be used as cationic adsorbent under acidic conditions due to protonation. Table 3 shows the synergistic effect of carboxymethyl lignin and some metal ions on the adsorption of dye from wastewater, which proves that the presence of metal ions can significantly improve the wastewater treatment capacity of carboxymethyl lignin. However, carboxymethyl compounds often exhibit inferior temperature resistance, which severely limits the commercial application of carboxymethyl lignin.

The cationic etherified lignin prepared by reacting lignin with cationic etherification agent is one of the most popular wastewater treatment agents. Common cationic etherification agents include tertiary amine salts and quaternary ammonium salts. However, since the tertiary amine salts have a strong dependence on pH and show electropositivity only under acidic conditions, the lignin-based wastewater treatment agent with the tertiary amine salts only has an efficient flocculation effect within a narrow pH range (*119*). However, the quaternary ammonium salts have stronger electropositivity than tertiary amine salts. The 3-chloro-2-hydroxypropyl trimethyl ammonium chloride (CTA) and 2, 3-epoxypropyl trimethyl ammonium chloride (ETA) are two commonly used quaternary ammonium etherification agents. In one study, kraft lignin was etherified using ETA to prepare cationic lignin-based flocculant in alkaline solution (*120*). The lignin-based flocculant prepared under optimized conditions had 90 wt.% solubility and 1.10 meq/g charge density at 1 wt.% lignin concentration, which had excellent decolorization effect on Remazol Brilliant Violet, Reactive Black, and Direct Yellow simulative wastewaters. Notably, the thermogravimetry (TGA) analysis proved that the thermal stability of etherified lignin was reduced.

The lignin derivatives having quaternary ammonium salts can effectively remove the pollutants from wastewater within a wide range of pH value, and fewer doses are needed to achieve the desired flocculation effect due to stronger charge attraction, which are more practical. Moreover, the preparation of cationic etherified lignin generally has the advantages of simple operation and high conversion rate. Although etherified lignin containing quaternary ammonium salts has high charge density, its molecular weight and water solubility are inferior (*121*). In order to solve these problems, Laszlo first prepared quaternized lignin by the etherification of kraft lignin and CTA in alkaline solution, and the quaternized lignin was divided into pH 7-soluble quaternized lignin and pH 7-insoluble quaternized lignin (*122*). Subsequently, the pH 7-soluble quaternized lignin was treated with soybean peroxidase and hydrogen peroxide under nitrogen to increase molecular weight and improve water solubility, obtaining a cationic lignin-based flocculant. The results showed that the lignin-based flocculant could effectively decolorize in a wide range of pH value, and it was compared to commercial flocculants. However, the high price and less variety are inevitable problems for quaternary ammonium salt monomers at this stage. The application of cheap and active quaternary ammonium salt monomers is the focus of future research.

In general, the etherification reaction is widely used in the preparation of wastewater treatment agents because of simple operation, low cost, and superior product performance. However, commonly used etherification agents, such as chloroacetic acid and ethylene oxide, are highly toxic and cause serious harm to the human body and environment. Moreover, the cationic etherification agent is monotonous, and most researches are conducted around CTA, which is not conducive to the diversified development of lignin-based wastewater treatment agents. A novel, efficient and environmentally friendly etherification agent has yet to be discovered. Furthermore, the composite etherification modification of lignin has broad development prospects in the field of wastewater treatment.

Esterification

Lignin has strong hydrogen bonds within and between molecules caused by a large number of hydroxyl groups, resulting in inferior compatibility and interfacial adhesion (*123*). The active hydroxyl of lignin can be esterified with acyl halides, anhydrides, acids, etc (*113*). The biocompatibility and biodegradability of esterified lignin are significantly improved because of the destruction of hydrogen bonds. Due to remarkable biocompatibility, esterified lignin is widely used as filler to increase the mechanical strength of the mixture. The application of esterified lignin in wastewater treatment focuses on decolorization. Masaru et al. prepared a cationic lignin-based flocculant through the esterification reaction of alkali lignin with bromoisobutyryl groups, which exhibited predominant flocculation efficiency for dye wastewater (*124*). Moreover, the synergistic effect of esterified lignin and magnetic particles will have a strong decolorization ability. In one study, larch lignin was first esterified with maleic anhydride (*125*). The hydrophobicity of esterified lignin was significantly increased due to the shielding effect of the carbonyl group from maleic anhydride on the hydrophilic hydroxyl groups on the lignin, and the esterified lignin could form hollow microspheres under the action of anti-solvent. Subsequently, nano-magnetic particles (Fe_3O_4) were embedded in lignin hollow microsphere or attached to its surface, obtaining a magnetic lignin hollow microsphere (MLS). The MLS was used as an adsorbent to treat methylene blue and Rhodamine B dye wastewaters, and the adsorption capacity reached 31.23 mg/g and 17.62 mg/g, respectively. In addition, the removal efficiency could reach more than 98% after three cycles of desorption.

Ionic liquid, as environmentally friendly solvent and catalyst, has the advantages of stability, fast reaction rate, reusability, high conversion rate and selectivity, and it has been widely studied as a reaction medium for lignin modification (*126*). Lignin can be dissolved in ionic liquid without changing its spatial structure, weakening the reaction resistance caused by hydrogen bonding. Ionic liquid can provide homogeneous reaction conditions for the esterification of lignin, which speeds up the main reaction rate and inhibits side reactions, significantly increasing the degree of substitution (*127*). In one study, four industrial lignins (Indulin AT softwood kraft lignin, Wayagamack softwood kraft lignin, Windsor hardwood kraft lignin and Lignol softwood organosolv lignin) were esterified with maleic anhydride in an acidic ionic liquid (1-butyl-3-methyl imidazolium hydrogen sulfate) (*128*). The results showed that the esterification degrees of hydroxyl groups on the four industrial lignins were as high as 30-50%, and the aliphatic hydroxyl groups of three kinds of industrial lignin (Indulin AT, Windsor and Lignol) had preferential reaction regioselectivity. The solubilities of the four esterified lignins in polar and protic solutions were significantly increased without affecting the thermal stability. Moreover, the ionic liquid used can easily be recovered and reused. However, it is regrettable that ionic liquid as a reaction solvent has little research on the preparation of lignin-based wastewater treatment agents. The main reason is the high price of ionic liquids. With the advancement of technology, ionic liquid is expected to be applied in catalyzing the modification of lignin to prepare wastewater treatment agents.

Oxidation

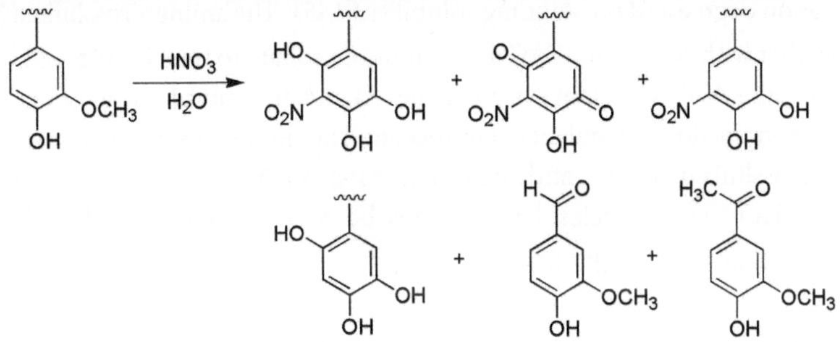

Figure 3. Oxidation reaction scheme and resulting products for softwood lignin (132). Reproduced with permission from reference (132). Copyright 2016 American Chemical Society.

The hydroxyl groups on lignin can be oxidized to aldehyde groups or carboxyl groups by oxidizing agent, and even the phenylpropanoid unit ring will be opened due to oxidative cleavage (*129*). After oxidation, the hydrogen bond of lignin decreases and the number of carboxyl groups increases, hence water solubility and metal chelating ability of lignin increase significantly (*130*). Dizhbite et al. oxidized lignin with polyoxometalate $H_3[PMo_{12}O_{40}]$, and the oxygen-containing adsorption-active sites of lignin were greatly increased (*131*). Under the optimal preparation conditions, the adsorption capacity of oxidized lignin was three times and twice than that of unmodified lignin for Cd(II) and Pb(II), respectively. However, mild oxidation reagents, such as hydrogen peroxide, metal oxides, nitrobenzene, and catalysts, can only oxidize the side chain of lignin, and cannot act on the phenylpropanoid skeleton. Therefore, stronger oxidation conditions are required. In one study, lignin extracted from the thermomechanical pulp was oxidized using nitric acid to prepare anionic lignin-based flocculant (*132*). Under the action of nitric acid, the aliphatic part of lignin generated numerous active sites, and lignin skeletons were oxidatively degraded. Figure

3 showed the scheme of lignin oxidation and possible products. The oxidation products of lignin mainly depended on the acid concentration and reaction temperature. The high acid concentration or reaction temperature helped lignin to introduce nitro groups. Otherwise, lignin could more easily introduce carboxyl groups. The anionic lignin-based flocculant synthesized by 30 wt % nitric acid concentration at 80 °C for 1.5 h, had 2.9 mmol/g carboxyl content and 97% solubility, which could remove 77% dyes and 70% COD for ethyl violet, and 88% dye and 86% COD for basic blue, respectively.

Crosslinking

Lignin can crosslink with compounds containing two or more functional groups by soft bonds, making its molecular weight and surface area greatly increased. Crosslinked lignin can form loose soft macromolecules in water, which can capture and adsorb fine solid particles or colloidal particles in suspension (*133*). In one study, lignin was cross-linked with polyethylene glycol under alkaline conditions to obtain viscous polymers, which had prominent flocculation effect on fine solids particles (*134*). However, lignin-based wastewater treatment agents prepared by crosslinking modification alone have single function and limited wastewater treatment ability. They usually need to be used in conjunction with coagulants to achieve the desired flocculation effect. Fortunately, it is a feasible method to further modify the cross-linked lignin in order to increase the charge density and improve flocculation ability. In one study, an amine-crosslinked lignin was synthesized by the reaction of alkali lignin, epichlorohydrin, N, N-dimethylformamide and ethylenediamine, and it exhibited high cation degree and outstanding solubility (*135*). The amine-crosslinked lignin was used as flocculant together with coagulants PAC and aluminum sulfate to treat humic acid wastewater. The results showed that the dual-coagulant system achieved higher removal rate as compared with single coagulant system and the dual-coagulants of unmodified lignin. Furthermore, composite adsorbents based on the crosslinked lignin and nanomaterials, such as carbon nanotubes, magnetic nanoparticles and TiO_2 nanoparticles, have recently been extensively studied, and they are usually highly efficient and environmentally friendly. For example, Zhang et al. synthesized lignin-based hybrid magnetic nanoparticles by the cross-linking of carboxymethylated lignin and amino-functionalized magnetic nanoparticles using epichlorohydrin as a cross-linker (*136*). The hybrid nanoparticles were used to adsorb Pb^{2+} and Cu^{2+}, and the adsorption capacities were up to 150.33 mg/g and 70.69 mg/g, respectively. Importantly, the adsorption equilibrium of the hybrid nanoparticles could be achieved within 30 s, which was among the fastest lignin-based adsorbents.

In addition to the chemical crosslinking described above, some noncovalent interactions can also crosslink lignin and three-dimensional (3D) porous materials to prepare a lignin-based adsorbent with high adsorption capacity. Generally, lignin can provide adsorption sites, and the 3D porous materials can offer multileveled porous skeleton and large surface area (*137*). In one study, a one-step method was applied to the crosslinking of lignosulfonate (LS) and graphene oxide (GO) in order to prepare lignosulfonate-modified graphene hydrogel (LS-GH) linked by hydrogen bonding and π-π conjugation (*138*), as is shown in Figure 4. The adsorption capacity of LS-GH for Pb(II) was up to 1210 mg/g, which was among the adsorbents with the highest adsorption capacity as reported by now. It could be used as a column-packed device to continuously adsorb Pb(II) from wastewater, with an ultra-high adsorption capacity (1308 mg/g within 40 min). Moreover, LS-GH retained 90% adsorption capacity after 5 times desorption and 82% adsorption capacity after 10 times desorption.

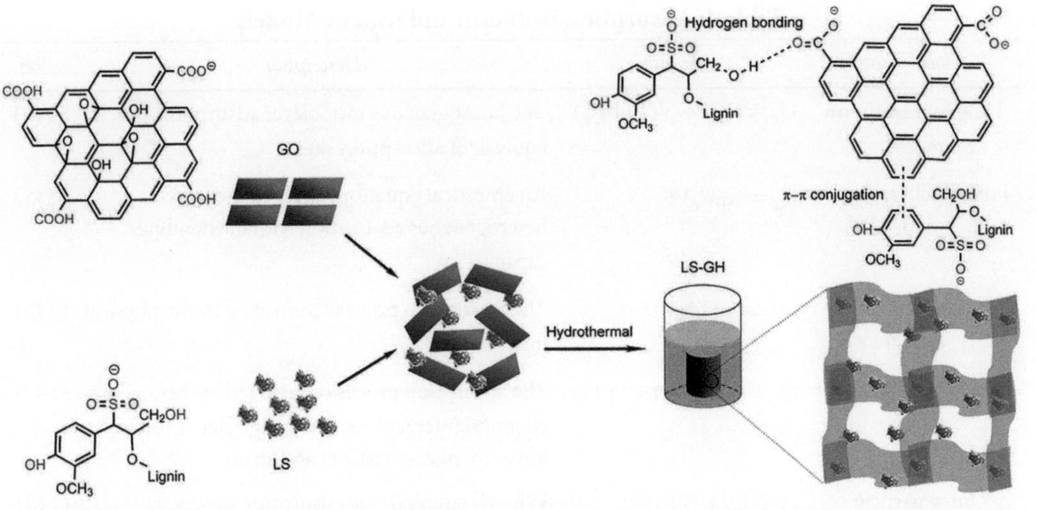

Figure 4. The preparation scheme of LS-GH and the interaction between GO and lignin (138). Reproduced with permission from reference (138). Copyright 2016 The Royal Society of Chemistry.

Application in Wastewater Treatment

Absorbent

A large number of studies have shown that lignin and its derivatives, as adsorbents, can effectively remove heavy metal ions, dyes, chemicals, and proteases from wastewaters (*139–141*). Adsorption is a complex endothermic process, which includes physical adsorption, chemical adsorption, and even supramolecular adsorption (*142*). The adsorption isotherm and kinetics are important parameters for understanding the adsorption process. Table 4 describes several generally accepted adsorption isotherms and kinetic models, which are helpful for the study of adsorption type and adsorption mechanism. Where Q_e, C_e and Q_m are the adsorption capacity, equilibrium concentration, saturated adsorption capacity, respectively, and b is the Langmuir constant. K_F and n are the adsorption capacity and energy constants, respectively. The k_1, k_2, and k_{int} are the pseudo-first-order, pseudo-second-order and intra-particle diffusion equations constants, respectively. The θ is the intercept, and Q_t is the adsorption amount at t minutes. Most adsorption processes of lignin-based adsorbents are included in these five models. However, it is worth noting that the adsorption model of the lignin-based adsorbent is not constant, which is related to the type of wastewater. In one study, a lignin-based resin prepared by the polymerization of lignosulfonate and glucose was used as adsorbent to remove five metal ions Pb(II), Cu(II), Cd(II), Ni(II), and Cr(III) from wastewater (*143*). The fitting results showed that the adsorption processes of lignin-based resin for Cu(II), Pb(II), and Cr(III) were consistent with the Langmuir isotherm model, while the adsorption processes for Ni(II) and Cd(II) were in accordance with Freundlich isotherm model. Moreover, the adsorption processes for the five metal ions were applicable to the Pseudo-second-order model, which proved that these adsorption processes belong to chemical adsorption.

Table 4. Adsorption Isotherm and Kinetic Models

Model Type	*Equation*	*Description*	*Ref*
Langmuir isotherm model	$Q_e = Q_m bC_e/(1+bC_e)$	The homogeneous monolayer adsorption and equivalent adsorption sites	(*144*)
Freundlich isotherm model	$Q_e = K_F C_e^{1/n}$	An empirical equation, apply to the heterogeneous adsorption with outstanding capacity	(*145*)
Pseudo-first-order model	$Q_t = Q_e(1-e^{k1t})$	The adsorption process controlled by the physical interactions	(*146*)
Pseudo-second-order model	$Q_t = Q_e^2 k_2 t/(1+ Q_e k_2 t)$	The adsorption process controlled by the chemical interactions, involving valence forces, ion exchange, chelation, and so on	(*147*)
Intra-particle diffusion model	$Q_t = k_{int} t^{1/2} + \theta$	When θ equals 0, the adsorption process is dominated by the intraparticle diffusion	(*148*)

Unmodified lignin can be used as an adsorbent to treat wastewater, especially for heavy metal ion wastewater. Lignin contains oxygen-containing groups, such as hydroxyl and carboxyl groups, and the unshared electron pairs on the oxygen atoms of oxygen-containing groups can form coordination bonds with the outer empty electron orbitals of metal ions, forming lignin-metal chelates. The adsorption capacity of lignin to various heavy metal ions is different. In one study, alkali lignin extracted from papermaking black liquor was subjected to adsorption tests for Pb(II), Cu(II), Cd(II), Zn(II), and Ni(II). The results showed that the static saturation adsorption capacity of the lignin for Pb(II), Cu(II), Cd(II), Zn(II), and Ni(II) was 89.51, 22.88, 25.41, 11.25 and 5.99 mg/L, respectively (*149*). In another study, the kraft lignin obtained from the black liquor by sulfuric acid precipitation adsorbed Cr(IV), Cd(II), Cu(II) and Zn(II) from wastewater, and the maximum adsorption capacity was 11.08, 8.21, 1.77 and 3.37 mg/L, respectively (*150*). A comparison of the two studies shows that the adsorption capacity of industrial lignin has a crucial relationship with the separation method. The conformation and functional groups of industrial lignin are affected by the separation process, which determines its adsorption capacity. Generally, alkali lignin and organosolv lignin have better adsorption capacity.

However, there is still a large gap between the adsorption capacity of unmodified lignin and commercial adsorbents. The chemical modifications mentioned above can improve the adsorption capacity and application range of lignin. In addition to the treatment of the wastewaters with metal ion or cationic dye, the modified lignin can also purify food wastewater containing organic matter, chemicals, enzymes, and so on. It can even be used as a gas adsorbent for natural gas storage. It is necessary to fully understand the structural factors that affect modified lignin adsorption capacity, which is of great significance for the design, preparation and application of lignin-based adsorbents. Because adsorption usually occurs on the surface of the adsorbent, the specific surface area is critical. The rough surface, micro- and mesoporosity can provide greater specific surface area for lignin. In one study, alkaline lignin was first Mannich modified using polyethylenimine, and then was esterified by CS_2 to obtain a lignin-based adsorbent with a porous structure (*151*). This lignin-based adsorbent was rich in mesoporous structure, and its specific surface area was up to 22.3 m^2/g, which was 12 times that of original lignin. Its adsorption capacity for Pb(II) was 188 mg/g, which was 13 times that of alkaline lignin. Moreover, the amorphous branches of lignin will form strong hydrogen bonds, making it difficult for lignin to expand in water, reducing specific surface area. Li et al. used graphene oxide to modify lignin, reducing the hydrogen bonding between lignin molecules and providing porous skeleton and large specific surface area derived from graphene oxide (*138*). The specific surface area of lignin-graphene oxide increased to 459.3 m^2/g, and its adsorption capacity for Pb(II) was as high as 1210 mg/g.

In addition, the content of oxygen-containing functional groups is a key factor for adsorption capacity of lignin. Oxidation, phenolization, and the introduction of functional groups with carboxyl group or hydroxyl group are commonly used methods to increase the content of oxygen-containing functional groups. For example, in one study, lignin was oxidized using polyoxometalate $H_3[PMo_{12}O_{40}]$ and the co-oxidant O_2 or H_2O_2 (*131*). The content of carboxyl groups and aliphatic hydroxyl groups in oxidized lignin was increased significantly, and its adsorption capacity for Cd(II) and Pb(II) increased by three times (35.9 mg/g) and twice (155.4 mg/g), respectively. Additionally, some studies had shown that carboxyl groups and phenolic hydroxyl groups were more inclined to adsorb Pb(II) than other metal ions, and the surface complexation analysis showed that the adsorption capacity of hydroxyl groups to metal ions was higher than that of carboxyl groups (*143*, *149*). Moreover, in addition to oxygen-containing functional groups, the introduction of sulfur- and nitrogen-containing functional groups can also increase the lignin adsorption capacity. Sulfomethylation and sulfonation are commonly used methods for introducing sulfur-containing functional groups. Amination and Mannich reactions are used to introduce nitrogen-containing functional groups to the lignin skeleton (*140*). Furthermore, the increase in charge density or molecular weight, and the introduction of hydrophobic groups are also feasible ways to increase lignin adsorption capacity. The electrostatic attraction between lignin-based adsorbents and contaminations having opposite charges is enhanced by increasing the surface charge density of the adsorbents. The large molecular weight can provide larger dynamic diameter, and the introduction of hydrophobic groups helps the solid-liquid separation of the adsorbed particles, thereby speeding up the adsorption rate.

There are many external factors that affect the effectiveness of lignin-based adsorbents in purifying wastewater, such as the adsorption time, temperature, wastewater pH, stirring speed, and initial concentration of contaminants. Among them, temperature and wastewater pH value are the main factors affecting the adsorption effectiveness (*152*). Since adsorption is an endothermic process, high temperature can increase the adsorption rate and adsorption capacity (*153*). In addition, the kinetic energy and surface activity of lignin-based adsorbents have been improved at high temperatures, thereby increasing the chance of collision and adsorption capacity. However, excessively high temperature will hydrolyze part of functional groups on the lignin-based adsorbent, reducing its adsorption capacity (*154*). The applicable temperature range of the lignin-based adsorbent is one of the important parameters to measure its commercial value. The lignin-based adsorbent with wider applicable temperature range can be used in many regions and seasons, making it more universal. Moreover, the ionization degree and surface charge of the lignin-based adsorbent are seriously affected by the wastewater pH (*155*). In terms of anionic adsorbents, the anion groups may be hydrolyzed in low pH solution, and excess H^+ will compete for adsorption sites with cationic contaminants, which causes decrease in the adsorption effect (*156*). Determining the optimal application pH is essential for the use of lignin-based adsorbents, which will not only improve the adsorption effect, but also reduce the dose of adsorbent and save the cost.

Flocculant

Flocculation, as the front step of wastewater treatment, is often used for solid-liquid separation, and the outstanding flocculation effect is of great significance for the whole wastewater treatment system (*157*). Different from adsorbents, flocculants usually need a certain degree of water solubility, which can form flexible long chains in water, thereby aggregating and precipitating colloidal particles in wastewater. The actual flocculation process is complex and variable, which involves van der Waals forces, thermodynamic equilibrium, hydrogen bonding, electrostatic interaction, and so on (*158*). In the following, several generally accepted flocculation mechanisms, such as charge neutralization, bridging, charge patching and sweeping, are briefly introduced.

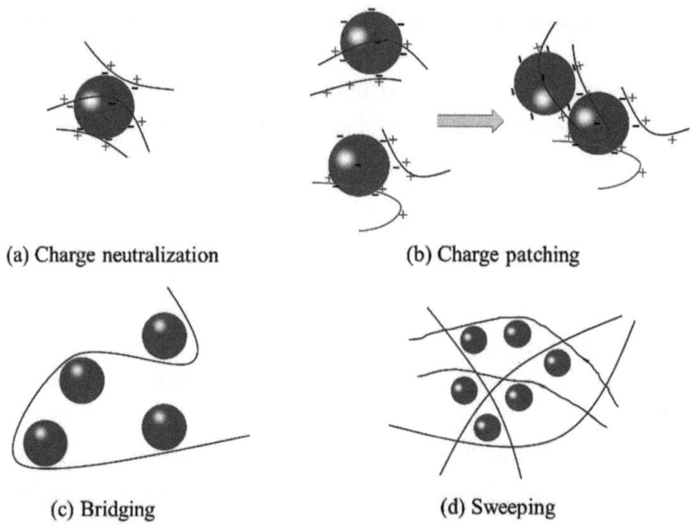

Figure 5. Schematic diagram of flocculation mechanisms of lignin-based flocculants.

Charge attraction is based on the diffused double layer model and Derjguin-Landau-Verwey-Overbeek (DLVO) theory (*159*). Normally, the contaminants maintain dynamic stability in the wastewater due to the strong electrostatic repulsion caused by the surface charge. When the flocculants with opposite charges are added, the ionized flocculants and the counterions in the solution gather around the diffused double layer of contaminants, causing the electric double layer to be compressed. At this time, the electrostatic repulsion between the particles is greatly reduced, and they are easily aggregated to form flocs due to the significant increase in electrostatic attraction and collision frequency between particles, which is defined as charge attraction. Charge attraction can be divided into charge neutralization and charge patching.

The flocculation process dominated by charge neutralization is that when the dose of the flocculant reaches the optimum, the surface charge of the particles is completely neutralized and the zeta potential of the solution approaches zero, achieving the best flocculation effect (Figure 5a) (*160*). At this juncture, the diffused double layer of the particles is compressed to the thinnest and the mutual attraction between the particles is maximized. However, when excessive flocculants are added, the electrostatic repulsion between the particles is increased again because their surfaces are covered by an excessive amount of opposite charges. The zeta potential of the solution is far from zero in the opposite direction, which causes the re-stabilization phenomenon and inferior flocculation. Therefore, it is necessary to accurately calculate the optimal dose. In addition, another feature of the flocculation process dominated by charge neutralization is that the optimal dose of flocculant is positively correlated with the initial concentration of contaminants in wastewater (*161*).

However, in the actual wastewater treatment process, the excessive dose of flocculants with opposite charges is difficult or even does not cause the re-stabilization phenomenon, which is inconsistent with the charge neutralization theory. Therefore, the charge patching is proposed, which is described in Figure 5b (*162*). The charge patching theory is that when flocculants with the opposite charge are added, part of the surface of the contaminants quickly adsorbs ionized flocculation, not all. As a result, various regions of a particle surface exhibit different charge types. Under rapid stirring, regions with opposite charges from different particles collide with each other, eventually forming large flocs and precipitating. It is worth noting that the flocculation process dominated by charge patching is insensitive to excessive flocculant doses, and the zeta potential of the solution is not usually near zero when the flocculation effect is optimal. Furthermore, rapid stirring and high charge density of flocculants may promote the effect of charge patching (*163*).

Bridging is a unique flocculation mechanism for polymeric flocculants and is shown in Figure 5c. Generally, polymeric flocculants with long-chain conformation are absorbed onto the surface of the suspended colloid forming loops and the end dangles in the water. During flocculation, the loops and dangles contact with and are attracted by other particles (*164*). The polymeric chain will adhere to two or more particles at the same time to connect the dispersed particles in a "bridge" manner, which is the bridging mechanism (*165*). The prerequisite for bridging action is that the surface of the particle is not completely wrapped by the polymer chain, and there are still some blank sites that can be adhered by other polymeric chains. When the concentration of the polymeric flocculants is excessive and the surface of particle is completely covered by polymeric flocculants, a thicker adsorption film is formed on the surface of the particle. Therefore, the repulsive force between particles is greatly enhanced, and the polymeric flocculant plays a positive role in stabilizing the contaminations in the wastewater (*166*). In general, short chain flocculants are difficult to from "bridge" between the particles. The chain conformation and state of the polymeric flocculant in water significantly affect the bridging effect. The polymeric chains with a more extended conformation have stronger bridging. The hydrodynamic size and bridging effect of polymeric flocculant can be improved by increasing the molecular weight. As demonstrated by the Flory-Krigbaum equation, the scaling exponent between the hydrodynamic size of a polymeric flocculant (R) and its molecular weight (MW) in a good solvent is 0.6, as shown in Eq. (*167*):

$$R \propto MV^{0.6}$$

The sweeping is originally found in inorganic flocculants, and the small colloidal contaminations can be enmeshed and entrapped by insoluble large metal hydroxide (*168*). For some lignin-based flocculants with poor solubility, they can also capture some small colloidal particles through physical action, which is known as the sweeping of polymeric flocculants (Figure 5d) (*169*). In the actual wastewater treatment process, flocs with a three-dimensional network structure will first be formed under the action of other flocculation mechanisms. These flocs retain a certain adsorption capacity and have a large surface area, which can enmesh and entrap the small contamination particles remaining in the wastewater, forming larger flocs. Interestingly, even if the contaminant particles are not destabilized, they can also be removed under the action of sweeping. However, it is worth noting that the destabilization of contaminant particles can accelerate the rate of sweeping. The main factor that affects the sweeping effect is the initial concentration of contaminants in the wastewater. Generally, a higher initial concentration is favorable for the sweeping effect (*170*). Moreover, the influence of the flocculant dose on the sweeping effect is also related to the initial concentration of contaminants. When the initial concentration of contaminants is low, a large amount of flocculant is needed, and the flocculation effect is poor. When the initial concentration of contaminants is higher, the required dose of flocculant is lower and the superior flocculation effect is realized (*171*).

In brief, lignin-based flocculants first destabilize contaminations under the action of charge attraction, and then aggregate the unstable particles into large flocs by bridging or sweeping (*172*). From the above description of these flocculation mechanisms, it can be known that charge density is the main factor affecting the flocculation performance of the lignin derivatives. For non-ionic or weak ionic lignin-based flocculants, the flocculation process mainly relies on bridging. The flocs formed in this case are loose and have inferior shear resistance (*48*, *68*). These flocculants generally need to be used with coagulants in order to achieve the desired flocculation effect. Therefore, the current research of lignin-based flocculants mainly focuses on ionic type, and lignin-based flocculants with high charge density can exhibit excellent flocculation efficiency without coagulants. Hasan et al. prepared two lignin-based flocculants with different charge densities by the polymerization of kraft lignin with AM and DMC (*44*). The results of flocculation tests showed that lignin-based flocculant with higher charge density exhibited stronger flocculation efficiency for simulated clay suspensions. Moreover, flocculants with high charge density have a wider pH application range.

It is worth noting that the lignin-based flocculant is generally regarded as polyelectrolyte, but the ionizing groups on its side chain cannot be uniformly dispersed in the solution due to the limitation of the lignin skeleton and the charge unit distance (*173*). Therefore, there is no infinite dilution for lignin-based flocculant with high charge density. According to the counter-ion condensation theory, the actual charge density of lignin-based flocculants in solution is limited (*174*). The actual charge density of lignin-based flocculant with low charge density in solution may be equal to its maximum charge density. However, when the charge density on the macromolecule exceeds a certain level, the actual charge density of the lignin-based flocculant in solution is less than its maximum charge density. That is, when the distance between adjacent charges on the side chain of the lignin-based flocculant is smaller than the Bjerrum length determined by the solution and temperature, a large number of counter ions in solution will gather near the ionizing group, thereby reducing its actual charge density in the solution (*175*, *176*). Therefore, in the preparation of lignin-based flocculants, the ionizing group introduced to the lignin should be appropriate in consideration of cost and flocculation efficiency.

In addition, the molecular weight of the lignin-based flocculant also has a profound influence on its flocculation capacity. Generally, the lignin derivatives with larger molecular weight have stronger bridging effect and flocculation capacity. The molecular weight of the lignin derivative is determined by the chain length and chain density. Kajihara et al. prepared a series of lignin-based flocculants with different chain lengths and different chain densities via esterification with bromoisobutyryl groups, atom transfer radical polymerization with 2-(dimethylamino) ethyl methacrylate, and quaternization with the amino groups, which were used to treat simulated dye wastewater (*124*). The nitrogen content of the lignin-based flocculants remained constant. When the side chain density of the lignin-based flocculant was low, the flocculation capacity was positively correlated with the side chain length. Conversely, the flocculation capacity decreased with the increase of side chain length at high side chain density.

Other Applications

Membrane filtration technology is also a commonly used way to purify wastewater (*177*). Nanofiber membrane is widely welcomed because of its simple operation, the large surface area and high performance (*178*). In one study, lignin-based composite nanofiber membrane was prepared by the electrospinning of lignin/poly vinyl alcohol (PVA) blend (*178*). The strong hydrogen bond between lignin and PVA provided sufficient strength for the nanofiber membrane. In addition, the addition of lignin could reduce the viscosity of blend, which made the blend easier to be electrospun and produce finer fibers. However, when there were excessive lignins in the blend, the strength of the electrospun fibers decreased severely, and bead fibers might appear. The experimental results showed that the proportion of lignin in the blend could be up to 50%, and the nanofiber membrane prepared under the conditions had no bead fibers, which exhibited excellent removal effect on cationic dye (Safranine T). Moreover, the nanofiber membrane could maintain efficient adsorption performance after five cycles of desorption. With the in-depth study of high value of lignin, lignin-based membranes used in the field of wastewater treatment will receive more attention.

The rapid spread of COVID-19 has seriously jeopardized the safety of human life and property (*179*). How to effectively prevent the spread of pathogens such as viruses, bacteria and parasites is receiving unprecedented attention. Wastewater is one of the main routes of pathogen transmission (*180*). Therefore, the removal and inactivation of pathogens in wastewater are vital to the human health. In one study, cationic lignin particles were prepared by coating cationic sulfate lignin containing ETA on spherical lignin prepared by nanoprecipitation (*181*). The cationic lignin particles had an excellent virus-affinity, which could effectively remove nano-scale cowpea chlorotic mottle virus (CCMV) from wastewater. In addition, the antiviral and antibacterial properties of lignin derivatives have been reported in some studies (*182, 183*). Moreover, it is well known that quaternary ammonium salts with high cationicity have strong inactivation ability against pathogens and are often used to prepare disinfectants (*184, 185*). Moreover, several studies have confirmed that some natural polysaccharide derivatives with quaternary ammonium salts not only effectively removed bacteria from wastewater, but also inactivated them (*186–188*). For example, in one study, when the Escherichia coli (*E. coli*) wastewater was treated using excess lignin-based flocculant (LBF) with quaternary ammonium salt groups, large flocs were first formed and precipitated under the action of charge attraction and bridging (*189*). Subsequently, the quaternary ammonium salts group adsorbed on *E. coli* surface acted on the cell wall and cell membrane, thereby rupturing and killing *E. coli*, as shown in Figure 6.

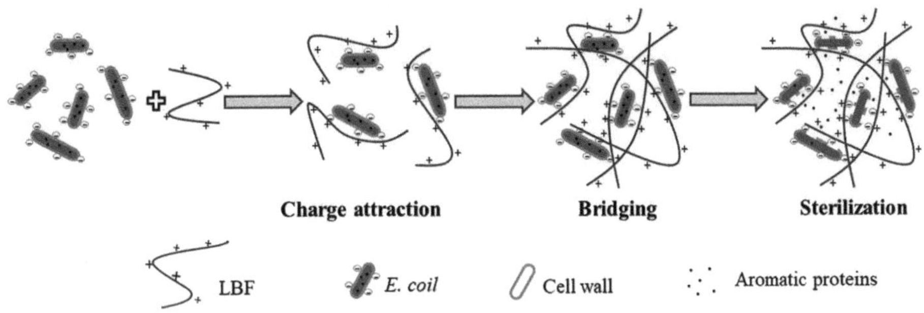

Figure 6. Schematic diagram of lignin-based flocculant treating E. coli wastewater (189). Reproduced with permission from reference (189). Copyright 2014 American Chemical Society.

Conclusions and Future Perspectives

Lignin and its derivatives show outstanding potential in the field of wastewater purification. However, the current researches on lignin-based wastewater treatment agents are still at the laboratory stage, and there is still a long way to go before commercial application. The heterogeneous structure and the complicated influencing factors, such as tree species and extraction methods, have seriously hindered the commercialization of lignin derivatives. Moreover, although the biodegradability of lignin-based wastewater treatment agents can reduce the risk of secondary pollution, it will cause unstable performance and short shelf life of the product. The chemical modifications such as cross-linking with epichlorohydrin, glutaraldehyde and the addition of stabilizers are effective methods to improve product stability and shelf life (*190*, *191*). In addition, the composition of actual wastewater from living or industry is generally quite complex while most of the current researches on lignin-based wastewater treatment agent only focus on the purification of simulated wastewater with single component, which is insufficient to deal with actual complex wastewater. The performance test of lignin-based wastewater treatment agents should be developed from static to dynamic, from simulated wastewater to actual wastewater, from individual cases to universal cases. Furthermore, it is necessary to further study the effect of lignin derivative structure on its wastewater treatment capacity and the purification mechanisms. It is of great significance for the design and application of lignin-based wastewater treatment agents to establish scientific theoretical relations among structure, purification capacity and treatment effect. The preparation of lignin-based wastewater treatment agents having multiple functions, such as adsorption, flocculation, sterilization, scale inhibition and corrosion inhibition, is the future development direction.

Acknowledgments

This research was supported by the National Natural Science Foundation of China (51961125207) and Beijing Forestry University Outstanding Young Talent Cultivation Project (2019JQ03005).

References

1. Shannon, M. A.; Bohn, P. W.; Elimelech, M.; Georgiadis, J. G.; Marinas, B. J.; Mayes, A. M. Science and technology for water purification in the coming decades. *Nature.* **2008**, *452*, 301–310.
2. Shakir, E.; Zahraw, Z.; Al-Obaidy, A. H. M. Environmental and health risks associated with reuse of wastewater for irrigation. *Egypt. J. Petrol.* **2017**, *26*, 95–102.
3. Raheem, A.; Sikarwar, V. S.; He, J.; Dastyar, W.; Dionysiou, D. D.; Wang, W.; Zhao, M. Opportunities and challenges in sustainable treatment and resource reuse of sewage sludge: a review. *Chem. Eng. J.* **2018**, *337*, 616–641.
4. Wang, C.; Hou, Y.; Xue, Y. Water resources carrying capacity of wetlands in Beijing: Analysis of policy optimization for urban wetland water resources management. *J. Clean Prod.* **2017**, *161*, 1180–1191.
5. Regkouzas, P.; Diamadopoulos, E. Adsorption of selected organic micro-pollutants on sewage sludge biochar. *Chemosphere.* **2019**, *224*, 840–851.

6. Wei, H.; Gao, B.; Ren, J.; Li, A.; Yang, H. Coagulation/flocculation in dewatering of sludge: a review. *Water Res.* **2018**, *143*, 608–631.
7. Tipnis, N. P.; Burgess, D. J. Sterilization of implantable polymer-based medical devices: A review. *Int. J. Pharm.* **2018**, *544*, 455–460.
8. Kumar, P.; Pournara, A.; Kim, K.-H.; Bansal, V.; Rapti, S.; Manos, M. J. Metal-organic frameworks: challenges and opportunities for ion-exchange/sorption applications. *Prog. Mater. Sci.* **2017**, *86*, 25–74.
9. Martínez-Huitle, C. A.; Panizza, M. Electrochemical oxidation of organic pollutants for wastewater treatment. *Curr. Opin. Electroche.* **2018**, *11*, 62–71.
10. Luo, T.; Abdu, S.; Wessling, M. Selectivity of ion exchange membranes: A review. *J. Membr. Sci.* **2018**, *555*, 429–454.
11. Grenda, K.; Arnold, J.; Gamelas, J. A.; Cayre, O. J.; Rasteiro, M. G. Flocculation of silica nanoparticles by natural, wood-based polyelectrolytes. *Sep. Purif. Technol.* **2020**, *231*, 115888.
12. Toczyłowska-Mamińska, R. Limits and perspectives of pulp and paper industry wastewater treatment–A review. *Renew. Sust. Energ. Rev.* **2017**, *78*, 764–772.
13. Li, X.; Xu, R.; Yang, J.; Nie, S.; Liu, D.; Liu, Y.; Si, C. Production of 5-hydroxymethylfurfural and levulinic acid from lignocellulosic biomass and catalytic upgradation. *Ind. Crop. Prod.* **2019**, *130*, 184–197.
14. He, W.; Zhang, Y.; Fatehi, P. Sulfomethylated kraft lignin as a flocculant for cationic dye. Colloid Surf. A-Physicochem. *Eng. Asp.* **2016**, *503*, 19–27.
15. Zeng, D.; Wu, J.; Kennedy, J. F. Application of a chitosan flocculant to water treatment. *Carbohydr. Polym.* **2008**, *71*, 135–139.
16. Lan, G.; Zhang, M.; Liu, Y.; Qiu, H.; Xue, S.; Zhang, T.; Xu, Q. Synthesis and Swelling Behavior of Super-Absorbent Soluble Starch-g-poly (AM-co-NaAMC14S) Through Graft Copolymerization and Hydrolysis. *Starch-Starke* **2019**, *71*, 1800272.
17. Du, H.; Liu, W.; Zhang, M.; Si, C.; Zhang, X.; Li, B. Cellulose nanocrystals and cellulose nanofibrils based hydrogels for biomedical applications. *Carbohydr. Polym.* **2019**, *209*, 130–144.
18. Chen, J.; Eraghi Kazzaz, A.; AlipoorMazandarani, N.; Hosseinpour Feizi, Z.; Fatehi, P. Production of flocculants, adsorbents, and dispersants from lignin. *Molecules.* **2018**, *23*, 868.
19. Wang, Z.; Huang, W.; Yang, G.; Liu, Y.; Liu, S. Preparation of cellulose-base amphoteric flocculant and its application in the treatment of wastewater. *Carbohydr. Polym.* **2019**, *215*, 179–188.
20. Wei, H.; Ren, J.; Li, A.; Yang, H. Sludge dewaterability of a starch-based flocculant and its combined usage with ferric chloride. *Chem. Eng. J.* **2018**, *349*, 737–747.
21. Sun, R. C. Lignin Source and Structural Characterization. *ChemSusChem.* **2020**.
22. Tribot, A.; Amer, G.; Alio, M. A.; de Baynast, H.; Delattre, C.; Pons, A.; Mathias, J.-D.; Callois, J.-M.; Vial, C.; Michaud, P. Wood-lignin: Supply, extraction processes and use as bio-based material. *Eur. Polym. J.* **2019**, *112*, 228–240.
23. Okolie, J. A.; Nanda, S.; Dalai, A. K.; Kozinski, J. A. Chemistry and Specialty Industrial Applications of Lignocellulosic Biomass. *Waste Biomass Valor.* **2020**, *272*.
24. Aro, T.; Fatehi, P. Production and application of lignosulfonates and sulfonated lignin. *ChemSusChem.* **2017**, *10*, 1861–1877.

25. Chio, C.; Sain, M.; Qin, W. Lignin utilization: a review of lignin depolymerization from various aspects. *Renew. Sust. Energ. Rev.* **2019**, *107*, 232–249.
26. Shamaei, L.; Khorshidi, B.; Islam, M. A.; Sadrzadeh, M. Industrial waste lignin as an antifouling coating for the treatment of oily wastewater: Creating wealth from waste. *J. Clean Prod.* **2020**, *256*, 120304.
27. Bouffard, S. C.; Tshilombo, A.; Westsells, P. Use of lignosulfonate for elemental sulfur biooxidation and copper leaching. *Miner. Eng.* **2009**, *22*, 100–103.
28. Adhikari, B. B.; Gurung, M.; Alam, S.; Tolnai, B.; Inoue, K. Kraft mill lignin–a potential source of bio-adsorbents for gold recovery from acidic chloride solution. *Chem. Eng. J.* **2013**, *231*, 190–197.
29. Laurichesse, S.; Avérous, L. Chemical modification of lignins: Towards biobased polymers. *Prog. Polym. Sci.* **2014**, *39*, 1266–1290.
30. Grossman, A.; Vermerris, W. Lignin-based polymers and nanomaterials. *Curr. Opin. Biotechnol.* **2019**, *56*, 112–120.
31. Liu, H.; Chung, H. Lignin-based polymers via graft copolymerization. *J. Polym. Sci. Pol. Chem.* **2017**, *55*, 3515–3528.
32. Jin, C.; Yang, M.; Shuang, E.; Liu, J.; Zhang, S.; Zhang, X.; Sheng, K.; Zhang, X. Corn stover valorization by one-step formic acid fractionation and formylation for 5-hydroxymethylfurfural and high guaiacyl lignin production. *Bioresour. Technol.* **2020**, *299*, 122586.
33. Ganewatta, M. S.; Lokupitiya, H. N.; Tang, C. Lignin biopolymers in the age of controlled polymerization. *Polymers.* **2019**, *11*, 1176.
34. Maradur, S. P.; Kim, C. H.; Kim, S. Y.; Kim, B.-H.; Kim, W. C.; Yang, K. S. Preparation of carbon fibers from a lignin copolymer with polyacrylonitrile. *Synth. Met.* **2012**, *162*, 453–459.
35. Tobimatsu, Y.; Schuetz, M. Lignin polymerization: how do plants manage the chemistry so well? *Curr. Opin. Biotechnol.* **2019**, *56*, 75–81.
36. Kubo, S.; Kadla, J. Lignin-based carbon fibers: Effect of synthetic polymer blending on fiber properties. *J. Polym. Environ.* **2005**, *13*, 97–105.
37. Thielemans, W.; Can, E.; Morye, S.; Wool, R. Novel applications of lignin in composite materials. *J. Appl. Polym. Sci.* **2002**, *83*, 323–331.
38. Victor, P. A.; Gonçalves, S. B.; Machado, F. Styrene/lignin-based polymeric composites obtained through a sequential mass-suspension polymerization process. *J. Polym. Environ.* **2018**, *26*, 1755–1774.
39. Zahran, M. A.; El-Sherbiny, I. M.; El-Raheem, H.; Basuni, M. M. Synthesis, characterization and aqueous properties of a newly prepared lignin-based graft copolymer as a drilling mud additive. *Int. J. Thin. Fil. Sci. Tec.* **2017**, *6*, 1–7.
40. Xiao-hong, W. Study on Grafting Modification of Lignosulfonate and Acrylamide [J]. *J. Anhui Agr. Sci.* **2010**, *16*.
41. Santhosh, P.; Gopalan, A.; Lee, K.-P. Gold nanoparticles dispersed polyaniline grafted multiwall carbon nanotubes as newer electrocatalysts: preparation and performances for methanol oxidation. *J. Catal.* **2006**, *238*, 177–185.
42. Guldi, D. M. Capped fullerenes: stabilization of water-soluble fullerene monomers as studied by flash photolysis and pulse radiolysis. *J. Phys. Chem. A* **1997**, *101*, 3895–3900.

43. Hasan, A.; Fatehi, P. Cationic kraft lignin-acrylamide as a flocculant for clay suspensions: 1. Molecular weight effect. *Sep. Purif. Technol.* **2018**, *207*, 213–221.
44. Hasan, A.; Fatehi, P. Cationic kraft lignin-acrylamide copolymer as a flocculant for clay suspensions:(2) Charge density effect. *S Sep. Purif. Technol.* **2019**, *210*, 963–972.
45. Aldajani, M. *Flocculation performance of lignin-acrylamide derivatives*. 2019.
46. Friedman, M. Chemistry, biochemistry, and safety of acrylamide. A review. *J. Agric. Food Chem.* **2003**, *51*, 4504–4526.
47. Liu, R.; Dong, A.; Fan, X.; Wang, Q.; Yu, Y.; Cavaco-Paulo, A. HRP-mediated polyacrylamide graft modification of raw jute fabric. *J. Mol. Catal. B-Enzym.* **2015**, *116*, 29–38.
48. Hasan, A.; Fatehi, P. Stability of kaolin dispersion in the presence of lignin-acrylamide polymer. *Appl. Clay Sci.* **2018**, *158*, 72–82.
49. Li, R.; Gao, B.; Guo, K.; Yue, Q.; Zheng, H.; Wang, Y. Effects of papermaking sludge-based polymer on coagulation behavior in the disperse and reactive dyes wastewater treatment. *Bioresour. Technol.* **2017**, *240*, 59–67.
50. Wang, Y. P.; Zhou, P.; Luo, S. Z.; Guo, S.; Lin, J.; Shao, Q.; Guo, X.; Liu, Z.; Shen, J.; Wang, B. In situ polymerized poly (acrylic acid)/alumina nanocomposites for Pb^{2+} adsorption. *Adv. Polym. Technol.* **2018**, *37*, 2981–2996.
51. Wang, X.; Zhou, Z.; Guo, X.; He, Q.; Hao, C.; Ge, C. Ultrasonic-assisted synthesis of sodium lignosulfonate-grafted poly (acrylic acid-co-poly (vinyl pyrrolidone)) hydrogel for drug delivery. *RSC Adv.* **2016**, *6*, 35550–35558.
52. Kong, F.; Wang, S.; Gao, W.; Fatehi, P. Novel pathway to produce high molecular weight kraft lignin–acrylic acid polymers in acidic suspension systems. *RSC Adv.* **2018**, *8*, 12322–12336.
53. Rong, H.; Gao, B.; Zhao, Y.; Sun, S.; Yang, Z.; Wang, Y.; Yue, Q.; Li, Q. Advanced lignin-acrylamide water treatment agent by pulp and paper industrial sludge: Synthesis, properties and application. *J. Environ. Sci.* **2013**, *25*, 2367–2377.
54. Fang, R.; Cheng, X.-S.; Fu, J.; Zheng, Z.-B. Research on the graft copolymerization of EH-lignin with acrylamide. *Nat. Sci.* **2009**, *1*, 17.
55. Hasan, A.; Fatehi, P. Self-assembly of kraft lignin-acrylamide polymers. Colloid Surf. A-Physicochem. *Eng. Asp.* **2019**, *572*, 230–236.
56. Liu, F.; Zhang, Z.; Wang, Z.; Dai, X.; Chen, M.; Zhang, J. Novel lignosulfonate/N, N-dimethylacrylamide/γ-methacryloxypropyl trimethoxy silane graft copolymer as a filtration reducer for water-based drilling fluids. *J. Appl. Polym. Sci.* **2020**, *137*, 48274.
57. Kong, F.; Wang, S.; Price, J. T.; Konduri, M. K.; Fatehi, P. Water soluble kraft lignin–acrylic acid copolymer: Synthesis and characterization. *Green Chem.* **2015**, *17*, 4355–4366.
58. Lin, X.; Zhang, J.; Luo, X.; Zhang, C.; Zhou, Y. Removal of aniline using lignin grafted acrylic acid from aqueous solution. *Chem. Eng. J.* **2011**, *172*, 856–863.
59. Zhao, T.; Zhang, K.; Chen, J.; Shi, X.; Li, X.; Ma, Y.; Fang, G.; Xu, S. Changes in heavy metal mobility and availability in contaminated wet-land soil remediated using lignin-based poly (acrylic acid). *J. Hazard. Mater.* **2019**, *368*, 459–467.
60. Chen, R.; Kokta, B. V. Graft Copolymerization of Lignosulfonate with Methacrylic Acid and Acrylate Monomers. *Acs Symposium Series American Chemical Society* **1982**.
61. Stanzione, J., III; Giangiulio, P.; Sadler, J.; La Scala, J.; Wool, R. *ACS Sust. Chem. Eng* **2013**, *1*, 419–426.

62. Guo, K.; Gao, B.; Yue, Q.; Xu, X.; Li, R.; Shen, X. Characterization and performance of a novel lignin-based flocculant for the treatment of dye wastewater. *Int. Biodeterior. Biodegrad.* **2018**, *133*, 99–107.
63. Wang, S.; Kong, F.; Gao, W.; Fatehi, P. Novel process for generating cationic lignin based flocculant. *Ind. Eng. Chem. Res.* **2018**, *57*, 6595–6608.
64. Wang, S.; Kong, F.; Fatehi, P.; Hou, Q. Cationic high molecular weight lignin polymer: a flocculant for the removal of anionic azo-dyes from simulated wastewater. *Molecules.* **2018**, *23*, 2005.
65. Price, J. T.; Gao, W.; Fatehi, P. Lignin-g-poly (acrylamide)-g-poly (diallyldimethyl-ammonium chloride): Synthesis, Characterization and Applications. *ChemistryOpen.* **2018**, *7*, 645–658.
66. Agha, H.; Pedram, F. Flocculation of kaolin particles with cationic lignin polymers. *Sci Rep.* **2019**, *9*.
67. Yang, Z.; Degorce-Dumas, J.-R.; Yang, H.; Guibal, E.; Li, A.; Cheng, R. Flocculation of Escherichia coli using a quaternary ammonium salt grafted carboxymethyl chitosan flocculant. *Environ. Sci. Technol.* **2014**, *48*, 6867–6873.
68. Hasan, A.; Fatehi, P. Synthesis and characterization of lignin–poly (acrylamide)–poly (2-methacryloyloxyethyl) trimethyl ammonium chloride copolymer. *J. Appl. Polym. Sci.* **2018**, *135*, 46338.
69. Alsulami, Q. A.; Albukhari, S. M.; Hussein, M. A.; Tay, G.; Rozman, H. Biodegradable lignin as a reactive raw material in UV curable systems. *Polym.-Plast. Tech. Mat.* **2020**, 1–20.
70. Kovalenko, E.; Popova, O.; Aleksandrov, A.; Galikyan, T. Electrochemical modification of lignins. *Russ. J Electrochem.* **2000**, *36*, 706–711.
71. Hamada, T.; Yamashita, S.; Omichi, M.; Yoshimura, K.; Ueki, Y.; Seko, N.; Kakuchi, R. Multicomponent-reaction-ready biomass-sourced organic hybrids fabricated via the surface immobilization of polymers with lignin-based compounds. *ACS Sustain. Chem. Eng.* **2019**, *7*, 7795–7803.
72. Qian, Y.; Zhong, X.; Li, Y.; Qiu, X. Fabrication of uniform lignin colloidal spheres for developing natural broad-spectrum sunscreens with high sun protection factor. *Ind. Crop. Prod.* **2017**, *101*, 54–60.
73. Hassan, M. M.; Islam, M. R.; Shehrzade, S.; Khan, M. A. Influence of Mercerization Along with Ultraviolet (UV) and Gamma Radiation on Physical and Mechanical Properties of Jute Yarn by Grafting with 3-(Trimethoxysilyl) Propylmethacrylate (Silane) and Acrylamide Under UV Radiation. *Polym.-Plast. Technol. Eng.* **2003**, *42*, 515–531.
74. Zheng, X.; Wu, D.; Su, T.; Bao, S.; Liao, C.; Wang, Q. Magnetic nanocomposite hydrogel prepared by ZnO-initiated photopolymerization for La (III) adsorption. *ACS Appl. Mater. Interfaces* **2014**, *6*, 19840–19849.
75. Ma, J.; Fu, K.; Shi, J.; Sun, Y.; Zhang, X.; Ding, L. Ultraviolet-assisted synthesis of polyacrylamide-grafted chitosan nanoparticles and flocculation performance. *Carbohydr. Polym.* **2016**, *151*, 565–575.
76. Breitenbach, J.; Srna, C. Electrochemical initiation of polymerization. *Pure Appl. Chem.* **1962**, *4*, 245–254.
77. Zhifa, Z.; Guohua, D.; Yaohua, Y.; Zhuomin, I.; Handao, W. Electrochemical graft of lignosulfonates with acrylic acid. *J. Cell. Sci. Technol.* **1998**, *8*.

78. Hüttermann, A.; Mai, C.; Kharazipour, A. Modification of lignin for the production of new compounded materials. *Appl. Microbiol. Biotechnol.* **2001**, 55, 387–394.
79. Carenza, M. Recent achievements in the use of radiation polymerization and grafting for biomedical applications. International Journal of Radiation Applications and Instrumentation. Part C. *Radiat. Phys. Chem.* **1992**, *39*, 485–493.
80. Greene, F. D.; Berwick, M. A.; Stowell, J. C. Stereochemistry of free-radical recombination reactions. Cage effect in decomposition of SS-(-)-azobis-.alpha.-phenylethane. *J. Am. Chem. Soc.* **1970**, 92, 867–874.
81. Chen, J.; Fan, X.; Zhang, L.; Chen, X.; Sun, S.; Sun, R. C. Research Progress in Lignin-Based Slow/Controlled Release Fertilizer. *ChemSusChem.* **2020**.
82. Matsushita, Y.; Yasuda, S. Reactivity of a condensed–type lignin model compound in the Mannich reaction and preparation of cationic surfactant from sulfuric acid lignin. *J. Wood Sci.* **2003**, *49*, 166–171.
83. Chen, X.; Peng, X.; Ma, X.; Wang, J. Investigation of Mannich reaction during co-liquefaction of microalgae and sweet potato waste. *Bioresour. Technol.* **2019**, *284*, 286–292.
84. Ibrahim, M. N. M.; Zakaria, N.; Sipaut, C. S.; Sulaiman, O.; Hashim, R. Chemical and thermal properties of lignins from oil palm biomass as a substitute for phenol in a phenol formaldehyde resin production. *Carbohydr. Polym.* **2011**, *86*, 112–119.
85. Tramontini, M.; Angiolini, L. , *Mannich bases: Chemistry and Uses*; CRC Press Inc.: Boca Raton, 1994.
86. Jiao, Y. Synthesis and application of the cationic lignin amine flocculant. *Tenside Surfactants Deterg.* **2010**, *47*, 381–384.
87. Cummings, T. F.; Shelton, J. R. Mannich reaction mechanisms. *J. Org. Chem.* **1960**, *25*, 419–423.
88. Wang, M.; Sjöholm, E.; Li, J. Fast and reliable quantification of lignin reactivity via reaction with dimethylamine and formaldehyde (Mannich reaction). *Holzforschung.* **2017**, *71*, 27–34.
89. Meng, X.; Scheidemantle, B.; Li, M.; Wang, Y.-y.; Zhao, X.; Toro-González, M.; Singh, P.; Pu, Y.; Wyman, C. E.; Ozcan, S. Synthesis, Characterization, and Utilization of a Lignin-Based Adsorbent for Effective Removal of Azo Dye from Aqueous Solution. *ACS omega.* **2020**, *5*, 2865–2877.
90. Chiu, C. W.; Huang, T. K.; Wang, Y. C.; Alamani, B. G.; Lin, J.-J. Intercalation strategies in clay/polymer hybrids. *Prog. Polym. Sci.* **2014**, *39*, 443–485.
91. Zhao, Z.; Li, L.; Shao, X.; Liu, X.; Zhao, S.; Xie, S.; Xin, Z. Tannic acid-assisted green fabrication of functionalized graphene towards its enhanced compatibility in NR nanocomposite. *Polym. Test* **2018**, *70*, 396–402.
92. Guo, Q. X.; Liu, H.; Guo, C.; Luo, S. W.; Gu, Y.; Gong, L. Z. Chiral Brønsted acid-catalyzed direct asymmetric Mannich reaction. *J. Am. Chem. Soc.* **2007**, *129*, 3790–3791.
93. Jiang, Y.; Schaus, S. E. Asymmetric Petasis Borono-Mannich Allylation Reactions Catalyzed by Chiral Biphenols. *Angew. Chem.-Int. Edit.* **2017**, *56*, 1544–1548.
94. Jo, H.; Hassan, A. H.; Jung, S. Y.; Lee, J. K.; Cho, Y. S.; Min, S.-J. Construction of 8-Azabicyclo [3.2. 1] octanes via Sequential DDQ-Mediated Oxidative Mannich Reactions of N-Aryl Pyrrolidines. *Org. Lett.* **2018**, *20*, 1175–1178.

95. Gul, H. I.; Tugrak, M.; Gul, M.; Mazlumoglu, S.; Sakagami, H.; Gulcin, I.; Supuran, C. T. New Phenolic Mannich Bases with Piperazines and Their Bioactivities. *Bioorganic Chem.* **2019**, 103057.
96. Trulli, L.; Sciubba, F.; Fioravanti, S. Chiral trans-carboxylic trifluoromethyl 2-imidazolines by a Ag2O-catalyzed Mannich-type reaction. *Tetrahedron.* **2018**, *74*, 572–577.
97. Al Hamouz, O. C. S.; Akintola, O. S. Removal of lead and arsenic ions by a new series of aniline based polyamines. *Process Saf. Environ. Protect.* **2017**, *106*, 180–190.
98. Goni, M. A.; Rosenberg, E.; Meregude, S.; Abbott, G. A methods study of immobilization of PONOP pincer transition metal complexes on silica polyamine composites (SPC). *J. Organomet. Chem.* **2016**, *807*, 1–10.
99. Ünlü, C. H.; Kutlu, M.; Atıcı, O. G. Mannich reaction of polysaccharides: Xylan functionalization in aqueous basic medium. *Carbohydr. Polym.* **2015**, *127*, 19–27.
100. Santos, D. E.; Li, D.; Ramstedt, M.; Gautrot, J. E.; Soares, T. A. Conformational Dynamics and Responsiveness of Weak and Strong Polyelectrolyte Brushes: Atomistic Simulations of Poly (dimethyl aminoethyl methacrylate) and Poly (2-(methacryloyloxy) ethyl trimethylammonium chloride). *Langmuir.* **2019**, *35*, 5037–5049.
101. Chen, Y.; Cass, S. L.; Kutty, S. K.; Yee, E. M.; Chan, D. S.; Gardner, C. R.; Vittorio, O.; Pasquier, E.; Black, D. S.; Kumar, N. Synthesis, biological evaluation and structure–activity relationship studies of isoflavene based Mannich bases with potent anti-cancer activity. *Bioorg. Med. Chem. Lett.* **2015**, *25*, 5377–5383.
102. Lian, X.; Yan, B. Phosphonate MOFs composite as off–on fluorescent sensor for detecting purine metabolite uric acid and diagnosing hyperuricuria. *Inorg. Chem.* **2017**, *56*, 6802–6808.
103. Rajasekharan, S. K.; Kamalanathan, C.; Ravichandran, V.; Ray, A. K.; Satish, A. S.; Mohanvel, S. K. Mannich base limits Candida albicans virulence by inactivating Ras-cAMP-PKA pathway. *Sci Rep.* **2018**, *8*, 14972.
104. Gmeiner, C.; Klose, D.; Mileo, E.; Belle, V.; Marque, S. R.; Dorn, G.; Allain, F. d. r. H.; Guigliarelli, B.; Jeschke, G.; Yulikov, M. Orthogonal tyrosine and cysteine site-directed spin labeling for dipolar pulse EPR spectroscopy on proteins. *J. Phys. Chem. Lett.* **2017**, *8*, 4852–4857.
105. Li, Z.; Toh, Z. X.; Xiang, Y.; Ye, E.; Loh, X. J. Chemical Modification of Lignin. *Funct. Mater. Lignin.* **2018**, 81–120.
106. Li, R.; Wang, X.; Lin, Q.; Yue, F.; Liu, C.; Wang, X.; Ren, J. Structural Features of Lignin Fractionated From Industrial Furfural Residue Using Alkaline Cooking Technology and Its Antioxidant Performance. *Front. Energy Res.* **2020**, *8*, 83.
107. Zhou, M.; Wang, D.; Peng, R.; Yang, D.; Qiu, X.; Qian, Y. Synthesis of a hindered amine-grafted lignin-based emulsifier and its application in a green emulsifiable concentrate. *J. Agric. Food Chem.* **2019**, *67*, 11129–11136.
108. Ang, A. F.; Ashaari, Z.; Lee, S. H.; Tahir, P. M.; Halis, R. Lignin-based copolymer adhesives for composite wood panels–A review. *Int. J. Adhes. Adhes.* **2019**, *95*, 102408.
109. Lai, C.; Tu, M.; Xia, C.; Shi, Z.; Sun, S.; Yong, Q.; Yu, S. Lignin alkylation enhances enzymatic hydrolysis of lignocellulosic biomass. *Energy Fuels* **2017**, *31*, 12317–12326.
110. Wang, B.; Wen, J.-L.; Sun, S.-L.; Wang, H.-M.; Wang, S.-F.; Liu, Q.-Y.; Charlton, A.; Sun, R.-C. Chemosynthesis and structural characterization of a novel lignin-based bio-sorbent and its strong adsorption for Pb (II). *Ind. Crop. Prod.* **2017**, *108*, 72–80.

111. Fang, R.; Cheng, X.; Xu, X. Synthesis of lignin-base cationic flocculant and its application in removing anionic azo-dyes from simulated wastewater. *Bioresour. Technol.* **2010**, *101*, 7323–7329.
112. Chen, C.; Zhu, M.; Li, M.; Fan, Y.; Sun, R.-C. Epoxidation and etherification of alkaline lignin to prepare water-soluble derivatives and its performance in improvement of enzymatic hydrolysis efficiency. *Biotechnol. Biofuels* **2016**, *9*, 1–15.
113. Liu, L.-Y.; Hua, Q.; Renneckar, S. A simple route to synthesize esterified lignin derivatives. *Green Chem.* **2019**, *21*, 3682–3692.
114. Chen, P.; Zhang, L.; Peng, S.; Liao, B. Effects of nanoscale hydroxypropyl lignin on properties of soy protein plastics. *J. Appl. Polym. Sci.* **2006**, *101*, 334–341.
115. Da Silva, L. G.; Ruggiero, R.; Gontijo, P. d. M.; Pinto, R. B.; Royer, B.; Lima, E. C.; Fernandes, T. H.; Calvete, T. Adsorption of Brilliant Red 2BE dye from water solutions by a chemically modified sugarcane bagasse lignin. *Chem. Eng. J.* **2011**, *168*, 620–628.
116. Adebayo, M. A.; Prola, L. D.; Lima, E. C.; Puchana-Rosero, M.; Cataluña, R.; Saucier, C.; Umpierres, C. S.; Vaghetti, J. C.; da Silva, L. G.; Ruggiero, R. Adsorption of Procion Blue MX-R dye from aqueous solutions by lignin chemically modified with aluminium and manganese. *J. Hazard. Mater.* **2014**, *268*, 43–50.
117. Cerrutti, B.; De Souza, C.; Castellan, A.; Ruggiero, R.; Frollini, E. Carboxymethyl lignin as stabilizing agent in aqueous ceramic suspensions. *Ind. Crop. Prod.* **2012**, *36*, 108–115.
118. Lee, J.-Y.; Lee, Y.-J.; Kang, P.-H.; Jeun, J.-P. Synthesis and Adsorption Properties of Carboxymethyl Lignin with Irradiated Lignin by Electron Beam. *Polym.-Korea* **2016**, *40*, 70–76.
119. Preparation and characterization of high boiling solvent lignin-based polyurethane film with lignin as the only hydroxyl group provider. *RSC Adv.* **2015**, *5*, 53949–53955.
120. Kong, F.; Parhiala, K.; Wang, S.; Fatehi, P. Preparation of cationic softwood kraft lignin and its application in dye removal. *Eur. Polym. J.* **2015**, *67*, 335–345.
121. Ma, Y.; Dai, J.; Wu, L.; Fang, G.; Guo, Z. Enhanced anti-ultraviolet, anti-fouling and anti-bacterial polyelectrolyte membrane of polystyrene grafted with trimethyl quaternary ammonium salt modified lignin. *Polymer.* **2017**, *114*, 113–121.
122. Laszlo, J. Solubility and dye-binding properties of quaternized and peroxidase-polymerized kraft lignin. *Environ. Technol.* **1999**, *20*, 607–615.
123. Dehne, L.; Babarro, C. V.; Saake, B.; Schwarz, K. U. Influence of lignin source and esterification on properties of lignin-polyethylene blends. *Ind. Crop. Prod.* **2016**, *86*, 320–328.
124. Kajihara, M.; Aoki, D.; Matsushita, Y.; Fukushima, K. Synthesis and characterization of lignin-based cationic dye-flocculant. *J. Appl. Polym. Sci.* **2018**, *135*, 46611.
125. Li, Y.; Wu, M.; Wang, B.; Wu, Y.; Ma, M.; Zhang, X. Synthesis of magnetic lignin-based hollow microspheres: a highly adsorptive and reusable adsorbent derived from renewable resources. *ACS Sustain. Chem. Eng.* **2016**, *4*, 5523–5532.
126. Qu, Y.; Luo, H.; Li, H.; Xu, J. Comparison on structural modification of industrial lignin by wet ball milling and ionic liquid pretreatment. *Biotechnol. Rep.* **2015**, *6*, 1–7.
127. Younesi-Kordkheili, H.; Pizzi, A. Properties of plywood panels bonded with ionic liquid-modified lignin–phenol–formaldehyde resin. *J. Adhes.* **2018**, *94*, 143–154.

128. Sarazin, C.; Husson, E.; Hulin, L.; Hadad, C.; Stevanovic, T. Acidic ionic liquid as both solvent and catalyst for fast chemical esterification of industrial lignins: performances and regioselectivity. *Front. Chem.* **2019**, 7, 578.
129. An, L.; Si, C.; Wang, G.; Sui, W.; Tao, Z. Enhancing the solubility and antioxidant activity of high-molecular-weight lignin by moderate depolymerization via in situ ethanol/acid catalysis. *Ind. Crop. Prod.* **2019**, *128*, 177–185.
130. Wang, Z.; Xue, J.; Liu, W. Nitrogen fixation and chelating property of wheat ammonium sulfite pulping spent liquor. *BioResources.* **2012**, 7, 0777–0788.
131. Dizhbite, T.; Jashina, L.; Dobele, G.; Andersone, A.; Evtuguin, D.; Bikovens, O.; Telysheva, G. Polyoxometalate (POM)-aided modification of lignin from wheat straw biorefinery. *Holzforschung.* **2013**, *67*, 539–547.
132. Couch, R. L.; Price, J. T.; Fatehi, P. Production of flocculant from thermomechanical pulping lignin via nitric acid treatment. *ACS Sustain. Chem. Eng.* **2016**, *4*, 1954–1962.
133. Parajuli, D.; Inoue, K.; Ohto, K.; Oshima, T.; Murota, A.; Funaoka, M.; Makino, K. Adsorption of heavy metals on crosslinked lignocatechol: a modified lignin gel. *React. Funct. Polym.* **2005**, *62*, 129–139.
134. Ludwig, C. Process for flocculation of solids from aqueous suspensions. *Google Patents* **1974**.
135. Li, R.; Gao, B.; Sun, S.; Yue, Q.; Li, M.; Yang, X.; Song, W.; Jia, R. Amine-Cross-Linked lignin-based polymer: modification, characterization, and flocculating performance in humic acid coagulation. *ACS Sustain. Chem. Eng.* **2015**, 3, 3253–3261.
136. Zhang, Y.; Ni, S.; Wang, X.; Zhang, W.; Lagerquist, L.; Qin, M.; Willför, S.; Xu, C.; Fatehi, P. Ultrafast adsorption of heavy metal ions onto functionalized lignin-based hybrid magnetic nanoparticles. *Chem. Eng. J.* **2019**, *372*, 82–91.
137. Li, Z.; Chen, J.; Ge, Y. Removal of lead ion and oil droplet from aqueous solution by lignin-grafted carbon nanotubes. *Chem. Eng. J.* **2017**, *308*, 809–817.
138. Li, F.; Wang, X.; Yuan, T.; Sun, R. A lignosulfonate-modified graphene hydrogel with ultrahigh adsorption capacity for Pb (II) removal. *J. Mater. Chem. A* **2016**, *4*, 11888–11896.
139. Fang, R.; Lin, Y.; Cheng, X.-S. Preparation of lignin derivatives and their application as protease adsorbents. *Nat. Sci.* **2009**, *1*, 185.
140. Ge, Y.; Li, Z. Application of lignin and its derivatives in adsorption of heavy metal ions in water: a review. *ACS Sustain. Chem. Eng.* **2018**, *6*, 7181–7192.
141. Wang, X.; Jiang, C.; Hou, B.; Wang, Y.; Hao, C.; Wu, J. Carbon composite lignin-based adsorbents for the adsorption of dyes. *Chemosphere.* **2018**, *206*, 587–596.
142. Ge, Y.; Li, Z.; Kong, Y.; Song, Q.; Wang, K. Heavy metal ions retention by bi-functionalized lignin: Synthesis, applications, and adsorption mechanisms. *J. Ind. Eng. Chem.* **2014**, *20*, 4429–4436.
143. Liang, F.-B.; Song, Y.-L.; Huang, C.-P.; Li, Y.-X.; Chen, B.-H. Synthesis of novel lignin-based ion-exchange resin and its utilization in heavy metals removal. *Ind. Eng. Chem. Res.* **2013**, *52*, 1267–1274.
144. Mathias, P. M.; Kumar, R.; Moyer, J. D.; Schork, J. M.; Srinivasan, S. R.; Auvil, S. R.; Talu, O. Correlation of multicomponent gas adsorption by the dual-site Langmuir model. Application to nitrogen/oxygen adsorption on 5A-zeolite. *Ind. Eng. Chem. Res.* **1996**, *35*, 2477–2483.
145. Freundlich, H. Über die adsorption in lösungen. *Ztschrift Fr Physikalische Chemie.* **1906**, *57U* (1), 385–470.

146. Azizian, S. Kinetic models of sorption: a theoretical analysis. *J Colloid Interfaces Sci*, *276*, 47–52.

147. Ho, Y.-S. Review of second-order models for adsorption systems. *J. Hazard. Mater.* **2006**, *136*, 681–689.

148. Annadurai, G.; Juang, R.-S.; Lee, D.-J. Use of cellulose-based wastes for adsorption of dyes from aqueous solutions. *J. Hazard. Mater.* **2002**, 92, 263–274.

149. Guo, X.; Zhang, S.; Shan, X.-q. Adsorption of metal ions on lignin. *J. Hazard. Mater.* **2008**, *151*, 134–142.

150. Šćiban, M. B.; Klašnja, M. T.; Antov, M. G. Study of the biosorption of different heavy metal ions onto Kraft lignin. *Ecol. Eng.* **2011**, *37*, 2092–2095.

151. Li, W.; Li, M. International Journal of Software Science and Computational Intelligence. *Intelligence.* **2016**, *8*.

152. Fu, Y.; Liu, X.; Chen, G. Adsorption of heavy metal sewage on nano-materials such as titanate/ TiO2 added lignin. *Results in Physics* **2019**, *12*, 405–411.

153. Hameed, B.; Ahmad, A.; Aziz, N. Isotherms, kinetics and thermodynamics of acid dye adsorption on activated palm ash. *Chem. Eng. J.* **2007**, *133*, 195–203.

154. Zhang, P.; Li, Y.; Cao, Y.; Han, L. Characteristics of tetracycline adsorption by cow manure biochar prepared at different pyrolysis temperatures. *Bioresour. Technol.* **2019**, *285*, 121348.

155. Demirbas, A. Adsorption of lead and cadmium ions in aqueous solutions onto modified lignin from alkali glycerol delignication. *J. Hazard. Mater.* **2004**, *109*, 221–226.

156. Batool, S.; Idrees, M.; Al-Wabel, M. I.; Ahmad, M.; Hina, K.; Ullah, H.; Cui, L.; Hussain, Q. Sorption of Cr (III) from aqueous media via naturally functionalized microporous biochar: mechanistic study. *Microchem J.* **2019**, *144*, 242–253.

157. Bratby, J. Coagulation and flocculation in water and wastewater treatment-Third Edition. *Water intell. online* **2016**.

158. Sharma, B.; Dhuldhoya, N.; Merchant, U. Flocculants—an ecofriendly approach. *J. Polym. Environ* **2006**, *14*, 195–202.

159. Schmitz, K. S. Macroion clustering in solutions and suspensions: the roles of microions and solvent. *J. Phys. Chem. B* **2009**, *113*, 2624–2638.

160. Hjorth, M.; Jørgensen, B. U. Polymer flocculation mechanism in animal slurry established by charge neutralization. *Water Res.* **2012**, *46*, 1045–1051.

161. Guibal, E.; Roussy, J. Coagulation and flocculation of dye-containing solutions using a biopolymer (Chitosan). *React. Funct. Polym.* **2007**, *67*, 33–42.

162. Liu, Z.; Huang, M.; Li, A.; Yang, H. Flocculation and antimicrobial properties of a cationized starch. *Water Res.* **2017**, *119*, 57–66.

163. Saveyn, H.; Meersseman, S.; Thas, O.; Van der Meeren, P. Influence of polyelectrolyte characteristics on pressure-driven activated sludge dewatering. *Colloid Surf. A-Physicochem. Eng. Asp.* **2005**, *262*, 40–51.

164. Swenson, J.; Smalley, M.; Hatharasinghe, H. Mechanism and strength of polymer bridging flocculation. *Phys. Rev. Lett.* **1998**, *81*, 5840.

165. Ries, H. E.; Meyers, B. L. Flocculation mechanism: charge neutralization and bridging. *Science.* **1968**, *160*, 1449–1450.

166. Gregory, J. Polymer adsorption and flocculation in sheared suspensions. *Colloid. Surface.* **1988**, *31*, 231–253.

167. Dannhauser, W. Polymers: chemistry and physics of modern materials (Cowie, J.M.G.). *J Chem Educ.* **1974**, *51*, A550.
168. Tang, H.; Luan, Z.; Wang, D.; Gao, B. Composite inorganic polymer flocculants. *In Chem. Water Wastewater Treat. V.* **1998**, 25–34.
169. Gao, Z.-M.; Yang, X.-Q.; Wu, N.-N.; Wang, L.-J.; Wang, J.-M.; Guo, J.; Yin, S.-W. Protein-based pickering emulsion and oil gel prepared by complexes of zein colloidal particles and stearate. *J. Agric. Food Chem.* **2014**, *62*, 2672–2678.
170. Tanneru, C. T.; Rimer, J. D.; Chellam, S. Sweep flocculation and adsorption of viruses on aluminum flocs during electrochemical treatment prior to surface water microfiltration. *Environ. Sci. Technol.* **2013**, *47*, 4612–4618.
171. Renault, F.; Sancey, B.; Badot, P.-M.; Crini, G. Chitosan for coagulation/flocculation processes–an eco-friendly approach. *Eur. Polym. J.* **2009**, *45*, 1337–1348.
172. Wang, B.; Wang, S.-F.; Lam, S. S.; Sonne, C.; Yuan, T.-Q.; Song, G.-Y.; Sun, R.-C. A review on production of lignin-based flocculants: Sustainable feedstock and low carbon footprint applications. *Renew. Sust. Energ. Rev.* **2020**, *134*, 110384.
173. Yang, R.; Li, H.; Huang, M.; Yang, H.; Li, A. A review on chitosan-based flocculants and their applications in water treatment. *Water Res.* **2016**, *95*, 59–89.
174. Zimm, B. H.; Bret, M. L. Counter-ion condensation and system dimensionality. *J. Biomol. Struct. Dyn.* **1983**, *1*, 461–471.
175. Muthukumar, M. Theory of counter-ion condensation on flexible polyelectrolytes: Adsorption mechanism. *J. Chem. Phys.* **2004**, *120*, 9343–9350.
176. Manning, G. S. Limiting laws for equilibrium and transport properties of polyelectrolyte solutions. *Polyelectrolytes.* **1974**, 9–37.
177. Zagklis, D. P.; Vavouraki, A. I.; Kornaros, M. E.; Paraskeva, C. A. Purification of olive mill wastewater phenols through membrane filtration and resin adsorption/desorption. *J. Hazard. Mater.* **2015**, *285*, 69–76.
178. Zhang, W.; Yang, P.; Li, X.; Zhu, Z.; Chen, M.; Zhou, X. Electrospun lignin-based composite nanofiber membrane as high-performance absorbent for water purification. *Int. J. Biol. Macromol.* **2019**, *141*, 747–755.
179. Helmy, Y. A.; Fawzy, M.; Elaswad, A.; Sobieh, A.; Kenney, S. P.; Shehata, A. A. The COVID-19 pandemic: a comprehensive review of taxonomy, genetics, epidemiology, diagnosis, treatment, and control. *J. Clin. Med.* **2020**, *9*, 1225.
180. García-Aljaro, C.; Blanch, A.; Campos, C.; Jofre, J.; Lucena, F. Pathogens, faecal indicators and human-specific microbial source-tracking markers in sewage. *J. Appl. Microbiol.* **2019**, *126*, 701–717.
181. Rivière, G. N.; Korpi, A.; Sipponen, M. H.; Zou, T.; Kostiainen, M. A.; O�sterberg, M. Agglomeration of Viruses by Cationic Lignin Particles for Facilitated Water Purification. *ACS Sustain. Chem. Eng.* **20** derivatives with beneficial effects on human health. *Int. J. Mol. Sci.* **2017**, *18*, 1219.
182. Vinardell, M. P.; Mitjans, M.; Ikarashi, T.; Shimizu, A.; Yano, K.; Kanamoto, T.; Terakubo, S.; Nakashima, H.; Hasegawa, H. Anti-HIV and vitamin C-synergized radical scavenging activity of cacao husk lignin fractions. *Vivo* **2008**, *22*, 327–332.
183. Sakagami, H.; Satoh, K.; Fukamachi, H.; 20, 8, 4167-4177.

184. Gupta, A.; Ahmad, I.; Summerbell, R. Fungicidal activities of commonly used disinfectants and antifungal pharmaceutical spray preparations against clinical strains of Aspergillus and Candida species. *Med. Mycol.* **2002**, *40*, 201–208.
185. Liu, W.; Du, H.; Zhang, M.; Liu, K.; Liu, H.; Xie, H.; Zhang, X.; Si, C. Bacterial Cellulose-Based Composite Scaffolds for Biomedical Applications: A Review. *ACS Sustain. Chem. Eng.* **2020**, *8*, 7536–7562.
186. Zhang, S.; Zheng, H.; Tang, X.; Zhao, C.; Zheng, C.; Gao, B. Sterilization by flocculants in drinking water treatment. *Chem. Eng. J.* **2020**, *382*, 122961.
187. Li, X.; Zheng, H.; Wang, Y.; Sun, Y.; Xu, B.; Zhao, C. Fabricating an enhanced sterilization chitosan-based flocculants: Synthesis, characterization, evaluation of sterilization and flocculation. *Chem. Eng. J.* **2017**, *319*, 119–130.
188. Huang, M.; Wang, Y.; Cai, J.; Bai, J.; Yang, H.; Li, A. Preparation of dual-function starch-based flocculants for the simultaneous removal of turbidity and inhibition of Escherichia coli in water. *Water Res.* **2016**, *98*, 128–137.
189. Wang, B.; Wang, H.-M.; Sun, D.; Yuan, T.-Q.; Song, G.-Y.; Shi, Q.; Zheng, L.; Wang, S.-F.; Sun, R.-C. Chemosynthesis, characterization and application of lignin-based flocculants with tunable performance prepared by short-wavelength ultraviolet initiation. *Ind. Crop. Prod.* **2020**, *157*, 112897.
190. Mendes, A. A.; Giordano, R. C.; de LC Giordano, R.; de Castro, H. F. Immobilization and stabilization of microbial lipases by multipoint covalent attachment on aldehyde-resin affinity: application of the biocatalysts in biodiesel synthesis. *J. Mol. Catal. B-Enzym.* **2011**, *68*, 109–115.
191. Chen, K.; Lei, L.; Qian, Y.; Yang, D.; Qiu, X. Development of anti-photo and anti-thermal high internal phase emulsions stabilized by biomass lignin as a nutraceutical delivery system. *Food Funct.* **2019**, *10*, 355–365.

Chapter 8

Recent Strategies for Lignin-Based Thermosets

Martin Lawoko[*,1,2] **and Claudio Gioia**[3]

[1]**Wallenberg Wood Science Center (WWSC), Department of Fiber and Polymer Technology, School of Engineering Sciences in Chemistry, Biotechnology and Health, Royal Institute of Technology, KTH, Teknikringen 56-58, 100 44 Stockholm, Sweden**

[2]**Division of Wood Chemistry and Pulp Technology, Department of Fiber and Polymer Technology, School of Engineering Sciences in Chemistry, Biotechnology and Health, Royal Institute of Technology, KTH, Teknikringen 56-58, 100 44 Stockholm, Sweden**

[3]**Department of Civil, Chemical, Environmental and Materials Engineering, University of Bologna, Via Terracini 28, 40131 Bologna, Italy**

[*]**Email: lawoko@kth.se.**

Lignin has gained interest in academia as a raw material for advanced materials such as thermosets. Attractive features include its natural and technical abundances, its structural compositions, distribution of functionalities and most importantly its origin from renewable sources, i.e. lignocellulosic biomass. In this chapter, we will focus on lignin-based thermosets, specifically reviewing the recent strategies for such applications in polyurethanes, phenol-formaldehyde resins and epoxy thermosets. In these contexts, the advantages and competitiveness of lignin over conventional petrol-based will be discussed based on comparisons with benchmark properties. The chapter emphasizes the unique polymer chemistries adopted by different researchers, the effects of the chemical and physical properties of lignin on both reactivities during synthesis and thermomechanical properties of materials. Furthermore, the direct use of lignin in the unmodified state is contrasted with the use of selectively modified lignins in these polymer systems. The use of multifunctional lignin oligomers as raw material is also contrasted with the use of lignin platform monomers obtained through de-polymerization approaches. Finally, the challenges of lignin or monomers thereof, as a raw material in these applications are discussed to direct the future outlook of this research area.

Polyurethanes

Introduction on Polyurethane Thermosets

Polyurethane (PU) thermosets are traditionally synthesized from petroleum-derived polyols and isocyanates (Figure 1). They are a very versatile family of materials, available in various forms such as fibers, foams, coatings and elastomers. In 2018, the global polyurethane market size was estimated at 65.5 billion dollars, with an annual growth of 7.0% from 2019 to 2025 (*1*). Such success is mainly due to their tremendously tunable properties made possible by customized formulations constituted by mixtures of soft and hard blocks. PU market increment will be driven in the upcoming period by the increasing demand for lightweight and durable products in the automotive, construction and electronics industries. Polyurethane common applications range from flexible foam in upholstered furniture, rigid foam for insulation in walls and roofs to coatings, adhesives, sealants, and elastomers used on floors and automotive interiors (*2*).

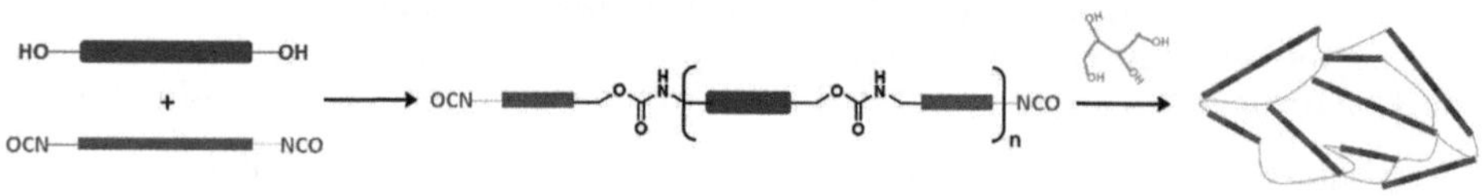

Figure 1. Schematic representation of the fabrication of a generic polyurethane network.

Lignin is potentially a perfect candidate as a component in PU polymer synthesis: it is bio-based, renewable, abundant, cheap, multifunctional in phenolic and aliphatic hydroxyls and its aromatic framework can impart a great rigidity when introduced into a crosslinked architecture (*3*, *4*). In this subsection, the most recent advances in two main strategies for the creation of lignin-based polyurethane architectures will be outlined. The first one concerns the direct use of lignins as a partial replacement of petrol-based polyols to understand the potential applications and performances range of lignin PU resins, while the second proceeds through a chemical modification of lignin to introduce additional functionalities and to obtain elevated properties.

Unmodified Lignin in Polyurethane Thermosets

The main drawbacks of the use of lignin in PU networks are related to its moderate solubility in the reaction medium, often enhanced by the high molecular weight, and the sterical hindrance of some of the hydroxy groups by the bulky lignin structures. For these reasons, one challenge is to achieve a stoichiometric reaction of lignin hydroxyls with the urethane "-NCO" moieties.

Annamalai and co-workers addressed the issue of the poor solubility of lignin investigating a procedure to disperse Kraft lignin into a polyether polyol (Voranol RH360 from DOW Chemicals) at 120°C using glycerol as compatibilizer (*5*). The obtained mixtures were cured with polymeric methylene diphenyl diisocyanate and the obtained materials fully characterized. The results demonstrated that the high-temperature processing induced a process of disaggregation of lignin microparticle enabling the solubility of lignin in the system. This simple process had a beneficial impact on material properties such as thermal insulation and compressive strength.

The main approach to counterbalance the poor reactivity of hindered lignin hydroxyls is to use an excess of the isocyanate component to enhance the advancement of the addition reaction.

This method, however, deeply affects the crosslinking density and the properties of the material and therefore requires a fine optimization. An example of this approach was reported in a study by Pan et al. on the fabrication of rigid polyurethane foams (*6*). Their aim was to understand the impact of several factors such as the lignin content, the molecular weight of the chain extenders and the -NCO/-OH ratio on the thermo-mechanical properties of lignin-based PU. Specifically, the authors demonstrated that hardwood organosolv lignin performed much better than Kraft lignin due to its higher solubility with the polyol. As a consequence, organosolv lignin could substitute a higher amount of polyol maintaining unaltered strength.

The impressive versatility of the polyurethane family is the result of the contribution of several factors such as the chemical structure of the hard and soft segment, the structure of the crosslinker, the ratio of each main component, possible excesses of specific components to favor the crosslinking as well as the presence of catalysts, initiators, chain extenders and foaming additives. Clearly, the involvement of a complex system such as lignin introduces additional variables. To optimize the formulation of the resins and their curing conditions, several research groups adopted mathematical methods such as the response surface methodology (RSM). Avelino and co-workers, for example, adopted a central composite design of the experiments with two levels and two independent variables (7). They manage to optimize a three-component system consisting of lignin (in this case a coconut shell ethanosolv lignin), polyethylene glycol (PEG 400) and toluene diisocyanate, through a solvent-free polymerization route. The authors reported that with the optimized formulations, the introduction of lignin was associated with superior thermal and thermo-oxidative stabilities. Contrarily, in other formulations, an excess of lignin would produce brittle materials unable to undergo compression. Xu et al., designed a single factor experiment to study the effect of enzymatic hydrolysis lignin content on the foams morphology and mechanical properties (8). The formulation consisted of lignin ranging from 2.5 to 15wt% and it was reported that higher lignin contents were able to produce irregular, inhomogeneous, and thick cell walls in the foams. However, outstanding mechanical performances were achieved for the foam containing 15% of lignin. Specifically, about a ten-fold increase of the compressive modulus and an impressive ninety eight-fold increase in the compressive strength were obtained.

A final example of highly performing PU systems based on unmodified lignin is provided by the work of He and co-workers (9). They synthesized new elastomeric materials from Kraft lignin as a hard segment and poly-(propylene glycol) tolylene 2,4-diisocyanate terminated as a soft segment (Figure 2). Particularly, they studied the effect of lignin molecular weight (600 and 3600 Da) and weight fraction (up to 40%) on the thermomechanical properties of the obtained elastomers. It was shown that the glass transition temperature (Tg) increases with the lignin content, and that low molecular weight lignin could impart excellent Young's modulus (176.4 MPa), tensile strength (33.0 MPa) and an outstanding strain at break of 1394% due to an efficient dispersion of low molecular weight lignin in the material. Besides, the authors unveiled that the contribution from lignin hydrogen bonding, especially when used in a higher amount, enhanced thermomechanical properties by improving intermolecular lignin-lignin and lignin-resin interactions, acting as a physical cross-linking (Figure 2).

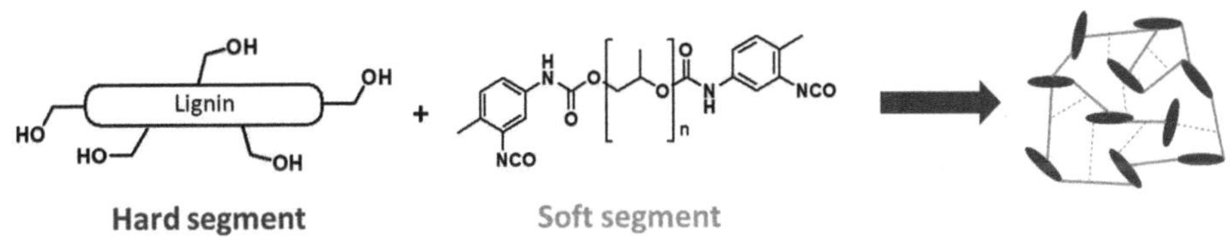

Figure 2. Fabrication of highly performing PU systems: dashed line represents hydrogen bonds (9).

Chemically Modified Lignin in Polyurethane Thermosets

Chemical modification introduces specific functional groups into the lignin structure with the objective of enhancing its reactivity or imparting additional properties to the final thermoset. Since the mechanical properties of traditional polyurethane resins are heavily controlled by the crosslinking density, many efforts have in recent years been devoted to increasing the number of reactive sites on lignin.

Phenolic hydroxyls in lignin are known to have a lower reactivity than aliphatic hydroxyls during crosslinking towards isocyanates. For this reason, oxypropylation was shown to be a valid method to obtain aliphatic polyols from lignin presenting enhanced reactivity (*10, 11*). It is achieved by reacting lignin from different sources with propylene oxide under basic conditions. As a matter of fact, lignin hydroxyls react with propylene oxide producing a fully aliphatic -OH lignin (Figure 3). Additionally, the oxypropylation is a reliable method for lignin liquefaction, therefore favoring the processability and lignin distribution in the material (*12*).

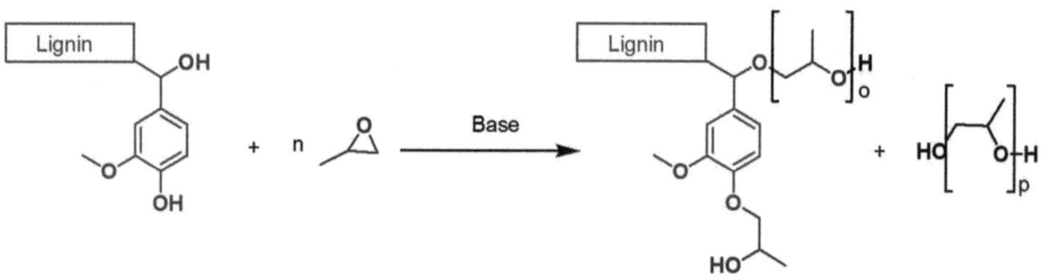

Figure 3. Oxypropylation of lignin.

A most recent example of this approach was reported by Cao and co-workers (*13*). Remarkably, they refined lignin by sequential precipitation in ethanol/water mixtures to obtain fractions characterized by decreasing molecular weight and low polydispersity. These fractions were then activated by oxypropylation to obtain homogeneous starting materials for rigid polyurethane foams. The results demonstrated that low Mw fractions produced the materials with better performances in terms of compressive strength (0.78 MPa), lower thermal conductivity (0.037 m^{-1} K^{-1}) and better thermal stability.

A completely different approach sees the introduction of isocyanate groups directly on lignin structure. Isocyanate lignin is produced by partially reacting lignin hydroxyls with di-isocyanate compounds (Figure 4). With such a method, the resulting modified lignin can be crosslinked with commercially available polyols, creating a new type of lignin-based polyurethanes.

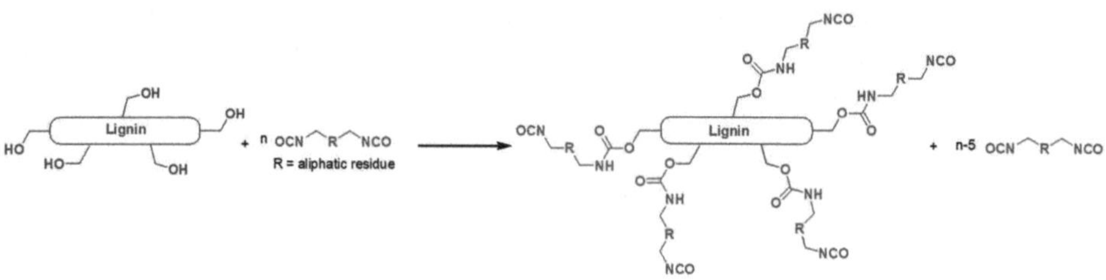

Figure 4. Isocyanate functionalized lignin.

For example, Eceiza et al. compared the influence of unmodified Kraft lignin and isocyanate lignin on the performances of flexible foams (*14*). The observed results demonstrated that modified lignin increased the reactivity of the system, decreased the foaming stages and required less catalyst in comparison with unmodified one. Furthermore, the flexibility of the resultant foams increased and chemical attachment of lignin in the resultant material improved as evinced by extraction experiments. In other work, Zhang and Fang synthesized high-performance polyurethanes based on isocyanate lignin and poly (caprolactone) (*15*). The authors modified alkali lignin with different amounts of hexamethylene diisocyanate and crosslinked the resulting products with biodegradable poly (caprolactone) as a soft segment. Interesting insights on the main parameters affecting the properties of the resins were unveiled. Specifically, the lignin content, the -NCO/-OH ratio and the molecular weight of PCL were found to be relevant parameters for the materials properties. As a result, an optimized material containing 37.3% of lignin demonstrated the highest performances in the tensile strength (19.35 MPa), elongation at break (188.36%), and tear strength (38.94 kN/m) as well as a remarkable solvent and thermal resistance at 340°C.

As previously mentioned, functionalization is also the pathway to introduce additional properties to lignin. A remarkable example was reported by Xing and coworkers who managed to obtain halogen-free, flame retardant rigid polyurethane foams by introducing phosphorous based functionalities (*16*). Wheat straw alkali lignin was chemically modified via a three-step reaction: the first one involved a hydroxymethylation procedure to increase the lignin aliphatic hydroxyls. This was followed by the grafting of POCl3 on the aliphatic hydroxyls, while the third step involves ethylene glycol to finally obtain a phosphorous functionalized polyol (Figure 5). The results demonstrated a clear effect of the modified lignin on the char formation at 700°C (42.7 % while lignin-free foams produced only 0.2% of char) as well as a relevant reduction of the heat released during the combustion process.

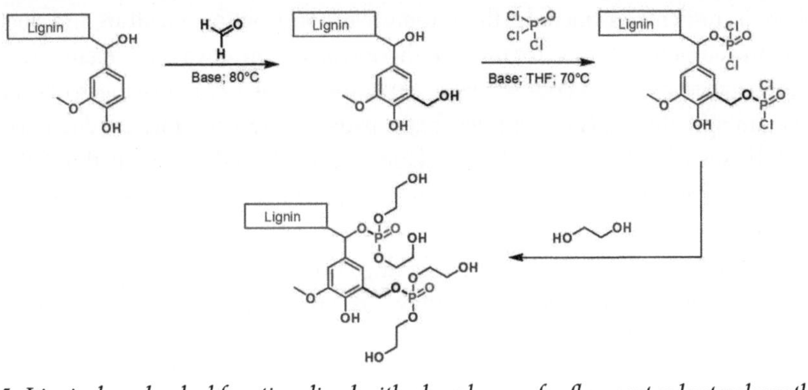

Figure 5. Lignin-based polyol functionalized with phosphorous for flame retardant polyurethanes.

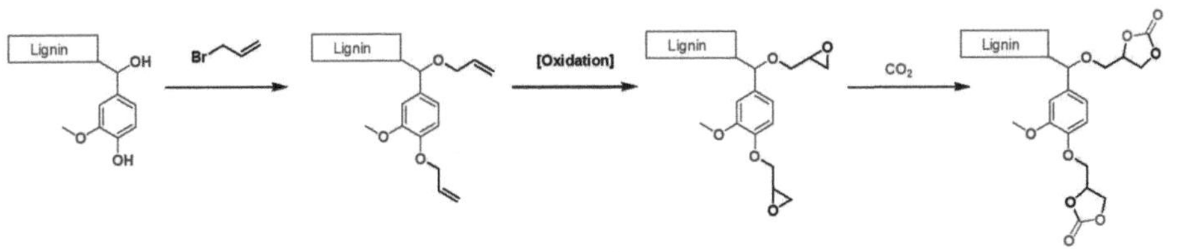

Figure 6. Carbonate-functionalized lignin for isocyanate-free polyurethane thermosets.

We caution, however, based on the adopted hydroxylation chemistry, that the method may be more suitable for softwood lignin dominated by guaiacyl units as opposed to hardwood lignins with syringyl units, where an additional methoxyl group at C5 would block the reaction.

A special mention of the work of Zoia et al. concerning their pioneering synthesis of lignin-based isocyanate-free polyurethanes is due. This class of new materials recently emerged to create an alternative route for the use of toxic isocyanate compounds in the polyurethane synthesis (*17*). For the first time, Zoia and coworkers develop a synthetic method to obtain lignin functionalized with cyclic carbonates able to react with polyamines to obtain new polyurethane systems (Figure 6) (*18*). Specifically, herbaceous lignin deriving from a soda pulping process was first cyclocarbonated and then reacted with 1,12- diaminododecane in the presence of poly(ethylene glycol) bis cyclic carbonate as a soft segment used to tune the material properties. The authors validated the synthetic method and demonstrated the possibility to obtain materials with tailored properties ranging from elastomers to thermosets.

Phenol-formaldehyde

Introduction on Phenol-formaldehyde Thermosets

Phenol-formaldehyde (PF) thermosets, commonly referred to as phenolic resins, represent one of the main classes of thermosetting materials used for a variety of applications. The phenolic resin market is expected to reach an estimated 14.53 billion dollars by 2024 with a constant growth of 5.1% from 2019 to 2024 (*19*). Phenolic thermosets are usually obtained by the multiple reactions of phenols with formaldehyde (Figure 7) and represents a family of materials presenting a wide range of desirable properties such as high heat resistance, high moldability, and high mechanical strength and are preferred over other synthetic thermosets due to their good chemical resistance, dimensional stability and resistance to moisture. They can be categorized as resol resins, novolac resins, and other subgroups according to their chemical structure and crosslinking density.

The phenolic resin market includes the automotive, construction, furniture and electrical sector. For example, the automotive industry uses PF thermosets to improve reinforcement, superior heat and friction resistance, as well as good adhesive performance for a variety of mechanical components such as brake linings, clutch plates, and disc brake pads. The low thermal conductivity of phenolic thermosets makes it suitable to be used as insulation materials in the construction field while water resistance is exploited in the electrical and furniture sector. Finally, PF thermosets find another main application in the field of wood adhesives.

Framed in this scenario, lignin represents a huge opportunity as a biobased substituent in PF thermosets due to its inherent phenolic nature, its enormous abundance and its cheap price (particularly in the case of technical sources). Here, we reported the most recent strategies adopted for lignin as a basic constituent in PF thermosets.

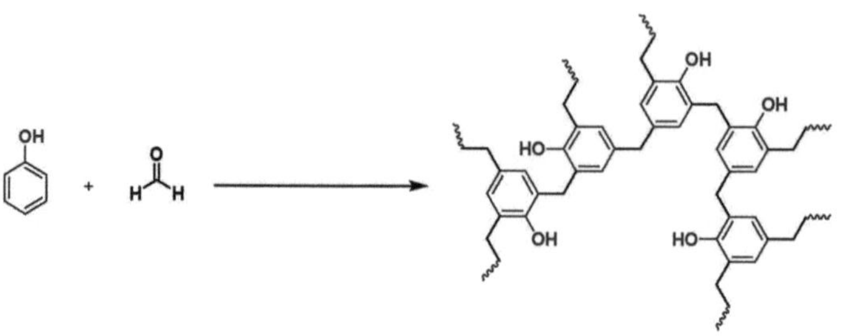

Figure 7. Schematic representation of the fabrication of a generic PF thermoset.

Lignin in PF Thermosets

One notable advantage of the involvement of lignin for the fabrication of PF thermosets is related to its polyether backbone, able to confer extra rigidity to the system once introduced in the network and often resulting in enhanced mechanical performances. Also, lignin phenolic moieties can be directly polymerized with common industrial procedures. One main issue hampering the widespread industrial production of lignin-based PF thermosets concerns the accessibility and the reactivity of its reactive groups in comparison with common industrially available phenols. Commonly, formaldehyde reacts with the 2-,4-, and 6- position of phenol due to the directing ability of the hydroxyl group for electrophilic substitution, in the case of lignin, the key reactive sites are located in the free phenolic structures of the lignin which can react only on the 3- and 5- position if not occupied by methoxy groups (Figure 8). Furthermore, only the lignin end-groups present reactive phenols while all the aromatic units within the chain cannot react. As a main consequence, the molecular weight and the chemical structure of the lignin have a tremendous effect on the final properties of the material.

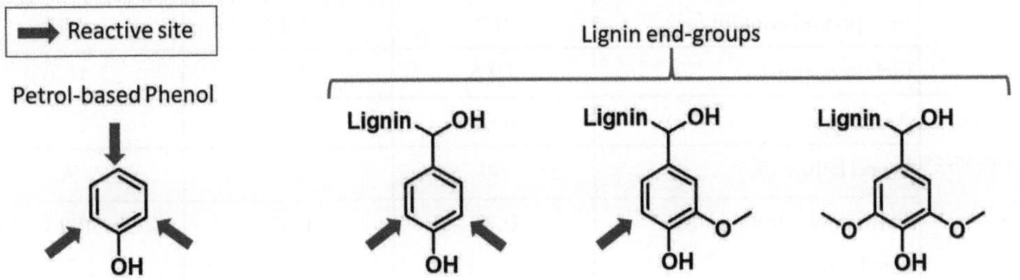

Figure 8. Reactivity comparison between petrol-based phenol and lignin phenolic structures.

Yang et al. synthesized adhesives from four different technical lignins (corn cob lignin, two different kinds of poplar wood lignin, and wheat straw lignin) for the fabrication of plywood panels (*20*). The authors created four adhesives substituting 50% of the phenol with the corresponding technical lignins. The physical-chemical analysis of the thermosets revealed that lignin-based materials generally present properties comparable with the commercial benchmark with some notable differences. In particular, the free phenol and free formaldehyde values were found to be far below the limit requirement of the Chinese National Standard (GB/T 14732-2006). Furthermore, the viscosity increased with the increase in the molecular weight of lignin. It was also shown that the solid content of the lignin-based adhesives increased by 50% in comparison with the commercial benchmark, and consequently favoring a continuous bond line between surfaces. Among all, corn cob based adhesive presented the highest bonding strength (1.18 MPa) while wheat straw presented the lowest (0.95 MPa). This aspect was postulated to be related to the higher molecular weight of wheat straw lignin, associated with a lower number of reactive sites and a higher viscosity. While the reactive sites directly influence the crosslinking density, a higher viscosity induces inhomogeneities in the adhesive and a reduction in the penetration of the adhesive in the veneer thus reducing the strength of the cured bond line of the plywood. Concerning the kinetic of the curing reaction, a higher molecular weight of lignin was found to require a higher curing temperature to overcome its lower reactivity.

Argyropoulos et al. made a relevant step forward in understanding the impact of lignin structures on the properties of lignin PF thermosets (*21*). Specifically, two distinct technical lignins were examined as obtained from an acidic and an alkaline organosolv pulping of bamboo. These samples were thoroughly characterized and demonstrated notable differences in the structural composition mainly related to the presence of aryl ether linkages, molecular weight, sidechain structures and presence of extractives. The resulting resins were used for the production of plywood and compared with a commercially available product. Table 1 reports the performances of the thermosets and the corresponding plywood samples.

Table 1. Performances Comparison between Lignin-Based PF Thermosets and a Commercially Available Product (*21*)

		Acidic Organosolv	*Aalkaline Organosolv*	*Commercial PF thermosets*
Properties	pH	11.7	11.5	12.6
	Viscosity (mPa·s)	1827	1119	133
	Solid content (%)	55.0	53.7	44.0
	Free formaldehyde content (%)	0.17	0.28	0.24
	Free phenol content (%)	0.9	1.13	0.11
	Gel time (min)	19.6	22.0	16.0
Plywood performances	Bonding strength (MPa)	0.94	0.98	1.53
	Wood failure (%)	30	10	90
	Formaldehyde emission ($mg\ L^{-1}$)	0.36	0.41	0.25

It was shown that parameters such as the purity (ash, carbohydrate, and extractive content), molecular weight, and chemical reactivity of the technical lignins mainly contributed to the LPF resin properties and thus the performance of the wood products prepared while structural units present a negligible impact on the bonding strength and formaldehyde emissions of plywood. Finally, they proposed two main strategies to boost the lignin impact on material properties. The first is related to chemical modification of technical lignin to improve reactivity while the second is based on formulation according to the effective number of reactive sites.

This latter approach was recently adopted by Lourençon and Tamminen focusing their research toward a less common hardwood lignin source (22). Lignin from grasses and softwood has been the objective of extensive studies as they present a higher number of theoretical reactive sites in comparison with hardwood. Syringyl units in hardwood lignin, in fact, do not react with formaldehyde since both 3- and 5- carbons are occupied by methoxy groups. The authors evaluated the maximum consumption of formaldehyde of hardwood and softwood lignins using hydroxymethylation and discovered that while hardwood lignin consumed formaldehyde close to the theoretical calculation, the reference softwood lignin consumed only about 25% of the theoretical value. The main reason is related to lower availability of the softwood lignin reactive sites due to a higher molar mass along with steric effects which hinder the crosslinking reaction as confirmed by a lower formation of methylene bridges. As a consequence, PF thermosets synthesized with a phenol substitution on 50% of lignin displayed comparable bonding strength irrespective of whether the resins were made with softwood or hardwood lignin (range of 3.4–4.8 N mm−2 at 150°C and 45–480 s of press time). Additionally, the resins obtained by this method demonstrated low values of free formaldehyde emissions (<0.23%) and higher thermal stability in comparison with commercially available PF resins.

Interestingly, a novel strategy exploiting lignin in the form of nanoparticles was reported (*23*). Nanolignin with a high specific surface area and a porous structure with an average particle size of about 300 nm were prepared and substituted to phenol at levels of up to 60% to produce wood adhesives. The results demonstrate an outstanding role of lignin contributing to the final properties of the material. With only 40% of lignin substitution, the bond strength of the resin reached 1.3 MPa outperforming grade 1 plywood standard requirements (0.7 MPa). Moreover, the formaldehyde emission of the plywood prepared from 40% of nanolignin was 0.40 mg L-1, reaching E0 grade and meeting the GB/T 14732-2006 standard. Finally, in comparison with normal lignin-based PF resins, the nanolignin significantly reduced the curing temperature from 186 °C to 145 °C. This new approach remarkably introduces a new perspective related to the role of the supramolecular assembly of lignin structures and their influence on the final properties of a new concept of thermosets.

Activated Lignin in PF Resin Architectures

As previously introduced, the low lignin reactivity constitutes one of the challenges for the fabrication of materials with reproducible properties due to the inherent variability in the number and accessibility of its reactive groups, especially when compared with phenol as a conventional reference. Kraft lignin, for example, is estimated to have only 0.3 reactive sites accessible for the formaldehyde reaction for every monomer. To address this problem, several approaches have been devised to enhance reactivity in lignin and produce reliable thermosets (*24*).

In one approach additional phenolic sites are created by simply depolymerizing the lignin structures by means of hydrolytic processes. Hydrolysis produces low molecular weight polyphenolic compounds through inter-unit ether bond cleavage. For example, hot compressed water (3.4 MPa and 260 °C) was demonstrated to effectively liquefy lignin with a liquefaction yield up to 62% without any kind of reactant or catalyst (*25*). The resulting mixture of phenols was successfully introduced in a PF resin, with a phenol substitution rate of up to 60%. The materials thus obtained demonstrated a remarkable adhesive strength (1.32 MPa) able to outperform reference materials such as a typical phenol-based resin (0.77 MPa) and an "unmodified" lignin-based resin (0.72 MPa). Additionally, the hydrolytic lignin resin demonstrated the lowest amount of free formaldehyde and a substantial decrease in the activation energy of the curing. Furthermore, hydrolyzed lignin can be further modified by additional hydrothermal liquefaction in ethanol/water co-solvents (50:50, v/v) at 350 °C in presence of hematite ore as the catalyst (*26*). As a result, the corresponding material demonstrated a remarkable boost in the dry bonding strength reaching 2.65 MPa.

A second method involves a demethylation process to eliminate the methoxy group and create new catechol moieties on lignin. Such treatments allow to activate the aromatic carbon 6- of the guaiacol structure thus increasing the number of reactive sites for the crosslinking reaction with formaldehyde. Several pathways were historically developed with the specific aim to remove the methoxy group such as sulfur mediated reactions (*27, 28*), reductions (*29*) and even oxidative processes (*30*). Recently, Pan et al. reported the demethylation of alkali lignin in presence of halogen acids at 130°C (*31*). Among all, HI produced the highest increment of phenolic functionalities due to its higher nucleophilic character. The thermosetting material obtained with this method met the bonding strength standard for exterior plywood with the modified lignin used to replace as much as 50 wt.% of phenol. The increased hydroxyl contents resulting from the lignin demethylations also imparted faster cure times for the lignin-containing resins and lower formaldehyde emissions.

Zhang and Li (27) studied the demethylation at room temperature of agricultural crop-based lignin in presence of sulfur- and sulfur-based reactants such as sodium hydrogen sulfide (NaSH), sodium sulfite (Na_2SO_3) and mercaptan (CH_3SH). Among all the reactants, Na_2SO_3 proved to be very effective in the demethylation process of lignin, increasing the phenolic content on the molecule and consequently enhancing molecular reactivity. Finally, it is worth mentioning the recent enzymatic demethylation carried out by fungal degradation and reported by Venkatesagowda and Dekker (32). Interestingly, a selected number of wood-decay fungi obtained from the Boreal Forest in northern Canada were screened for the lignin demethylation phenomenon, among which, Fomitopsis pinicola, Galerina autumnalis, and Aspergillus sp.3 BRI 270 displayed the highest lignin demethylation activity. With their result, the authors undoubtedly demonstrated that the mentioned fungi display the typical reactivity of specific enzymes such as O-demethylases, currently poorly explored and that biological processes can be potentially exploited for the synthesis of PF resins.

Phenolation represents the third main method to improve lignin reactivity toward crosslinking with phenol and formaldehyde. This process consists in the treatment of lignin with phenol in acid conditions to favor their condensation on the lignin aromatic rings and sidechains (Figure 9).

Ether bonds are also cleaved thus decreasing the molecular weight of lignin. As a result, pristine phenolic functionalities are added to the lignin structure favoring the subsequent curing reaction. Clearly, the structure/property relationship is a main contemporary target for academia with the specific aim to unravel the range of properties achievable with such an approach. Cheng at al., focused their research on a profound understanding of the reactivity of the different lignin substructures under the phenolation process and proposed specific mechanisms for each main connecting group (33). Specifically, the authors observed that phenol was incorporated onto both the side chains and aromatic nuclei of lignin while all lignin substructures of aryl ether (βO4), phenylcoumaran (β-5), benzyl ether (αO4), pinoresinol (ββ), α-carbonyl, etc. were reacted, resulting in an increase in the phenolic hydroxyl groups. Finally, the comprehensive characterization revealed that most of the ether linkages were cleaved during phenolation while the β-elimination of the γ-hydroxymethyl group as formaldehyde was the main reaction of side chains. To underline the great potential of the phenolation approach, PF thermosets were obtained from phenolated lignin, demonstrating a shorted gel time and a lower exothermic peak of curing (34). Such materials showed thermal stability comparable with commercial PF resins and were successfully employed for the fabrication of plywood fitting in the request of CNS 1349 standard for Type 1 plywood concerning technical performances such as dry, warm water-soaked, and repetitive boiling water soaked bonding strength.

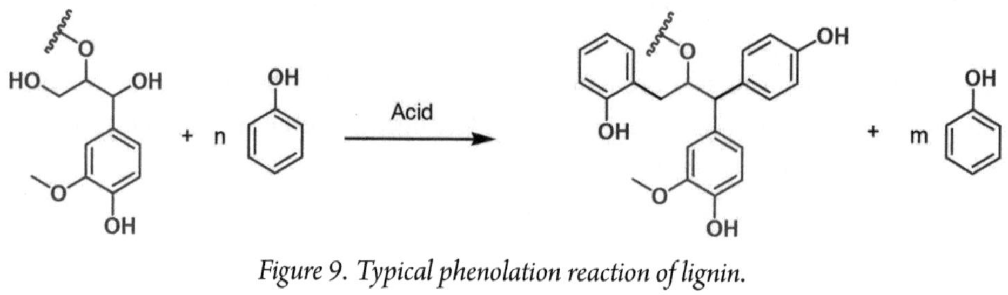

Figure 9. Typical phenolation reaction of lignin.

Epoxy Resins

Introduction on Epoxy Thermosets

Epoxy thermosetting materials represent another main family of thermosetting materials. In 2016, the global epoxy thermosets market reached 21.5 billion US$, with forecasted yearly growth estimated at 6.3% in the short-term and, up to 37.3 billion US$ in 2025. This is mainly due to the rapidly increasing demands of epoxy composite and adhesive markets (35).

The huge demand for epoxy resins is currently driven by key downstream markets such as composites materials, automobile, aerospace, electrical and electronic. For example, epoxies are used in the food industry as coatings for packaging, prolonging the shelf-life of food while in food-related machinery they protect food from contamination. In automotive and aerospace fields, epoxies are applied as coatings, providing water and corrosion resistance, and in structural parts where weight reduction is favorable, durability, heat resistance and mechanical strength. Furthermore, epoxy thermosets as constituents in adhesives, sealants, and fillers are essential in constructions. Finally, they are also used extensively in sports equipment such as tennis racquets, skis, surf-boards, etc.

With the recent advent of green chemistry, the urge for more sustainable epoxy resins is becoming more and more attractive and many efforts are currently directed toward the synthesis of epoxy thermosets from renewable materials (36). Specifically, due to its demonstrated acute toxicity on human health (37), the substitution of bisphenol A (BPA) based epoxidated pre-polymer, is an urgent need. In this context, the key challenge however is finding bio-based substitutes with matching thermomechanical performances. Here, we report the most recent strategies adopting lignin for the fabrication of epoxy thermosetting materials.

Lignin as a Filler for Epoxy Thermosets

In one strategy, lignin from different sources is physically blended with an epoxy pre-polymer and tested with several crosslinkers in the final curing step. In this case, lignin performs the role of the reinforcing agent due to its polyaromatic rigid structure, therefore little attention is devoted to its composition and functionality. Kong et al., for example, introduced only 5% of corn straw enzymatic hydrolyzed lignin into an epoxy-based adhesive (38). They demonstrated that such low concentrations of lignin could undoubtedly affect the properties of the adhesive. In particular, lignin considerably altered the curing temperature shifting it toward higher temperatures, while the adhesive demonstrated a higher crosslinking density and better bonding performances. The inclusion of lignin in polymer composite systems where both high-temperature shear strength and thermomechanical performance are desirable, such as carbon fiber/epoxy resin composites, is of interest.

Yin et al. prepared lignin/epoxy blends by mixing different amounts of corn straw enzymatic hydrolyzed lignin into an epoxy/polyamide system, and studied systematically the effect of the lignin content on bending strength, impact strength and density of the materials (39). The impact strength of the blends initially increased with the content of lignin reaching its maximum at 45% lignin content, while higher amounts negatively affected it. The bending strength gradually decreased with the increasing amount of lignin while density constantly increased. The combination of the rigid structure of lignin and moieties suitable for hydrogen bond formations was beneficial at the optimized lower lignin content while, for higher concentrations, the higher viscosity and inhomogeneity dominated the system, with consequent negative impact. In addition, they demonstrated the possibility for lignin to react directly with the epoxy pre-polymer in absence of the polyamide crosslinker.

Further optimization of the described system was developed by the same authors by a hot press molding process (*39*). Materials containing up to 60% of lignin were fabricated and the effect of molding conditions on the final material was evaluated by SEM, FTIR, DSC, and DMA analyses. The optimal molding conditions were higher temperature and pressure and resulted in better interfacial adhesion. Under these conditions, the resulting material was homogeneous as manifested in the lack of obvious interfaces upon SEM analysis.

Synthesis of Epoxidated Lignin

Undoubtedly, the chemical introduction of epoxy groups on lignin´s structure opens new possibilities, for compatibility with classical epoxy systems and progressively paving way for a consistent substitution of bisphenol A (BPA) in material science. An in-depth analysis of lignin structure and functionalities is critical for the subsequent modification with the controlled selectivity required for materials with tailored properties. Functional groups such as phenols, aliphatic alcohols, and carboxylic acids constitute the reactive sites that allow the insertion of oxirane moieties. Typically, such functional groups are introduced by reacting lignin from different sources with epichlorohydrin in an alkaline environment (Figure 10). The typical pathway of epoxidation is well known and involves two steps: initially, epichlorohydrin reacts with the available nucleophilic groups, grafting a chlorohydrin moiety on the structure. Subsequently, the alkaline environment promotes the formation of the oxirane ring by means of the elimination of HCl (*40*).

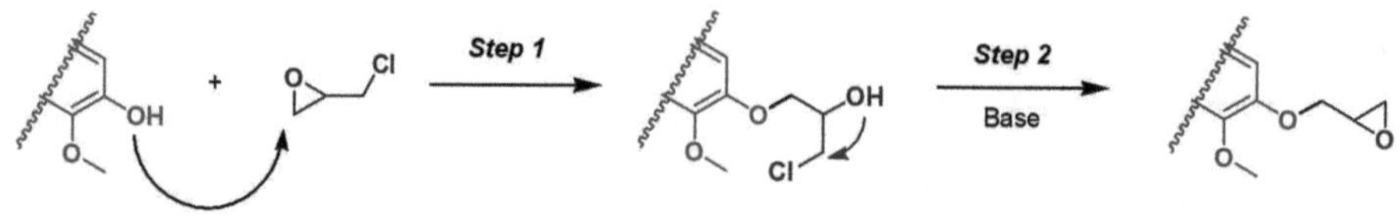

Figure 10. General mechanism of reaction of lignin´s phenol with epichlorohydrin.

The effect of parameters such as reaction time, temperature and amount of base on the reaction conditions were studied by Xu et al. on depolymerized organosolv lignin (*41*). They also designed a set of experiments by utilizing the central composite design (CCD) to achieve the optimal number of epoxy groups on the lignin structure (*42*). The results demonstrated that mild conditions such as 55°C, a 6.3 molar ratio of NaOH/lignin for 8h of reaction would result in the complete functionalization of the reactive sites. The mathematical model applied agreed with the experimental results, demonstrating the possibility to rationally design the condition of reaction to maximize the functionalization.

Malutan et al., instead, exploited a more empirical approach toward lignin epoxidation (*43*). Other than considering the usual variabilities (temperature, time and amount of base) they investigated the reactivity of different kinds of lignin such as wheat straw and Sarkanda grass lignin. In addition, they evaluated hydroxymethylation of lignin as a method to increase the number of reactive sites before epoxidation.

The selectivity of the epoxidation reaction toward lignin functional groups was evaluated by applying different catalytic systems. The contribution of a basic promoter such as NaOH or KOH resulted in the selective reaction of phenols and carboxylic acids (*44*) while the combination of a base and a phase transfer catalyst allowed the additional functionalization of aliphatic hydroxyls (*45*). The presence of quaternary ammonium salts, for example, favors the deprotonation of aliphatic alcohols thus increasing the epoxy content on lignin.

Epoxy-Lignin/BPA-Based Epoxy Thermosets Co-polymeric Thermosets

The simple modification of lignin, introducing epoxy functionalities, favored the synthesis of new macromolecular structures aiming to partially replace petrol-based epoxy components (in particular containing BPA) with the rigid backbone of lignin. This results in hybrid materials with partial renewable character and novel desirable properties.

Xu et al., for example, applied low molar mass epoxy lignin, obtained through reductive depolymerization process (*42*), and a commercially available BPA-based epoxy pre-polymer in the fabrication of hybrid thermosets with different percentages of lignin (*46*). Such a study was directed toward the comprehension of the effect of lignin content on the curing kinetics and the mechanical properties of thermosets, glass fiber reinforced composite materials and metal coatings. Specifically, two different sources such as hardwood depolymerized organosolv lignin (DOL) and softwood depolymerized Kraft lignin (DKL) were subjected to epoxidation and co-polymerized with BPA-based components at different ratio. Interestingly, a low amount of lignin (<25%) resulted beneficial for the kinetic of the curing due to the positive influence of free hydroxyl groups on the structure of lignin-based epoxy thermosets. Higher content of lignin on the other hand results in higher viscosity mixtures, therefore retarding the curing. The curing of DKL required a lower reaction temperature compared to DOL due to a lower amount of epoxy number. Mechanical tests performed on the glass fiber composites evidenced that bio-based epoxy composites presented, superior or comparable tensile and flexural strengths in comparison with fully petrol-based composites when the lignin-based epoxy resin blending ratio is less than 50–75 wt%. Furthermore, the bio-based epoxy films comprising up to 50 wt% of DOL-based epoxy resin exhibited higher adhesion strength on a metal substrate. Additionally, the thermal behavior of the synthesized blends was evaluated (*47*). TGA-FTIR analysis was performed to assess thermal stability and degradation kinetic while the activation energy was calculated by Kissinger-Akahira-Sunoe (KAS) and Flynn-Wall-Ozawa (FWO) methods. Undoubtedly, the presence of lignin affects the thermal behavior of the materials initially decreasing their activation energy but increasing it in the final stage. All the lignin-containing materials demonstrated more effective fire retardancy than fully BPA-based materials as evinced by limiting oxygen index analysis. Notably, high lignin content materials (more than 75% w/w) demonstrated self-extinguishing properties.

The mechanical performances of organosolv beech lignin/BPA epoxy blended epoxies were studied by Meier et al. (*48*) They demonstrated that lignin contents between 8 and 33%, produced materials with higher crosslinking density, higher glass transitions, lower swelling and higher Young´s modulus when compared with fully petrol-based thermosets.

Another study by Yu et al. reported the beneficial effect of lignin on the corrosion performances of epoxy coatings (*49*). Specifically, Q235 steel samples were epoxy- coated with commercially available bisphenol A-based coatings, treated with a NaCl solution and analyzed by polarization curves and electrochemical impedance spectroscopy. Under such conditions, the authors demonstrated that even the addition of 2% of epoxidated lignin in the coating could remarkably enhance the corrosion resistance and such effect was ascribed to a lower penetration rate of water due to a high compatibility of lignin with the resin.

An elegant approach aims at chemically treating lignin to control and increase the number of phenolic groups. This would subsequently result in higher epoxy content upon epoxidation, which when combined with BPA based thermoset would produce a material with higher crosslinking density when cured. For example, the demethylation reaction performed by a thiol in alkaline conditions allows to transform the lignin´s guaiacol structure into a catechol (Figure 11A), while the acid treatment of lignin in presence of phenol results in the grafting functionalization in α and γ position (Figure 11B).

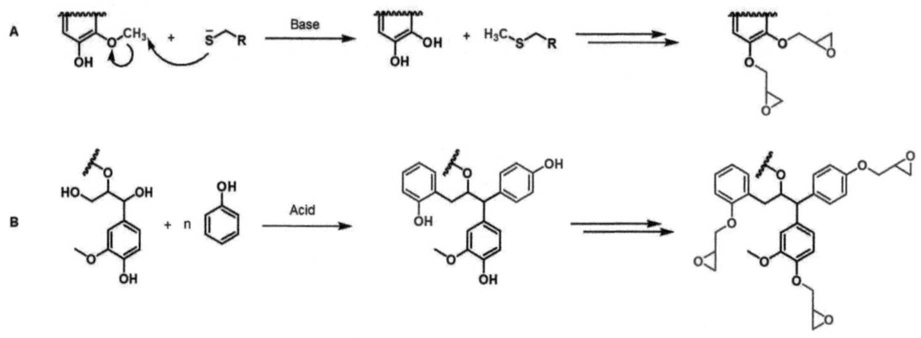

Figure 11. Demethylation (A) and phenolation (B) to obtain augmented epoxy-lignin.

As an example, an acetic acid produced eucalyptus lignin was phenolated, then epoxidated, and mixed with diglicidyl bisphenol A (DGEBA) in a ratio between 0 and 50%, cured with triethylenetetramine (TETA). It was then tested to evaluate adhesive properties (*50*). The results demonstrated a relevant impact of the lignin content on properties such as thermal degradation behavior, adhesive tensile shear strength, and water absorption. As a matter of fact, the introduction of lignin increased the mechanical performances of the adhesive (in particular when used 20% of lignin) and water absorption while resulted detrimental for the thermal resistance.

Epoxidated-demethylated enzymatic lignin from corncob was blended with industrially available epoxy-prepolymer E-51 and cure with polyamine 593 (*51*). The introduction of only 5% of epoxidated lignin substantially increased the mechanical properties of the thermoset. Specifically, tensile strength and elongation at break of the epoxy composites could be enhanced by 27% and 34%, respectively. Furthermore, rheological studies reported better elasticity and cohesiveness.

BPA-Free Lignin-Based Thermosets

All the recent advancements in lignin-based epoxy resins confirmed the great versatility of this system and created a solid background to explore the main aim of this field: the complete substitution of the toxic bisphenol-A and the creation of a reliable biobased alternative. Also in this case, Xu and co-workers gave a substantial contribution focusing their research on the curing reaction mechanism and the structure/property relationship of the obtained thermosets. For example, they epoxidized de-polymerized hydrolysis lignin and cured it with an aromatic diamine such as 4,4′-diaminodiphenyl methane and confronting it with a bisphenol-A reference material (*52*). An accurate thermal analysis of the curing reaction evinced that two main mechanisms compete in such conditions, the first occurs between the amine of the crosslinker and the epoxy group while the second consists of a self-curing promoted by the presence of some unreacted aliphatic hydroxy group of lignin. Interestingly, the first mechanism was predominant. Concerning performances, the lignin-based thermosets demonstrated good thermo-mechanical behavior comparable with bisphenol-A thermosets while the limiting oxygen index (LOI) resulted substantially improved.

The authors also investigated the performances of epoxies based on different lignins (organosolv and Kraft) cured with aliphatic and aromatic diamines. In particular, the dependency of activation energy of the curing reaction on the extent of curing was determined by model-free kinetics based on the DSC results (53). The results demonstrated that the aromatic amines have lower activation energy and therefore a better activity than an aliphatic amine for curing of the lignin-based epoxy thermosets. Furthermore, they showed that the activation energy of Kraft-based epoxy resins is lower than Organosolv based ones, and attributed this to higher availability of free hydroxyl groups in the Kraft-based epoxy resins able to promote the self-curing process. Finally and of significance, the thermal stability of all lignin-based samples was comparable to the thermal properties of petroleum-based epoxy resins.

A further step towards a better understanding of the effect of lignin molecular structures and molecular weight on epoxy thermosets was developed by Gioia and Lawoko (*44*, *54*). Their research aimed to understand how molecular weight and connecting units in lignin architecture may be used to design epoxy thermosets with customizable thermo-mechanical properties. Spruce Lignoboost Kraft lignin was refined by sequential solvent fractionation to obtain lignin fractions with low polydispersity and defined molecular weight. All the fractions were finely characterized in terms of main connecting units and functionalities and were crosslinked with a commercially available flexible diamine after proper epoxidation reaction. The materials were characterized in terms of thermal and mechanical properties. The obtained materials presented values of Young modulus ranging from 6 MPa for the low Mw lignins to 100 MPa and, for the first time, strong relationships between lignin structures and epoxy properties could be demonstrated. In order to develop deeper understand the effect of lignin chemical structure at the same molecular weight, refined Kraft lignin from spruce and eucalyptus with similar Mw were applied to fabricate epoxy resins for further studies. It was shown that softwood spruce Kraft lignin contributed to higher thermomechanical properties (1.7 GPa) compared with eucalyptus-based resins (1.4 GPa). These differences were attributed to molecular structure differences where the presence of unique guaiacyl units with several rigid C5- condensed aromatic units in spruce Kraft, reduced the molecular mobility and increased the aromatic density (Figure 12). In contrast, hardwood Kraft lignin is dominated by sinapyl units with lower content of condensed units but higher content of flexible methoxy units. Overall, this was the beginning of developing molecular-scale understanding in terms of the role of stiff and flexible segments on lignin molecule to material properties.

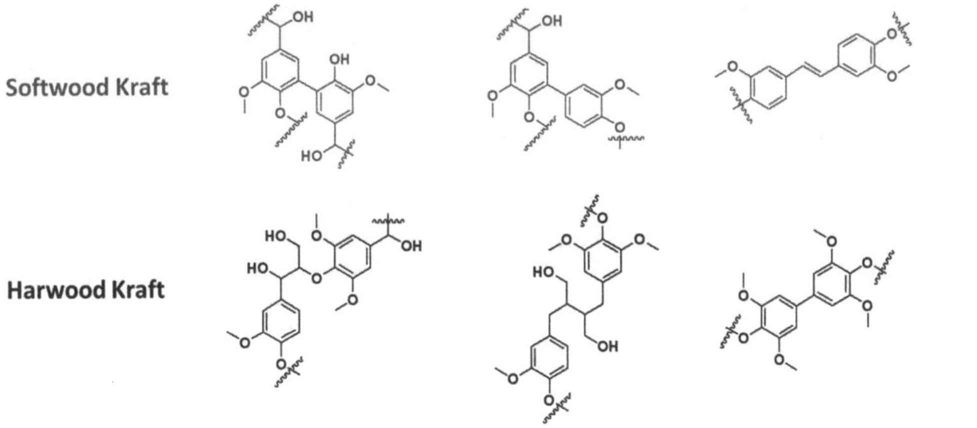

Figure 12. Common connecting units detected in Kraft lignins (54, 55, 56).

Another method to tune the properties of lignin-based epoxies adopts the complete control of lignin functionalities. To achieve this objective, subsequent chemical modifications can be arranged introducing additional phenolic groups deriving from natural compounds. Abu-Omar and coworkers reacted lignins with lignin-based phenols to obtain suitable biobased polyols (*57, 58*). As a matter of fact, the materials obtained from functionalized lignins exhibited improved cross-link density, α-relaxation temperature and storage modulus in a glassy region and increased thermal stability.

Monomers

Controlled Degradation of Lignin for Monomer Synthesis

The controlled depolymerization of lignin to obtain building blocks and monomers represents a complementary approach for the creation of lignin thermosetting materials. The approach adopts to degrade the complex lignin structure in several simple aromatic compounds suitable for further polymerization. However, the result is often a complex mixture of monomers, dimers and oligomers. Furthermore, the depolymerization of technical lignin often demonstrates low yields of monomers and complex mixtures of products, requiring harsh reaction conditions and producing undesired repolymerization reactions. Recent strategies have however yielded results in improving monomer yields and homogeneity. Consequently, the potential impact of such an approach on material science in the last decade is increasingly being investigated (*59, 60, 61*).

The main depolymerization pathways reported in the literature comprehend oxidative (*62*) and reductive depolymerization (*63*), acid (*64*) and base (*65, 66*) catalyzed depolymerization, sub- (*67*) and super- critical (*68*) fluid-based depolymerization, leading to the creation of a library of biobased aromatic building blocks (Figure 13) for the creation of new macromolecular structures.

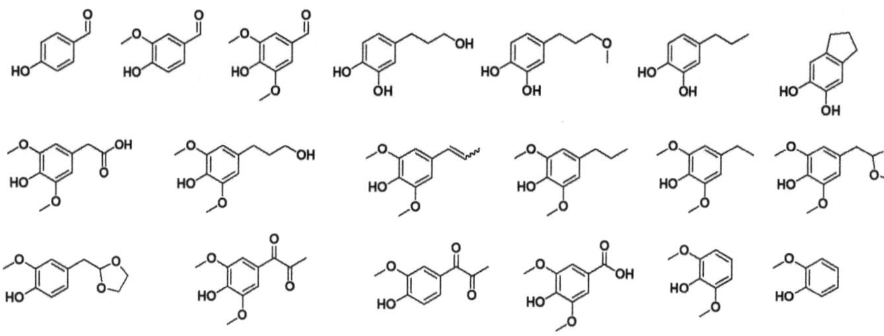

Figure 13. Examples of building blocks from lignin depolymerization (*59*).

It is worth mentioning the prominent role of vanillin as a key building block. Vanillin is currently the only monomer produced from lignin depolymerization on an industrial scale (*69*). It is produced mainly by the oxidative treatment of lignosulfonate liquor from the sulfite pulping of wood (Figure 14) (*70*).

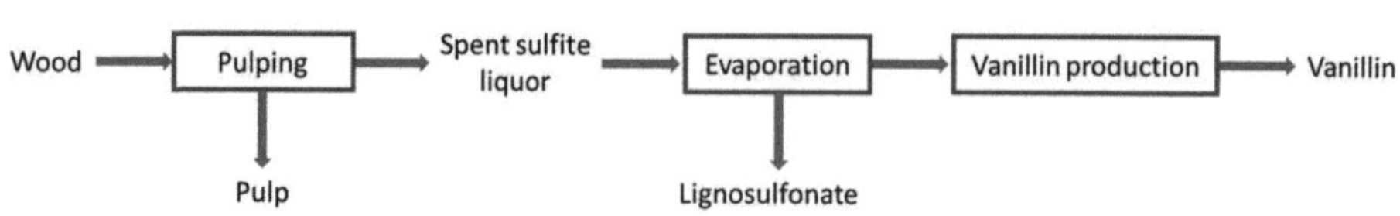

Figure 14. Generic scheme for the production of vanillin from wood (*71*).

The lignosulfonate liquor is treated with oxidants at high temperature and pressure. The proposed mechanism of degradation starts with dehydration of the lignin end group followed by oxidation processes and rearrangements (compounds 3, 4 and 5) to finally obtain the key intermediate 6. This compound degrades by the cleavage of C–C and C–O bonds to finally obtain the vanillate ion 7 (Figure 15) (*71*).

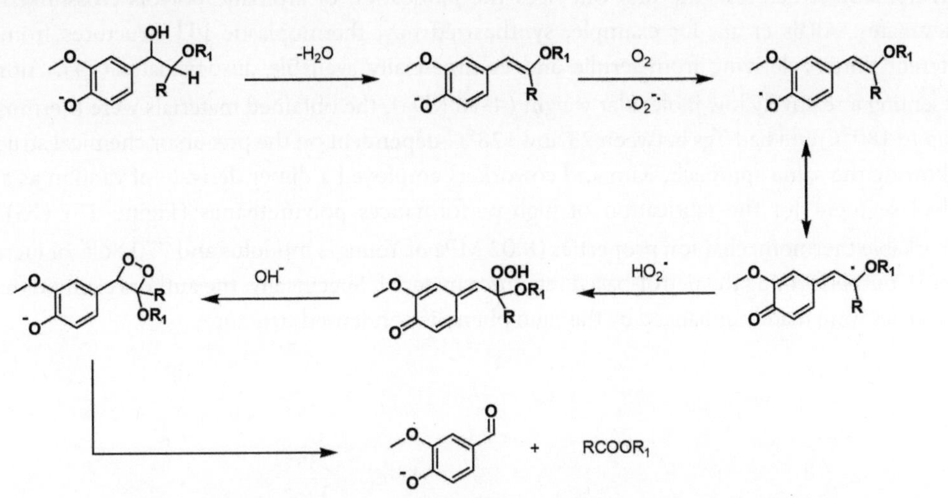

Figure 15. General reaction pathway for vanillin production from wood.

The great interest in vanillin is justified by the versatility of its aldehydic moiety which can be further modified to support the creation of a platform of vanillin based monomeric structures (72, 73). As a consequence, many of the most recent monomers reported in the literature are based on vanillin or its main derivatives reported in Figure 16.

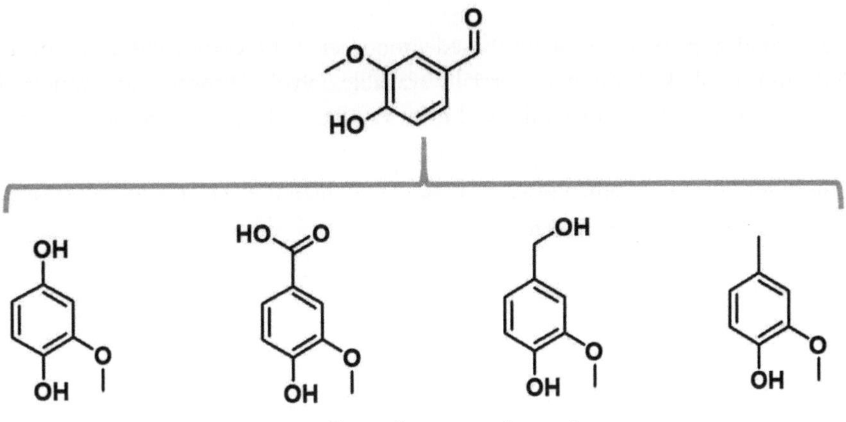

Figure 16. Vanillin and its most relevant derivatives.

Thermosetting Materials from Lignin Monomers

The development of the aforementioned processes directed to lignin degradation produced a variety of biobased building blocks creating the unprecedented opportunity to obtain novel macromolecular structures in material science. In this section, the latest development for the synthesis of polyurethanes, phenol-formaldehyde and epoxy resins from lignin-based monomers will be reported and discussed.

Polyurethanes

Two main approaches are adopted for the introduction of lignin-based aromatics into polyurethane structures. The first one sees the fabrication of aromatic polyols crosslinked with isocyanates. Allais et al., for example, synthesized new thermoplastic PU structures from four different dimers deriving from ferulic and commercially available diisocyanates (74). Although presenting a relatively low molecular weight (4-10 KDa), the obtained materials were thermostable at up to 180°C and had Tgs between 28 and 128°C dependent on the precursor chemical structure. Following the same approach, Kim and coworkers employed a dimer derivate of vanillin as a hard polyol segment for the fabrication of high-performances polyurethanes (Figure 17) (75) with remarkable thermomechanical properties (8.02 MPa of Young's modulus and 770.86% of increased strain), outperforming the petrol-based reference material. Specifically, the authors claimed that the properties were mainly enhanced by the rigid phenolic condensed structure.

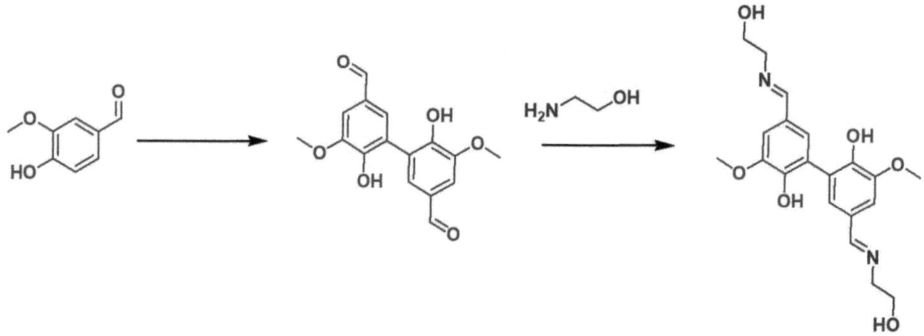

Figure 17. Synthesis of a vanillin dimer (75).

In the second approach, the lignin-based monomer is functionalized with the isocyanate functionality and crosslinked with commercially available polyols. Wadgaonkar synthesized a series of a bio-based aromatic diisocyanate, derived from vanillic acid and syringic acid, via the Curtius rearrangement (Figure 18) and subsequently polymerized with biobased aliphatic diols (76). The obtained thermoplastic polyurethanes were of reasonably high molecular weight and were stable at up to 308°C with Tgs ranging from 49 to 74°C.

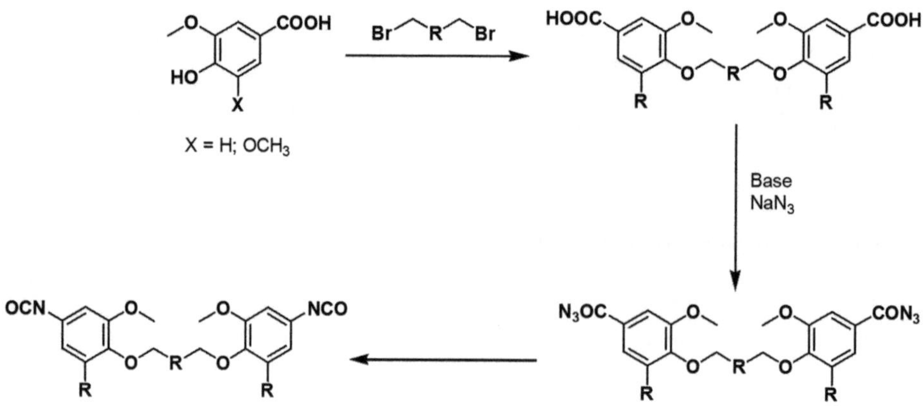

Figure 18. Synthesis of a biobased diisocyanate from vanillic acid.

Remarkably, D'Arrigo and co-workers used the very same system to crosslink different non-modified technical lignins to obtain coatings (77). With such an elegant approach the authors managed to create lignin-based materials crosslinked by lignin-derived monomers. In addition, the resulting materials demonstrated a distinct correlation between properties and the chemical–physical features of the parent lignins used suggesting the possibility to tailor the performances by choosing the lignin feedstock.

Finally, several authors are currently demonstrating the possibility to create isocyanate-free polyurethanes (NIPUs) exploiting lignin-based monomers. For example, Allais and Caillol synthesized, starting from ferulic acid, fully biobased bis- and tri-phenols (*78*) while Xie and Zhao (*79*) studied diphenols deriving from lignin-based cresol as raw material. In the second case, the diphenols were first epoxidated with epichlorohydrin then further reacted with CO_2 to obtain a bis(cyclic carbonate) and finally crosslinked by diamines (Figure 19).

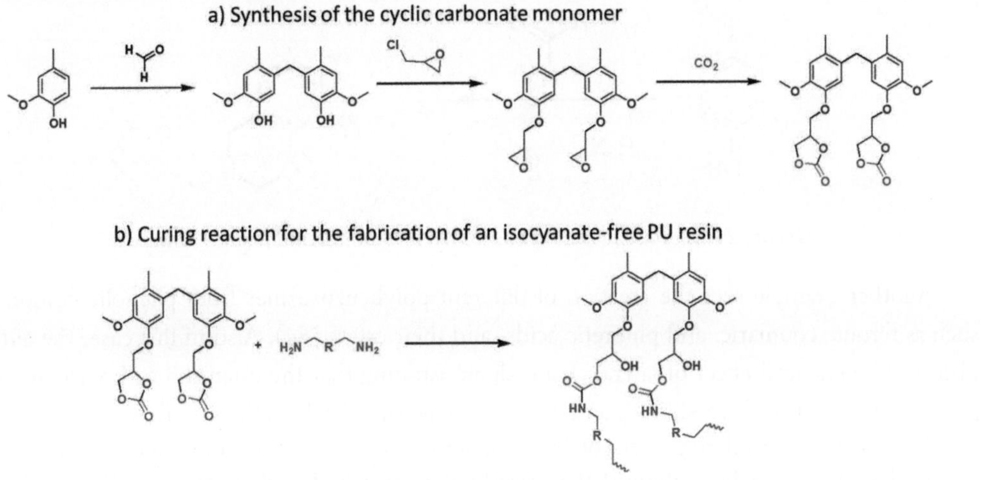

Figure 19. Synthesis of cyclic carbonate biobased monomers for isocyanate-free thermosets.

Phenol-formaldehyde

Lignin based monomeric phenols are the ideal replacement for phenol in phenolic resins (PF) from the viewpoint of chemical and physical characteristics. Interestingly, Biziks et al. studied the effect of different lignin-based mono phenols as a substitute for phenol in PF resins for wood treatment applications (*80*). Specifically, nine of the most common compounds derived from lignin liquefaction, namely ortho-, meta-, para cresol, guaiacol, catechol, 4-methylcatechol, resorcinol, syringol, 4-ethylphenol and resol, were reacted with formaldehyde to obtain the corresponding materials. The result undoubtedly demonstrated that compounds such as guaiacol or ortho-, or meta-, or para-cresol and/or 4-ethylphenol could be substituted effectively to phenol for wood treatment while resins made of catechol or 4-methylguaiacol and syringol demonstrated poor properties.

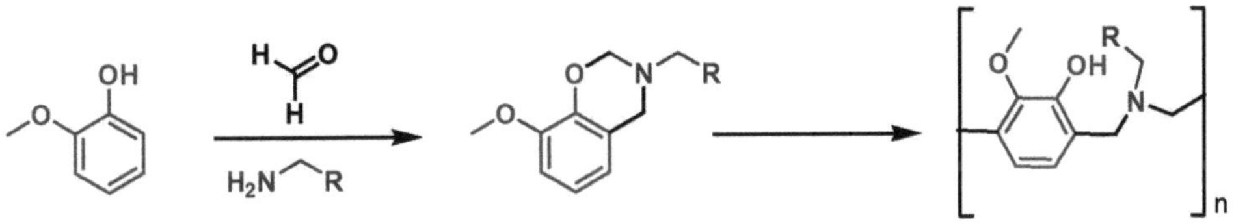

Figure 20. Synthesis of polybenzoxazines.

Caillol and co-workers, contributed to this line of research by developing sustainable thermosets based on green phenols such as resorcinol, guaiacol and tyrosol (*81*). Interestingly, the authors applied terephthalaldehyde as non-toxic replacement for formaldehyde.

A relatively novel approach in PF resin synthesis adopts the preparation of polybenzoxazines starting from phenolic derivates, formaldehyde and amines as reported in Figure 20 (*82*).

Varma et al., designed a fully biobased polybenzoxazine based on vanillin and biobased furfurylamine with enhanced thermal properties (Figure 21) (*83*). The authors demonstrated that the rigid aromatic backbone could confer a Tg of 270°C and elevated thermostability (T5% = 351°C). In addition, they concluded that the aldehydic group of vanillin could positively improve the curing process.

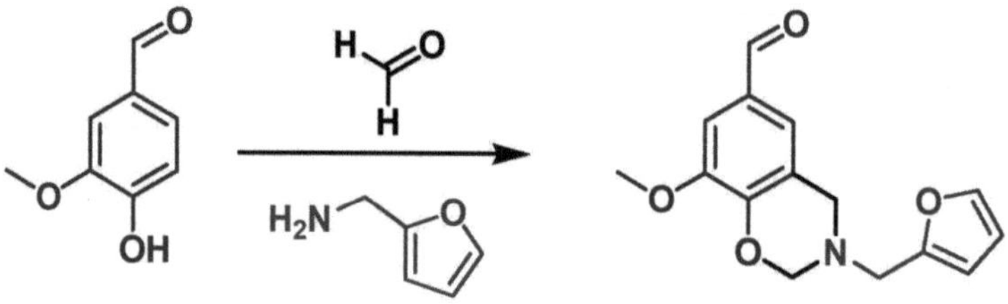

Figure 21. Biobased benzoxazine based on vanillin and furfurylamine.

Another example sees the creation of different polybenzoxazines from phenolic compounds such as ferulic, coumaric, and phloretic acids, and their esters (*84*). Also in this case, the authors observed a beneficial effect of lateral chains demonstrating that the ester and carboxylic moieties could confer additional crosslinking and therefore increasing the Tg of the materials.

Mhaske et al. started from guaiacol, but in their case, the benzoxazine monomer was synthesized using ethanolamine, which allowed the introduction of a hydroxyl group (*85*). This additional functionality was further crosslinked by an isocyanate hardener to prepare a poly(benzoxazine-urethane). The authors demonstrated that such material could be effectively used as a coating for corrosion protection of mild steel and that the concentration of urethane moieties in the coatings could be related to better the mechanical, chemical and thermal properties.

Finally, Lomonaco and coworkers designed an eco-friendly microwave-assisted synthesis to produce benzoxazine monomers from guaiacol and different amines. Remarkably, the reaction time was reduced from hours to 6 minutes in a solvent-free protocol (*86*).

Epoxy Resins

The main driving force for the creation of new monomers for epoxy resins is related to the necessity to find reliable alternatives to the toxic bisphenol A. An additional advantage in the use of monomeric compounds is related to better miscibility in the reaction mixture resulting in a more homogeneous material. In this field, several research groups developed a variety of new structures for epoxies by exploiting the rich chemistry of lignin-based phenols. Among all, vanillin derivates, guaiacol and eugenol represent the most exploited starting materials due to their versatile reactivity. Stanzione et al., for example, synthesized a bisphenolic compound through a condensation reaction between vanillyl alcohol and guaiacol (*87*). The resulting bisphenol was epoxidated and finally crosslinked with a commercially available diamine. Additionally, to better understand the effect of the phenolic structure on the viscoelastic properties of the final materials, the diglycidyl ethers of vanillyl alcohol, 4-hydroxymethylphenol and hydroquinone were studied (Figure 22). The results demonstrated that higher methoxy content was associated with lower values of the Tg, but at the same time to higher values of storage modulus at room temperature. Furthermore, the properties of the material can be piloted by simply mixing different ratios of monomers.

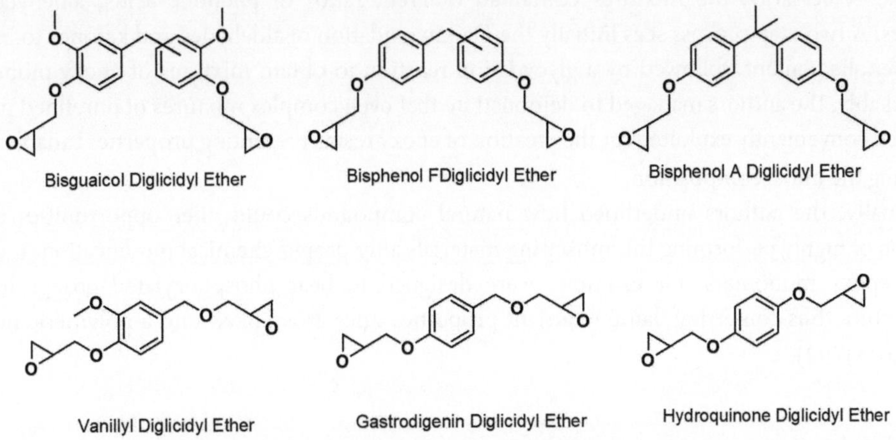

Figure 22. Phenolic and bisphenolic diglicidyl ethers exploited for a better understanding of the effect of structure on the viscoelastic properties of the final resins.

Another remarkable bisphenolic structure was reported by Zhu and coworkers (*88*). Interestingly, they created a fully biobased bisphenol bearing a labile acetal structure in the middle. The rigidity of this system, once crosslinked into a polymeric matrix, conferred exceptional thermomechanical performances with a Tg of 184°C, a modulus of 4.7 GPa and 0.30 GPa of hardness, able to outperform even the bisphenol A based reference. Furthermore, the authors proved that such material could be rapidly degraded by the selective cleavage of the acetal group.

As a matter of fact, Caillol's group highly contributed to the epoxy thermoset field by creating a solid line of research-based mainly on vanillin, ferulic acid (*89*) and their derivates. For example, methoxyhydroquinone, vanillic acid and vanillyl alcohol were glycidylated to obtain the corresponding epoxy monomers and then crosslinked with isophorone diamine (*90*). The materials demonstrated thermal properties comparable with bisphenol A based resins, ranging from 97°C and 152°C. The authors effectively demonstrated that such biobased structures are potentially suitable for commercial applications, however, they highlighted two main issues that would hamper their industrial applicability. The first is related to the fact that these pure monomers are solids requiring relatively high temperatures to react in a molten state. To overcome such issues the authors adapted a common industrial approach to obtain liquid oligomers suitable for subsequent crosslinking (*91*). The oligomers were obtained by reacting, in different ratios, methoxyhydroquinone with its epoxidized version (Figure 23). The researchers demonstrated that the thermomechanical properties were in line with commercially available resins, with Tg and Tα ranging from 80 and 110 °C. In addition, such properties could be tuned by changing the length of the starting oligomer.

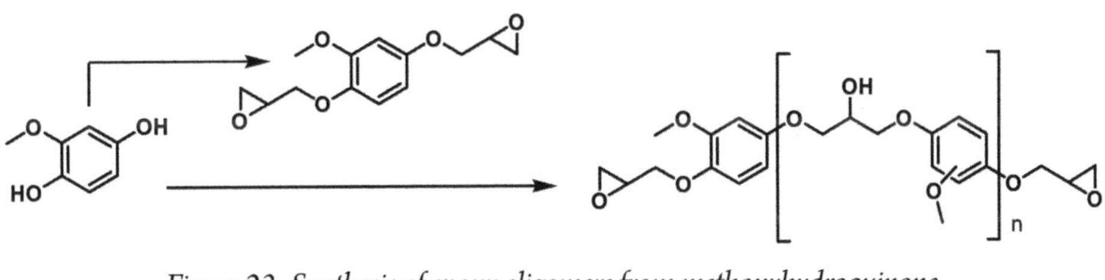

Figure 23. Synthesis of epoxy oligomers from methoxyhydroquinone.

The second main issue concerns the effective economic cost to obtain lignin-based pure structures from lignin depolymerization processes (92). The authors used as starting compounds mixtures of phenolics resembling those deriving from the product stream of the lignin to-vanillin process. Necessarily, the mixtures contained different ratios of phenolic acids, aldehydes and ketones. A two-step process sees initially the Daikin oxidation of aldehydes and ketones to increase the phenolic content, followed by a glycydation reaction to obtain mixtures of epoxy monomers. Remarkably, the authors managed to demonstrate that even complex mixtures of unrefined phenols could be conveniently exploited for the creation of epoxy resins presenting properties tunable by just adjusting the initial composition.

Finally, the authors underlined how natural compounds could offer opportunities for the creation of highly performing thermosetting materials after proper chemical modification. Eugenol-based epoxy monomers, for example, were designed to bear phosphorylated groups in their architecture thus conferring flame retardant properties once crosslinked into a polymeric network (Figure 24) (93).

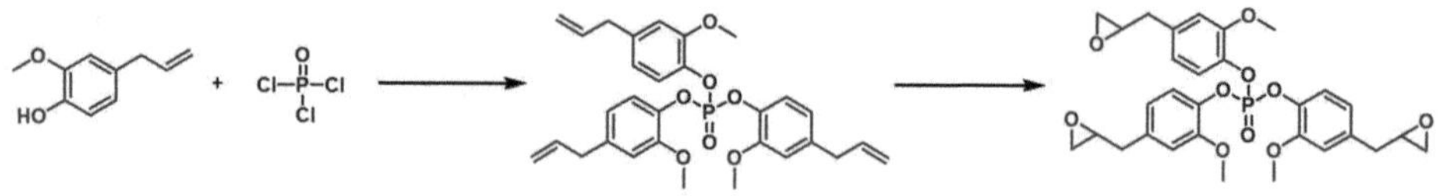

Figure 24. Synthesis of a phosphorylated eugenol-based epoxy monomer.

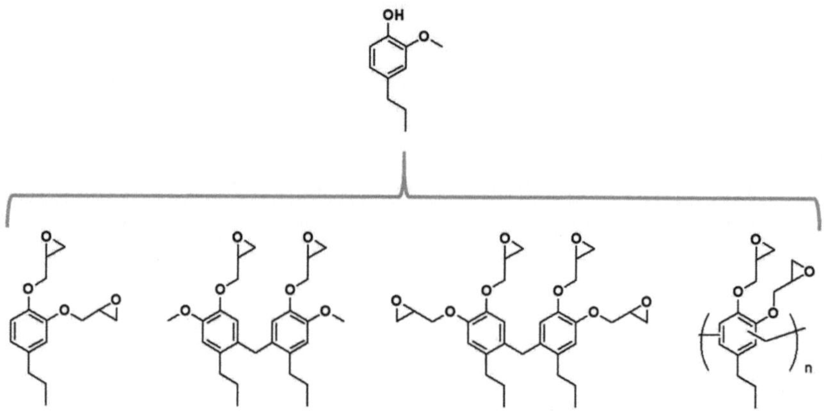

(1) (2) (3) (4)

Figure 25. Synthesis of Dihydroeugenol-based epoxidated building blocks.

Another remarkable line of research was developed by Abu-Omar and coworkers (94). They directed many efforts to widen the understanding of the effect of the chemical structure of the monomeric compounds on the final properties of the epoxy resins. For example, they synthesized a family of triphenols deriving from the condensation of phenols (phenol, guaiacol and 2,6-dimethoxyphenol) with different combinations of phenolic aldehydes (hydroxybenzaldehyde, vanillin, and syringaldehyde) (95). As a result, the authors obtained a library of triphenols presenting 0 to 6 methoxy groups. These compounds were subsequently epoxidized with epichlorohydrin and finally cured with diethylenetriamine. With this set of experiments, the authors managed to demonstrate that the higher content of the methoxy group was associated with a decrease of the Tg (from 132 to 118°C), glassy modulus (from 2.7 to 2.2 GPa) and onset degradation temperature (from 257 °C to 184°C). In another work, the same authors further widened the understanding of lignin-based materials by finely controlling their chemical structure (*96*). Dihydroeugenol, as a starting material, was modified through different synthetic pathways involving a demethylation process and phenol/formaldehyde reaction to obtain phenolic structures characterized by a controlled molecular weight, a defined number of hydroxy content other than their precise space disposition (Figure 25).

Remarkably, their epoxidation and crosslinking with commercially available diethylenetriamine produced a family of new resins whose structural features were as much as possible defined. The results revealed that mechanical and thermal performances of DHE-based networks improve with an increased cross-linking density as well as with the increasing molecular weight of the starting monomer. Interestingly, also the spatial distribution of the crosslinking sites has a huge impact on the performances as evinced confronting compound (2) and compound (3) in Table 2. Therefore, the highest performances were achieved by starting from the oligomerized system since both cross-linking density and molecular weight resulted significantly increased. Specifically, the $T\alpha$ rose from 40 to 139 °C, and the statistic heat-resistant index temperature (Ts) went from 125 to 153 °C. As a matter of fact, with their work, the authors demonstrated how finely it is possible to tune the performances with a high level of control on the chemical structures at play.

Table 2. Thermomechanical Properties of Dihydroeugenol-Based Materials

Epoxy network from	T_α *(°C)*	E_{30}' *(MPa)*	E_{30}' *at* $T_\alpha + 30°C$ *(MPa)*	*P (mpl/dm³)*
1	40	159	1.13	0.39
2	70	551	4.31	1.39
3	84	1441	10.6	3.28
4	139	1703	35.9	9.77

[a] T_α is the α-relaxation temperature; E_{30}' is the storage modulus at 30°C.

Challenges and Future Outlook

While lignin holds a promise as a bio-based source of aromatics in polymer applications, addressing lignin heterogeneity is an important milestone. Unlike conventional polymer precursors where homogeneity in structure, functionality and molecular weight is a given, inherent heterogeneity exists in lignin. The polydispersity in technical lignins is one issue that imposes that additional steps are required to produce narrow molecular weight dispersity. This is achieved through e.g. solvent fractionations (*97, 98*), and ultrafiltration techniques (*99, 100*). Although such refinings have been achieved to different levels of success, they would obviously confer additional cost. Nevertheless, for academic purposes, they have been very useful. The structural complexity of technical lignin is yet another challenge. As an example, the complete structure of Kraft lignins, which are used in several studies is still not known, however, recent efforts continue to shed more light on this thanks to the combined efforts of refining and advances in NMR characterization techniques (*101, 102, 103, 55, 56*). Lignin First Biorefinery approaches to obtain more homogeneous lignin oligomers are presently being studied by several groups. These new lignin sources are derived from e.g. organosolv processes such as co-solvent enhanced lignin fractionation (CELF) (*104*), which applies a THF solvent system, shown to molecularly dissolve lignin and selectively depolymerize lignin to produce narrow dispersity lignin oligomers with well characterized structures. Such approaches will be complementary to current technology if shown to be technically feasible.

A general challenge is that of a complete biorefinery concept taking into consideration the lignin, hemicellulose and fiber fractions. This is an essential milestone for the realization of a bio-economy based on sustainable lignocelluloses. Lawoko and coworkers (*105*), proposed a two-step strategy using green sustainable solvents, to obtain a hemicellulose fraction, pure native-like lignin fraction and a fiber residue rich in crystalline celluloses. The solvent systems used sequentially were sub-critical hot water to first extract the hemicelluloses, followed by sub-critical aqueous ethanol extraction with catalytic amounts of acid for lignin extraction. The latter step was performed using a static cycles approach that provided physical protection of lignin from detrimental lignin condensation reactions. Again here, narrow dispersity and well-characterized lignins with attractive features for further valorization were produced in high purity and good yield.

Lignin oligomers are very different from conventional polymer precursors, not only due to the heterogeneity in structure and functionality, but also the distribution of functional groups, which in turn would affect the crosslinking distribution in a thermoset. How exactly this affects the thermoset properties and morphology will require future studies. In this context, Johansson and coworkers (*106*) studied the morphology of Thiol-ene based thermosets using small and wide-angle X-ray scattering techniques and were able to identify Pi-stacking effects in the thermoset and related aggregate sizes, which were in turn related to specific structural components of molecular lignin. Such studies will be essential for the development of a deeper fundamental understanding of the properties of lignin-based thermosets. The diversity in types of lignin inter-unit linkages, their distribution along the polymer chain presents as well as substitutions on the aromatic ring all present an added complexity. Gioia and Lawoko (*54*), based on detailed lignin characterization, were able to show correlations between the degree of molecular stiffness with the material properties and postulated the role of stiff bonds in lignin, specifically direct couplings between aromatic rings (such as 5-5′ biphenyl couplings), in the enhancement of the thermomechanical properties of the thermosets. In contrast, substitutions such as methoxyl groups, more prominent in hardwood lignins than softwoods, where found to confer flexibility to the material. Such fundamental understanding should be further addressed in future studies, with benefits of enabling possibilities to molecularly tailor material properties by making the right choice of starting substrate.

References

1. Grand View Research. Polyurethane Market Analysis Report By Product (Rigid Foam, Flexible Foam, Coatings), By Application (Furniture & Interiors, Construction, Automotive), And Segment Forecasts, 2019 - 2025. *Rep. Mark.* **2019** (July), 1–16.
2. Wang, Y.; Cai, C. M.; Ragauskas, A. J. Recent Advances in Lignin-Based Polyurethanes. *TAPPI J.* **2017**, *16* (4), 203–207.
3. Karunarathna, M. S.; Smith, R. C. Valorization of Lignin as a Sustainable Component of Structural Materials and Composites: Advances from 2011 to 2019. *Sustain.* **2020**, *12* (734) https://doi.org/10.3390/su12020734.
4. Mahmood, N.; Yuan, Z.; Schmidt, J.; Charles, C. Depolymerization of Lignins and Their Applications for the Preparation of Polyols and Rigid Polyurethane Foams : A Review. *Renew. Sustain. Energy Rev.* **2016**, *60*, 317–329. https://doi.org/10.1016/j.rser.2016.01.037.
5. Nemati, A.; Alwyn, D.; Evans, C.; Laycock, B.; James, D. A Simple Methodology for Improving the Performance and Sustainability of Rigid Polyurethane Foam by Incorporating Industrial Lignin. *Ind. Crop. Prod.* **2018**, *117* (February), 149–158. https://doi.org/10.1016/j.indcrop.2018.03.006.

6. Pan, X.; Saddler, J. N. Effect of Replacing Polyol by Organosolv and Kraft Lignin on the Property and Structure of Rigid Polyurethane Foam. *Biotechnol. Biofuels* **2013**, *6* (1) https://doi.org/10.1186/1754-6834-6-12.
7. Avelino, F.; Almeida, S. L.; Duarte, E. B. Thermal and Mechanical Properties of Coconut Shell Lignin-Based Polyurethanes Synthesized by Solvent-Free Polymerization. *J. Mater. Sci.* **2018**, *53* (2), 1470–1486. https://doi.org/10.1007/s10853-017-1562-z.
8. Chen, K.; Xu, J.; Li, J. Bio-Based Polyurethane Foam Preparation Employing Lignin from Corn Stalk Enzymatic Hydrolysis Residues. *RSC Adv.* **2018**, *8*, 15754–15761. https://doi.org/10.1039/C8RA01089G.
9. Li, H.; Sun, J.; Wang, C.; Liu, S.; Yuan, D.; Zhou, X.; Tan, J.; Stubbs, L.; He, C. High Modulus, Strength, and Toughness Polyurethane Elastomer Based on Unmodified Lignin. *ACS Sustain. Chem. Eng.* **2017**, *5* (9), 7942–7949. https://doi.org/10.1021/acssuschemeng.7b01481.
10. Sadeghifar, H.; Cui, C.; Argyropoulos, D. S. Toward Thermoplastic Lignin Polymers. Part 1. Selective Masking of Phenolic Hydroxyl Groups in Kraft Lignins via Methylation and Oxypropylation Chemistries. *Ind. Eng. Chem. Res.* **2012**, *51* (51), 16713–16720. https://doi.org/10.1021/ie301848j.
11. Ahvazi, B.; Wojciechowicz, O.; Ton-That, T. M.; Hawari, J. Preparation of Lignopolyols from Wheat Straw Soda Lignin. *J. Agric. Food Chem.* **2011**, *59* (19), 10505–10516. https://doi.org/10.1021/jf202452m.
12. Cateto, C. A.; Barreiro, M. F.; Rodrigues, A. E.; Belgacem, M. N. Optimization Study of Lignin Oxypropylation in View of the Preparation of Polyurethane Rigid Foams. *Ind. Eng. Chem. Res.* **2009**, *48* (5), 2583–2589. https://doi.org/10.1021/ie801251r.
13. Li, B.; Zhou, M.; Huo, W.; Cai, D.; Qin, P.; Cao, H.; Tan, T. Fractionation and Oxypropylation of Corn-Stover Lignin for the Production of Biobased Rigid Polyurethane Foam. *Ind. Crops Prod.* **2020**, *143* (August 2019), 111887. https://doi.org/10.1016/j.indcrop.2019.111887.
14. Gómez-fernández, S.; Ugarte, L.; Calvo-correas, T.; Pe, C.; Corcuera, M. A.; Eceiza, A. *Properties of Flexible Polyurethane Foams Containing Isocyanate Functionalized Kraft Lignin.* **2017**, *100*, 51–64. https://doi.org/10.1016/j.indcrop.2017.02.005.
15. Zhang, Y.; Liao, J.; Fang, X.; Bai, F.; Qiao, K.; Wang, L. Renewable High-Performance Polyurethane Bioplastics Derived from Lignin-Poly(ε-Caprolactone). *ACS Sustain. Chem. Eng.* **2017**, *5* (5), 4276–4284. https://doi.org/10.1021/acssuschemeng.7b00288.
16. Xing, W.; Yuan, H.; Yang, H. Functionalized Lignin for Halogen-Free Flame Retardant Rigid Polyurethane Foam : Preparation , Thermal Stability , Fire Performance and Mechanical Properties. *J Polym Res* **2013**, *20*, 234. https://doi.org/10.1007/s10965-013-0234-1.
17. Maisonneuve, L.; Lamarzelle, O.; Rix, E.; Grau, E.; Cramail, H. Isocyanate-Free Routes to Polyurethanes and Poly(Hydroxy Urethane)S. *Chem. Rev.* **2015**, *115* (22), 12407–12439. https://doi.org/10.1021/acs.chemrev.5b00355.
18. Zoia, L.; Mauri, M. Utilization of Cyclocarbonated Lignin as a Bio-Based Cross-Linker for the Preparation of Poly(Hydroxy Urethane)S. *RSC Adv.* **2017**, 7, 25054–25065. https://doi.org/10.1039/C7RA03416D.
19. Lucintel. Phenolic Resin Market Report: Trends, Forecast and Competitive Analysis. *Lucintel* **2020** (June), 1–16.

20. Yang, S.; Zhang, Y.; Yuan, T. Q.; Sun, R. C. Lignin-Phenol-Formaldehyde Resin Adhesives Prepared with Biorefinery Technical Lignins. *J. Appl. Polym. Sci.* **2015**, *132* (36), 1–8. https://doi.org/10.1002/app.42493.
21. Pang, B.; Yang, S.; Fang, W.; Yuan, T.; Argyropoulos, D. S. Structure-Property Relationships for Technical Lignins for the Production of Lignin-Phenol-Formaldehyde Resins. *Ind. Crop. Prod.* **2017**, *108* (March), 316–326. https://doi.org/10.1016/j.indcrop.2017.07.009.
22. Lourençon, T. V.; Alakurtti, S.; Virtanen, T.; Jaä¨skelaïnen, A. S.; Liitiä, T.; Hughes, M.; Magalhães, W. L. E.; Muniz, G. I. B.; Tamminen, T. Phenol-Formaldehyde Resins with Suitable Bonding Strength Synthesized from "Less-Reactive" Hardwood Lignin Fractions. *Holzforschung* **2020**, *74* (2), 175–183. https://doi.org/10.1515/hf-2018-0203.
23. Chen, Y.; Gong, X.; Yang, G.; Li, Q.; Zhou, N. Preparation and Characterization of a Nanolignin Phenol Formaldehyde Resin by Replacing Phenol Partially with Lignin Nanoparticles. *RSC Adv.* **2019**, *9* (50), 29255–29262. https://doi.org/10.1039/c9ra04827h.
24. Hu, L.; Pan, H.; Zhou, Y.; Zhang, M. Methods to Improve Lignin's Reactivity as a Phenol Substitute and as Replacement for Other Phenolic Compounds: A Brief Review. *BioResources* **2011**, *6* (3), 3515–3525. https://doi.org/10.15376/biores.6.3.3515-3525.
25. Yan, L.; Cui, Y.; Gou, G.; Wang, Q.; Jiang, M.; Zhang, S.; Hui, D.; Gou, J.; Zhou, Z. Liquefaction of Lignin in Hot-Compressed Water to Phenolic Feedstock for the Synthesis of Phenol-Formaldehyde Resins. *Compos. Part B* **2017**, *112*, 8–14. https://doi.org/10.1016/j.compositesb.2016.10.094.
26. Paysepar, H.; Hu, Y.; Feng, S.; Yuan, Z.; Shui, H.; Xu, C. (Charles). Bio-Phenol Formaldehyde (BPF) Resoles Prepared Using Phenolic Extracts from the Biocrude Oils Derived from Hydrothermal Liquefaction of Hydrolysis Lignin. *React. Funct. Polym.* **2020**, *146* (September 2019), 104442. https://doi.org/10.1016/j.reactfunctpolym.2019.104442.
27. Li, J.; Wang, W.; Zhang, S.; Gao, Q.; Zhang, W.; Li, J. Preparation and Characterization of Lignin Demethylated at Atmospheric Pressure and Its Application in Fast Curing Biobased Phenolic Resins. *RSC Adv.* **2016**, *6* (71), 67435–67443. https://doi.org/10.1039/c6ra11966b.
28. Hu, L.; Pan, H.; Zhou, Y.; Hse, C. Y.; Liu, C.; Zhang, B.; Xu, B. Chemical Groups and Structural Characterization of Lignin via Thiol-Mediated Demethylation. *J. Wood Chem. Technol.* **2014**, *34* (2), 122–134. https://doi.org/10.1080/02773813.2013.844165.
29. Meister, J. J. Modification of Lignin. *J. Macromol. Sci. - Polym. Rev.* **2002**, *42* (2), 235–289. https://doi.org/10.1081/MC-120004764.
30. Bernini, R.; Barontini, M.; Mosesso, P.; Pepe, G.; Willför, S. M.; Sjöholm, R. E.; Eklund, P. C.; Saladino, R. A Selective De-O-Methylation of Guaiacyl Lignans to Corresponding Catechol Derivatives by 2-Iodoxybenzoic Acid (IBX). the Role of the Catechol Moiety on the Toxicity of Lignans. *Org. Biomol. Chem.* **2009**, *7* (11), 2367–2377. https://doi.org/10.1039/b822661j.
31. Wang, H.; Eberhardt, T. L.; Wang, C.; Gao, S.; Pan, H. Demethylation of Alkali Lignin with Halogen Acids and Its Application to Phenolic Resins. *Polymers (Basel).* **2019**, *11* (11), 1–16. https://doi.org/10.3390/polym11111771.
32. Venkatesagowda, B.; Dekker, R. F. H. Enzymatic Demethylation of Kraft Lignin for Lignin-Based Phenol-Formaldehyde Resin Applications. *Biomass Convers. Biorefinery* **2020**, *10* (2), 203–225. https://doi.org/10.1007/s13399-019-00407-3.

33. Jiang, X.; Liu, J.; Du, X.; Hu, Z.; Chang, H.; Jameel, H. Phenolation to Improve Lignin Reactivity toward Thermosets Application. *ACS Sustain. Chem. Eng. 2018* **2018**, *6*, 5504–5512. https://doi.org/10.1021/acssuschemeng.8b00369.
34. Lignin, P.; Lee, W.; Chang, K.; Tseng, I. Properties of Phenol-Formaldehyde Resins Prepared from Phenol-Liquefied Lignin. *J. Appl. Polym. Sci.* **2012**, *124*, 4782–4788. https://doi.org/10.1002/app.35539.
35. Acmite Market Intelligence. *Global Epoxy Resin Market* **2017**.
36. Caillol, S.; David, G.; Boutevin, B.; Pascault, J.; Lyon, U. De. Biobased Thermosetting Epoxy: Present and Future. *Chem. Rev.* **2014**, *114*, 1082–1115. https://doi.org/10.1021/cr3001274.
37. Rochester, J. R. Bisphenol A and Human Health : A Review of the Literature. *Reprod. Toxicol.* **2013**, *42*, 132–155. https://doi.org/10.1016/j.reprotox.2013.08.008.
38. Kong, X.; Xu, Z.; Guan, L.; Di, M. Study on Polyblending Epoxy Resin Adhesive with Lignin I-Curing Temperature. *Int. J. Adhes. Adhes.* **2014**, *48*, 75–79. https://doi.org/10.1016/j.ijadhadh.2013.09.003.
39. Yin, Q.; Yang, W.; Sun, C.; Di, M. Preparation and Properties of Lignin-Epoxy Resin Composite. *BioResources* **2012**, 7 (4), 5737–5748. https://doi.org/10.15376/biores.7.4.5737-5748.
40. Rowell, R. M.; Chen, G. C. Epichlorohydrin Coupling Reactions with Wood - Part 1. Reaction with Biologically Active Alcohols. *Wood Sci. Technol.* **1994**, *28* (5), 371–376. https://doi.org/10.1007/BF00195285.
41. Ferdosian, F.; Yuan, Z.; Anderson, M.; Xu, C. C. Chemically Modified Lignin Through Epoxidation and Its Thermal Properties. *J. Sci. Technol. For. Prod. Process.* **2016**, *2* (4), 11–15.
42. Ferdosian, F.; Yuan, Z.; Anderson, M.; Xu, C. Synthesis of Lignin-Based Epoxy Resins: Optimization of Reaction Parameters Using Response Surface Methodology. *RSC Adv.* **2014**, *4* (60), 31745–31753. https://doi.org/10.1039/c4ra03978e.
43. Malutan, T.; Nicu, R.; Popa, V. I. Lignin Modification by Epoxidation. *BioResources* **2008**, *3* (4), 1371–1376. https://doi.org/10.15376/biores.3.4.1371-13767.
44. Gioia, C.; Lo Re, G.; Lawoko, M.; Berglund, L. Tunable Thermosetting Epoxies Based on Fractionated and Well-Characterized Lignins. *J. Am. Chem. Soc.* **2018**, *140* (11) https://doi.org/10.1021/jacs.7b13620.
45. Jablonskis, A.; Arshanitsa, A.; Arnautov, A.; Telysheva, G.; Evtuguin, D. Evaluation of Ligno Boost Softwood Kraft Lignin Epoxidation as an Approach for Its Application in Cured Epoxy Resins. *Ind. Crops Prod.* **2018**, *112* (April 2017), 225–235. https://doi.org/10.1016/j.indcrop.2017.12.003.
46. Ferdosian, F.; Zhang, Y.; Yuan, Z.; Anderson, M.; Charles, C. Curing Kinetics and Mechanical Properties of Bio-Based Epoxy Composites Comprising Lignin-Based Epoxy Resins. *Eur. Polym. J.* **2016**, *82*, 153–165. https://doi.org/10.1016/j.eurpolymj.2016.07.014.
47. Ferdosian, F.; Yuan, Z.; Anderson, M.; Xu, C. C. Thermal Performance and Thermal Decomposition Kinetics of Lignin-Based Epoxy Resins. *J. Anal. Appl. Pyrolysis* **2016**, *119*, 124–132. https://doi.org/10.1016/j.jaap.2016.03.009.
48. Over, L. C.; Grau, E.; Grelier, S.; Meier, M. A. R.; Cramail, H. Synthesis and Characterization of Epoxy Thermosetting Polymers from Glycidylated Organosolv Lignin and Bisphenol A. *Macromol. Chem. Phys.* **2017**, *218* (4), 1–11. https://doi.org/10.1002/macp.201600411.

49. Ding, J.; Gu, L.; Dong, W.; Yu, H. Epoxidation Modification of Renewable Lignin to Improve the Corrosion Performance of Epoxy Coating. *Int. J. Electrochem. Sci.* **2016**, *11*, 6256–6265. https://doi.org/10.20964/2016.07.56.
50. Feng, P.; Chen, F. Preparation and Characterization of Acetic Acid Lignin-Based Epoxy Blends. *BioResources* **2012**, *7* (3), 2860–2870.
51. Wang, F.; Kuai, J.; Pan, H.; Wang, N.; Zhu, X. Study on the Demethylation of Enzymatic Hydrolysis Lignin and the Properties of Lignin–Epoxy Resin Blends. *Wood Sci. Technol.* **2018**, *52* (5), 1343–1357. https://doi.org/10.1007/s00226-018-1024-z.
52. Ferdosian, F.; Yuan, Z.; Anderson, M.; Charles, C. Synthesis and Characterization of Hydrolysis Lignin-Based Epoxy Resins. *Ind. Crop. Prod.* **2016**, *91*, 295–301. https://doi.org/10.1016/j.indcrop.2016.07.020.
53. Ferdosian, F.; Yuan, Z.; Anderson, M.; Charles, C. Sustainable Lignin-Based Epoxy Resins Cured with Aromatic and Aliphatic Amine Curing Agents : Curing Kinetics and Thermal Properties. *Thermochim. Acta* **2015**, *618*, 48–55. https://doi.org/10.1016/j.tca.2015.09.012.
54. Gioia, C.; Colonna, M.; Tagami, A.; Medina, L.; Sevastyanova, O.; Berglund, L. A.; Lawoko, M. Lignin-Based Epoxy Resins: Unravelling the Relationship between Structure and Material Properties. *Biomacromolecules* **2020**, *21* (5), 1920–1928. https://doi.org/10.1021/acs.biomac.0c00057.
55. Giummarella, N.; Lindén, P. A.; Areskogh, D.; Lawoko, M. Fractional Profiling of Kraft Lignin Structure: Unravelling Insights on Lignin Reaction Mechanisms. *ACS Sustain. Chem. Eng.* **2020**, *8* (2), 1112–1120. https://doi.org/10.1021/acssuschemeng.9b06027.
56. Giummarella, N.; Pylypchuk, I. V.; Sevastyanova, O.; Lawoko, M. New Structures in Eucalyptus Kraft Lignin with Complex Mechanistic Implications. *ACS Sustain. Chem. Eng.* **2020**, *8*, 10983–10994. https://doi.org/10.1021/acssuschemeng.0c03776.
57. Zhao, S.; Abu-Omar, M. M. Synthesis of Renewable Thermoset Polymers through Successive Lignin Modification Using Lignin-Derived Phenols. *ACS Sustain. Chem. Eng.* **2017**, *5* (6), 5059–5066. https://doi.org/10.1021/acssuschemeng.7b00440.
58. Zhao, S.; Huang, X.; Whelton, A. J.; Abu-omar, M. M. Formaldehyde-Free Method for Incorporating Lignin into Epoxy Thermosets. *ACS Sustain. Chem. Eng.* **2018**, *6* (2), 10628–10636. https://doi.org/10.1021/acssuschemeng.8b01962.
59. Sun, Z.; Fridrich, B.; De Santi, A.; Elangovan, S.; Barta, K. Bright Side of Lignin Depolymerization: Toward New Platform Chemicals. *Chem. Rev.* **2018**, *118* (2), 614–678. https://doi.org/10.1021/acs.chemrev.7b00588.
60. Zhang, C.; Wang, F. Catalytic Lignin Depolymerization to Aromatic Chemicals. *Acc. Chem. Res.* **2020**, *53*, 470–484. https://doi.org/10.1021/acs.accounts.9b00573.
61. Feghali, E.; Torr, K. M.; van de Pas, D. J.; Ortiz, P.; Vanbroekhoven, K.; Eevers, W.; Vendamme, R. Thermosetting Polymers from Lignin Model Compounds and Depolymerized Lignins. *Top. Curr. Chem.* **2018**, *376* (4), 1–25. https://doi.org/10.1007/s41061-018-0211-6.
62. Behling, R.; Valange, S.; Chatel, G. Heterogeneous Catalytic Oxidation for Lignin Valorization into Valuable Chemicals: What Results? What Limitations? What Trends? *Green Chem.* **2016**, *18* (7), 1839–1854. https://doi.org/10.1039/c5gc03061g.

63. Monsigny, L.; Feghali, E.; Berthet, J. C.; Cantat, T. Efficient Reductive Depolymerization of Hardwood and Softwood Lignins with Brookhart's Iridium(Iii) Catalyst and Hydrosilanes. *Green Chem.* **2018**, *20* (9), 1981–1986. https://doi.org/10.1039/c7gc03743k.

64. Deepa, A. K.; Dhepe, P. L. Solid Acid Catalyzed Depolymerization of Lignin into Value Added Aromatic Monomers. *RSC Adv.* **2014**, *4* (25), 12625–12629. https://doi.org/10.1039/c3ra47818a.

65. Katahira, R.; Mittal, A.; McKinney, K.; Chen, X.; Tucker, M. P.; Johnson, D. K.; Beckham, G. T. Base-Catalyzed Depolymerization of Biorefinery Lignins. *ACS Sustain. Chem. Eng.* **2016**, *4* (3), 1474–1486. https://doi.org/10.1021/acssuschemeng.5b01451.

66. Chaudhary, R.; Dhepe, P. L. Solid Base Catalyzed Depolymerization of Lignin into Low Molecular Weight Products. *Green Chem.* **2017**, *19* (3), 778–788. https://doi.org/10.1039/c6gc02701f.

67. Islam, M. N.; Taki, G.; Rana, M.; Park, J. H. Yield of Phenolic Monomers from Lignin Hydrothermolysis in Subcritical Water System. *Ind. Eng. Chem. Res.* **2018**, *57* (14), 4779–4784. https://doi.org/10.1021/acs.iecr.7b05062.

68. Huang, X.; Korányi, T. I.; Boot, M. D.; Hensen, E. J. M. Catalytic Depolymerization of Lignin in Supercritical Ethanol. *ChemSusChem* **2014**, *7* (8), 2276–2288. https://doi.org/10.1002/cssc.201402094.

69. Fache, M.; Boutevin, B.; Caillol, S. Vanillin Production from Lignin and Its Use as a Renewable Chemical. *ACS Sustain. Chem. Eng.* **2016**, *4* (1), 35–46. https://doi.org/10.1021/acssuschemeng.5b01344.

70. Belgacem, M. N.; Gandini, A. *Monomers , Polymers and Composites from Renewable Resources*; Belgacem, M. N., Gandini, A., Eds.; Elsevier Science, 2008. https://doi.org/10.1016/B978-0-08-045316-3.X0001-4.

71. Tarabanko, V. E.; Petukhov, D. V.; Selyutin, G. E. New Mechanism for the Catalytic Oxidation of Lignin to Vanillin. *Kinet. Catal.* **2004**, *45* (4), 569–577. https://doi.org/10.1023/B:KICA.0000038087.95130.a5.

72. Fache, M.; Darroman, E.; Besse, V.; Auvergne, R.; Caillol, S.; Boutevin, B. Vanillin, a Promising Biobased Building-Block for Monomer Synthesis. *Green Chem.* **2014**, *16* (4), 1987–1998. https://doi.org/10.1039/c3gc42613k.

73. Llevot, A.; Grau, E.; Carlotti, S.; Grelier, S.; Cramail, H. From Lignin-Derived Aromatic Compounds to Novel Biobased Polymers. *Macromol. Rapid Commun.* **2016**, *37* (1), 9–28. https://doi.org/10.1002/marc.201500474.

74. Oulame, M. Z.; Pion, F.; Allauddin, S.; Raju, K. V. S. N.; Ducrot, P. H.; Allais, F. Renewable Alternating Aliphatic-Aromatic Poly(Ester-Urethane)s Prepared from Ferulic Acid and Bio-Based Diols. *Eur. Polym. J.* **2015**, *63*, 186–193. https://doi.org/10.1016/j.eurpolymj.2014.11.031.

75. Gang, H.; Lee, D.; Choi, K. Y.; Kim, H. N.; Ryu, H.; Lee, D. S.; Kim, B. G. Development of High Performance Polyurethane Elastomers Using Vanillin-Based Green Polyol Chain Extender Originating from Lignocellulosic Biomass. *ACS Sustain. Chem. Eng.* **2017**, *5* (6), 4582–4588. https://doi.org/10.1021/acssuschemeng.6b02960.

76. Kuhire, S. S.; Nagane, S.; Wadgaonkar, P. P. Poly(Ether Urethane)s from Aromatic Diisocyanates Based on Lignin-Derived Phenolic Acids. *Polym. Int.* **2017**, *66*, 892–899. https://doi.org/10.1002/pi.5333.

77. De Haro, J. C.; Allegretti, C.; Smit, A. T.; Turri, S.; D'Arrigo, P.; Griffini, G. Biobased Polyurethane Coatings with High Biomass Content: Tailored Properties by Lignin Selection. *ACS Sustain. Chem. Eng.* **2019**, *7* (13), 11700–11711. https://doi.org/10.1021/acssuschemeng.9b01873.
78. Ménard, R.; Caillol, S.; Allais, F. Chemo-Enzymatic Synthesis and Characterization of Renewable Thermoplastic and Thermoset Isocyanate-Free Poly(Hydroxy)Urethanes from Ferulic Acid Derivatives. *ACS Sustain. Chem. Eng.* **2017**, *5* (2), 1446–1456. https://doi.org/10.1021/acssuschemeng.6b02022.
79. Chen, Q.; Gao, K.; Peng, C.; Xie, H.; Zhao, Z. K.; Bao, M. Preparation of Lignin/Glycerol-Based Bis(Cyclic Carbonate) for the Synthesis of Polyurethanes. *Green Chem.* **2015**, *17* (9), 4546–4551. https://doi.org/10.1039/c5gc01340b.
80. Biziks, V.; Fleckenstein, M.; Mai, C.; Militz, H. Suitability of a Lignin-Derived Mono-Phenol Mimic to Replace Phenol in Phenol-Formaldehyde Resin for Use in Wood Treatment. *Holzforschung* **2020**, *74* (4), 344–350. https://doi.org/10.1515/hf-2019-0061.
81. Granado, L.; Tavernier, R.; Henry, S.; Auke, R. O.; Foyer, G.; David, G.; Caillol, S. Toward Sustainable Phenolic Thermosets with High Thermal Performances. *ACS Sustain. Chem. Eng.* **2019**, *7* (7), 7209–7217. https://doi.org/10.1021/acssuschemeng.8b06286.
82. Matsumura, S.; Hlil, A. R.; Lepiller, C.; Gaudet, J.; Guay, D.; Shi, Z.; Holdcroft, S.; Hay, A. S. Stability and Utility of Pyridyl Disulfide Functionality in RAFT and Conventional Radical Polymerizations. *J. Polym. Sci. Part A Polym. Chem.* **2008**, *46*, 7207–7224. https://doi.org/10.1002/pola.
83. Sini, N. K.; Bijwe, J.; Varma, I. K. Renewable Benzoxazine Monomer from Vanillin: Synthesis, Characterization, and Studies on Curing Behavior. *J. Polym. Sci. Part A Polym. Chem.* **2014**, *52* (1), 7–11. https://doi.org/10.1002/pola.26981.
84. Comí, M.; Lligadas, G.; Ronda, J. C.; Galià, M.; Cádiz, V. Renewable Benzoxazine Monomers from "Lignin-like" Naturally Occurring Phenolic Derivatives. *J. Polym. Sci. Part A Polym. Chem.* **2013**, *51* (22), 4894–4903. https://doi.org/10.1002/pola.26918.
85. Phalak, G. A.; Patil, D. M.; Mhaske, S. T. Synthesis and Characterization of Thermally Curable Guaiacol Based Poly(Benzoxazine-Urethane) Coating for Corrosion Protection on Mild Steel. *Eur. Polym. J.* **2017**, *88*, 93–108. https://doi.org/10.1016/j.eurpolymj.2016.12.030.
86. Oliveira, J. R.; Kotzebue, L. R. V.; Ribeiro, F. W. M.; Mota, B. C.; Zampieri, D.; Mazzetto, S. E.; Ishida, H.; Lomonaco, D. Microwave-Assisted Solvent-Free Synthesis of Novel Benzoxazines: A Faster and Environmentally Friendly Route to the Development of Bio-Based Thermosetting Resins. *J. Polym. Sci. Part A Polym. Chem.* **2017**, *55* (21), 3534–3544. https://doi.org/10.1002/pola.28755.
87. Stanzione, J.; La Scala, J. Recent Advances in Bio-Based Epoxy Resins and Bio-Based Epoxy Curing Agents. *J. Appl. Polym. Sci.* **2016**, *133* (45) https://doi.org/10.1002/app.44212.
88. Yuan, W.; Ma, S.; Wang, S.; Li, Q.; Wang, B.; Xu, X.; Huang, K.; Chen, J.; You, S.; Zhu, J. Synthesis of Fully Bio-Based Diepoxy Monomer with Dicyclo Diacetal for High-Performance, Readily Degradable Thermosets. *Eur. Polym. J.* **2019**, *117*, 200–207. https://doi.org/10.1016/j.eurpolymj.2019.05.028.
89. Ménard, R.; Caillol, S.; Allais, F. Ferulic Acid-Based Renewable Esters and Amides-Containing Epoxy Thermosets from Wheat Bran and Beetroot Pulp: Chemo-Enzymatic Synthesis and

Thermo-Mechanical Properties Characterization. *Ind. Crops Prod.* **2017**, *95*, 83–95. https://doi.org/10.1016/j.indcrop.2016.10.016.

90. Fache, M.; Auvergne, R.; Boutevin, B.; Caillol, S. New Vanillin-Derived Diepoxy Monomers for the Synthesis of Biobased Thermosets. *Eur. Polym. J.* **2015**, *67*, 527–538. https://doi.org/10.1016/j.eurpolymj.2014.10.011.
91. Fache, M.; Viola, A.; Auvergne, R.; Boutevin, B.; Caillol, S. Biobased Epoxy Thermosets from Vanillin-Derived Oligomers. *Eur. Polym. J.* **2015**, *68*, 526–535. https://doi.org/10.1016/j.eurpolymj.2015.03.048.
92. Fache, M.; Boutevin, B.; Caillol, S. Epoxy Thermosets from Model Mixtures of the Lignin-to-Vanillin Process. *Green Chem.* **2016**, *18* (3), 712–725. https://doi.org/10.1039/c5gc01070e.
93. Faye, I.; Decostanzi, M.; Ecochard, Y.; Caillol, S. Eugenol Bio-Based Epoxy Thermosets: From Cloves to Applied Materials. *Green Chem.* **2017**, *19* (21), 5236–5242. https://doi.org/10.1039/c7gc02322g.
94. Zhao, S.; Abu-Omar, M. M. Biobased Epoxy Nanocomposites Derived from Lignin-Based Monomers. *Biomacromolecules* **2015**, *16* (7), 2025–2031. https://doi.org/10.1021/acs.biomac.5b00670.
95. Zhao, S.; Huang, X.; Whelton, A. J.; Abu-Omar, M. M. Renewable Epoxy Thermosets from Fully Lignin-Derived Triphenols. *ACS Sustain. Chem. Eng.* **2018**, *6* (6), 7600–7608. https://doi.org/10.1021/acssuschemeng.8b00443.
96. Zhao, S.; Abu-Omar, M. M. Renewable Epoxy Networks Derived from Lignin-Based Monomers: Effect of Cross-Linking Density. *ACS Sustain. Chem. Eng.* **2016**, *4* (11), 6082–6089. https://doi.org/10.1021/acssuschemeng.6b01446.
97. Duval, A.; Vilaplana, F.; Crestini, C.; Lawoko, M. Solvent Screening for the Fractionation of Industrial Kraft Lignin. *Holzforschung* **2016**, *70* (1), 11–20. https://doi.org/10.1515/hf-2014-0346.
98. Cui, C.; Sun, R.; Argyropoulos, D. S. Fractional Precipitation of Softwood Kraft Lignin: Isolation of Narrow Fractions Common to a Variety of Lignins. *ACS Sustain. Chem. Eng.* **2014**, 2 (4), 959–968. https://doi.org/10.1021/sc400545d.
99. Sevastyanova, O.; Helander, M.; Chowdhury, S.; Lange, H.; Wedin, H.; Zhang, L.; Ek, M.; Kadla, J. F.; Crestini, C.; Lindstr, M. E. Tailoring the Molecular and Thermo – Mechanical Properties of Kraft Lignin by Ultrafiltration. *J. APPL. POLYM. SCI.* **2014**, *40799* (April), 1–11. https://doi.org/10.1002/app.40799.
100. Toledano, A.; García, A.; Mondragon, I.; Labidi, J. Lignin Separation and Fractionation by Ultrafiltration. *Sep. Purif. Technol.* **2010**, *71* (1), 38–43. https://doi.org/10.1016/j.seppur.2009.10.024.
101. Crestini, C.; Lange, H.; Sette, M.; Argyropoulos, D. S. On the Structure of Softwood Kraft Lignin. *Green Chem.* **2017**, *19* (17), 4104–4121. https://doi.org/10.1039/c7gc01812f.
102. Lancefield, C. S.; Wienk, H. J.; Boelens, R.; Weckhuysen, B. M.; Bruijnincx, P. C. A. Identification of a Diagnostic Structural Motif Reveals a New Reaction Intermediate and Condensation Pathway in Kraft Lignin Formation. *Chem. Sci.* **2018**, *9* (30), 6348–6360. https://doi.org/10.1039/c8sc02000k.
103. Zhao, C.; Hu, Z.; Shi, L.; Wang, C.; Yue, F.; Li, S.; Zhang, H.; Lu, F. Profiling of the Formation of Lignin-Derived Monomers and Dimers from Eucalyptus Alkali Lignin. *Green Chem.* **2020**. https://doi.org/10.1039/d0gc01658f.

104. Meng, X.; Parikh, A.; Seemala, B.; Kumar, R.; Pu, Y.; Christopher, P.; Wyman, C. E.; Cai, C. M.; Ragauskas, A. J. Chemical Transformations of Poplar Lignin during Cosolvent Enhanced Lignocellulosic Fractionation Process. *ACS Sustain. Chem. Eng.* **2018**, *6* (7), 8711–8718. https://doi.org/10.1021/acssuschemeng.8b01028.
105. Karlsson, M.; Giummarella, N.; Lindén, P. A.; Lawoko, M. Toward a Consolidated Lignin Biorefinery: Preserving the Lignin Structure through Additive-Free Protection Strategies. *ChemSusChem* **2020**, *13*, 4666–4677. https://doi.org/10.1002/cssc.202000974.
106. Jawerth, M. E.; Brett, C. J.; Terrier, C.; Larsson, P. T.; Lawoko, M.; Roth, S. V.; Lundmark, S.; Johansson, M. Mechanical and Morphological Properties of Lignin-Based Thermosets. *ACS Appl. Polym. Mater.* **2020**, *2* (2), 668–676. https://doi.org/10.1021/acsapm.9b01007.

Chapter 9

Recent Advances in Lignin-Based Hydrogels and Its Synthesis and Applications

Da-feng Zheng,[1,*] Ling Hu,[1] and Xue-qing Qiu[2]

[1]School of Chemistry and Chemical Engineering, South China University of Technology, Guangzhou, Guangdong Province 510640, China

[2]School of Chemical Engineering and Light Industry, Guangdong University of Technology, Guangzhou, Guangdong Province 510640, China

[*]Email: zhengdf@scut.edu.cn.

Lignin is the second most abundant plant polymer in nature. However, lignin is often piled up as waste or directly burned to obtain low calorific value. This results not only in environmental issues but also fails to play the value of lignin itself. For several decades, researchers have devoted their efforts to the high-value utilization of lignin., but the use of lignin to prepare new materials still faces many challenges. Researchers have used lignin to synthesize excellent hydrogels to expand the application fields in recent years due to lignin has a natural network structure and contains various functional groups. This article first reviews the methods for synthesizing lignin-based hydrogels in the past five years, including chemical crosslinking and physical crosslinking. In addition, it also reviews the applications of lignin-based hydrogels in water purification, biomedicine, electronic devices and other fields.

Introduction of Lignin

Lignin is the second most abundant and renewable resources on earth and it is widely found in ferns and all higher plants. The content of lignin differs according to plant, 35% in softwood while 18% ~ 24% in hardwood. Moreover, the content of lignin differs according to the plant cell part. Plant cells are composed of cell walls and cell cavities, between the cell walls is an intercellular layer. The content of lignin is higher in the intercellular layer, about 60%~90%, and the content is lower as it goes to the inside of the plant cell, and it is 10%~20% near the cell cavities.

Lignin, as an important component of lignocellulose (Figure 1), is packaged on the surface of cellulose in cooperation with hemicellulose. It plays a dual role of binding fibers and making the fibers rigid. The plant cell wall is only composed of cellulose and hemicellulose in the juvenile phase, and then lignin is gradually formed in the cell wall to bind the cells together. Lignin not only greatly increases the mechanical strength of the tissue, but also makes the cell wall (water channel in the plant) hydrophobic, prevents water leakage, and ensures the long-distance transportation of water and nutrients in the plant.

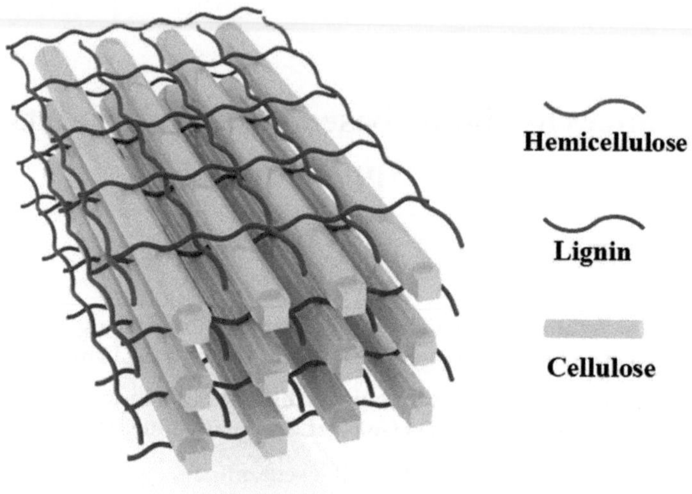

Figure 1. The structure of lignocellulose.

Structure of Lignin

Lignin is the second-most abundant natural aromatic polymer in nature. Compared with cellulose and hemicellulose, the structure of lignin is more complex and has not been characterized so far. Currently researchers generally believe that lignin consists of three phenylpropane structural units (Figure 2): p-hydroxyphenyl (H type), guaiacyl (G type) and syringyl (S type). They are randomly combined with different ether bonds and carbon-carbon bonds to form biopolymers with a three-dimension network structure. In recent years, through the two-dimensional heteronuclear single quantum coherence (2D-HSQC) nuclear magnetic technology, three more accurate connections of phenylpropane have been obtained (*1–3*) (Figure 3):β-O-4(the most common type),(β-5)+(α-O-4),(β-β)+(γ-O-α),4-O-5,(β-1)+(α-O-α), etc.

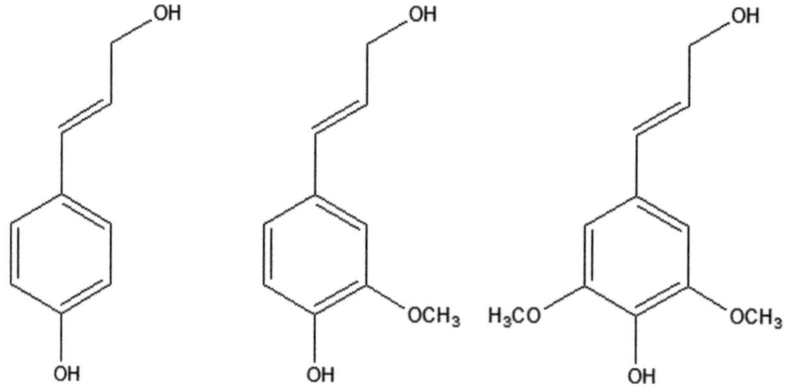

Figure 2. Three constituent units of lignin.

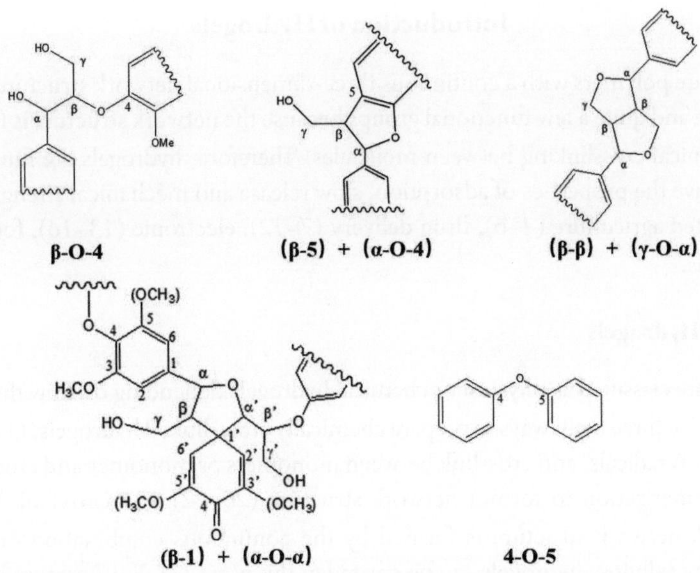

Figure 3. Accurate connections of phenylpropane. Reprinted with permission from (1). Copyright 2013 MDPI. Reprinted with permission from (2). Copyright 2016 American Chemical Society. Reprinted with permission from (3). Copyright 2003 American Chemical Society.

Commonly Used Industrial Lignin in Hydrogels

Alkali lignin (AL) and lignosulfonate (LS) are commonly used in hydrogel formulations.

Alkali Lignin (AL)

Alkali lignin comes from the waste liquid produced by the alkaline pulping process of the paper industry. The reaction conditions of alkaline pulping process are high temperature and pressure. Therefore, ether bonds and ester bonds of the lignin macromolecules are broken due to the violent reaction, forming smaller molecules (alkali lignin) which are rich in phenolic hydroxyl groups. The irregular breakage of chemical bonds in this process causes alkali lignin a wide molecular weight. Thus, alkali lignin usually needs to be purified by alkali dissolution and acid precipitation. Alkali lignin is rich in phenolic hydroxyl, alcoholic hydroxyl, carboxyl, and other hydrophilic functional groups and it can be completely ionized under alkaline conditions, showing a stable surface activity.

Lignosulfonate (LS)

Lignosulfonate comes from the waste liquid produced by the sulfite pulping process in the paper industry. Water-soluble lignosulfonate is produced during the pulping process because of sulfonylation occurred at high temperatures. Lignosulfonate usually needs to be purified through cation exchange resin due to sulfite residue. Lignosulfonate can be dissolved in aqueous solutions and has excellent adsorption, dispersibility and wettability.

About 50 million tons of lignin can be generated every year in the paper and biorefinery industries. However, most of them are generally treated as waste in the industrial, resulting in environmental pollution and resource wastage. Therefore, the high-value application of lignin has become a hot topic.

Introduction of Hydrogels

Hydrogels are polymers with a continuous three-dimensional network structure. It has a certain porous structure and quite a few functional groups because the network structure is formed based on physical or chemical crosslinking between molecules. Therefore, hydrogels are functional polymer materials that have the properties of adsorption, slow release and mechanical strength. Applications include penetrated agriculture (*4–6*), drug delivery (*7–12*), electronic (*13–16*), food (*17–19*) and many other fields.

Preparation of Hydrogels

Hydrogels are classified as physical or chemical hydrogels depending on how the materials were prepared. There are three main ways to prepare chemically cross-linked hydrogels: (1) Use light, heat, or initiator to form radicals, and crosslink between monomers or monomer and crosslinker through free radical polymerization to form a network structure (*20–22*). (2) Crosslink by the addition of crosslinker. A network structure is formed by the continuous combination of monomer and crosslinker. Many cellulose hydrogels can be created in this way (*23*). Furthermore, A great number of studies report on crosslinking two or more polymers (*24*, *25*). (3) Crosslink by the functional groups of monomers. Different from the chemical reaction between the monomer and the crosslinker, the monomer can generate a network through the chemical reaction of its own functional group, such as esterification reaction, isocyanate reaction and so on.

There are two common ways to prepare physically cross-linked hydrogels: (1) Crosslink through hydrogen-bonding interactions (*26*, *27*). A stable three-dimensional network structure is established relying on hydrogen bonds between monomers in this process. (2) Crosslink by electrostatic interaction (*28*, *29*). Oppositely charged polymers can be polymerized by electrostatic attraction to form a hydrogel in solution.

Properties of Hydrogels

Hydrogels are materials that stand out for their high-water content and high specific surface area. Some lignin-based hydrogels contain many hydrophilic groups and superior porous network structure, giving hydrogels excellent properties, such as swelling and mechanical properties. Swelling refers to the phenomenon that the volume of the hydrogel swells in a solvent. The swelling ratio is usually measured by the gravimetric method and is related to the nature of the solvent, the hydrophilic group and the network structure of the hydrogel, so it can be used to characterize the properties of a polymer network (*30*, *31*). Another essential feature of hydrogels is mechanical properties. Dense crosslinking network in hydrogels plays an important role in enhancing its mechanical stability. Rheological tests, stretching methods or compression methods are widely used to characterize the mechanical strength of hydrogels. Moreover, hydrogels with different properties and functions has been extensively explored in order to meet the needs of different application fields during the last decade, such as temperature-responsive hydrogels (*32*), pH-responsive hydrogels (*8*, *33*), conductive hydrogels (*34*, *35*) and so on.

Preparation of Lignin-Based Hydrogels

Using renewable materials (cellulose (*36*),chitosan (*37*) and pectin (*38*)) in preparation of hydrogels is one of the efficient ways to utilize biomass resources. The greatest advantage of this method, aside from the ability to reduce the consumption of fossil resources and meet the requirement of sustainable society, lies in improving the structure of traditional hydrogels and enhancing the biocompatibility and biodegradability. Recently, lignin-based hydrogels have attracted significant attention.

Chemical Crosslinking

Crosslinking by Free Radical Polymerization

Free radical polymerization usually contains monomers and crosslinkers. Acrylic acid and acrylamide are common monomers while N, N′-methylenebisacrylamide is used as a crosslinker. Generally, they can form polymer chains first, then the phenolic hydroxyl group of lignin can form free radicals in the presence of initiators and form a graft structure with polymer chains or monomers to obtain a lignin-based hydrogel. Much of the literature on lignin-based hydrogels focuses on this method. A lignin-based hydrogel (hardwood sulfate lignin) with high thermal stability and rheological stability was synthesized. N-isopropylacrylamide was used as a monomer and N,N′-methylenebisacrylamide was selected as a crosslinker in this system (*39*). In another preparation process, acrylamide and N,N′-methylenebisacrylamide was used to form polymer chains. Alkali lignin obtained from kraft pulp was served as raw material which can help to form smaller pore structure (*40*). Similarly, Wu Liu et al. (*41*) used lignin, acrylamide, and ethylene glycol dimethyl acrylate (EGDMA) as a crosslinker to prepare a pH-responsive and magnetic smart lignin-based hydrogel. In particular, purifying the lignin during the experiment will increase the experimental complexity. Yi Meng et al. (*42*) directly used the waste liquid (mainly containing polysaccharides and lignosulfonates) generated from the sulfite pulping process of the paper industry and N,N′-Methylenebisacrylamide to prepare super-swelled biopolymer hydrogel (Figure 4).FTIR showed that lignosulfonate and polysaccharide were grafted onto the polyacrylate network. Due to the abundant lignin and polysaccharides in the waste liquid, which is rich in hydroxyl, carboxylate, and sulfonic acid groups, the super-swelled hydrogel showed an excellent swelling rate. The swelling rate was increased to 280 g/g when adding only 0.4 g waste liquid to the hydrogel.

Researchers usually graft monomer with natural polymers (starch (*43*), cellulose (*44*), etc.) to form a grafting structure which results in high mechanical strength. Therefore, an emerging new type of lignin-based hydrogels preparation is forming covalent bonds between lignin and the monomer, and then create a network through free radical polymerization. Pan Jiang et al. (*45*) first synthesized phenolic lignin (PPAL) from phenol and soda lignin, and then thermally polymerized PPAL and N-isopropylacrylamide (NIPAAm) to obtain high-density hydrogels. Kalavathy Rajan et al. (*46*) used the organic solvent method to separate organic solvent lignin (OSL) from hybrid poplar, and then grafted 2-hydroxyethyl methacrylate (HEMA) to lignin through esterification to obtain different modified lignin (OSLH) with hydroxyl content(Figure 5), the OSLH and excess HEMA were free-radically polymerized to form a copolymer hydrogel with a lignin content up to 40 wt%. The material properties of the hydrogel were improved due to the graft copolymer generated a denser and more stable network. Studies have shown that the thermal stability of hydrogels was increased by 20% compared with hydrogels without lignin, and the storage modulus (G′) has increased by 3 orders of magnitude.

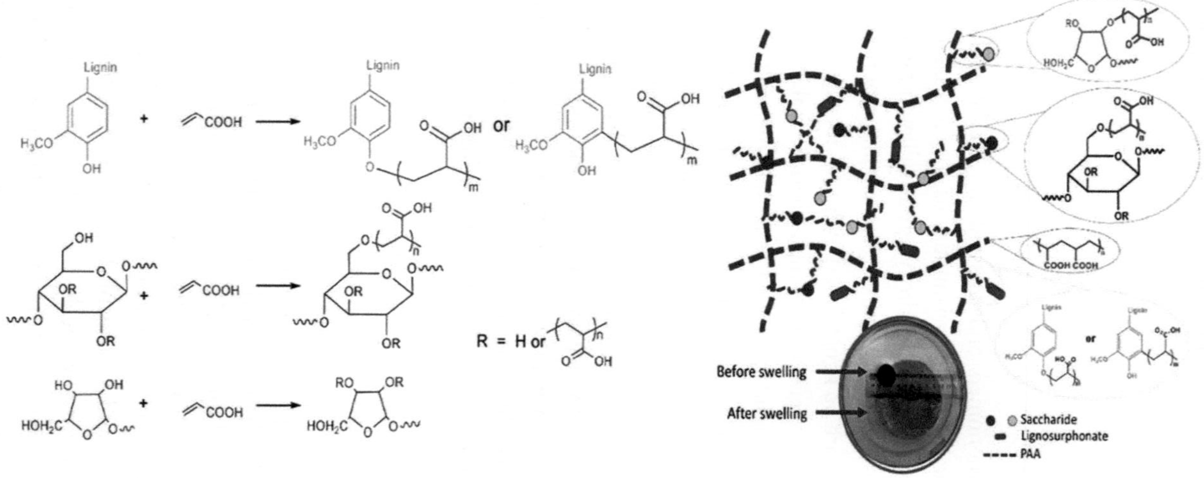

Figure 4. Schematic illustration of the hydrogel network and the chemical structures after the copolymerization of red liquor and acrylic acid. Reprinted with permission from (42). Copyright 2019 Elsevier.

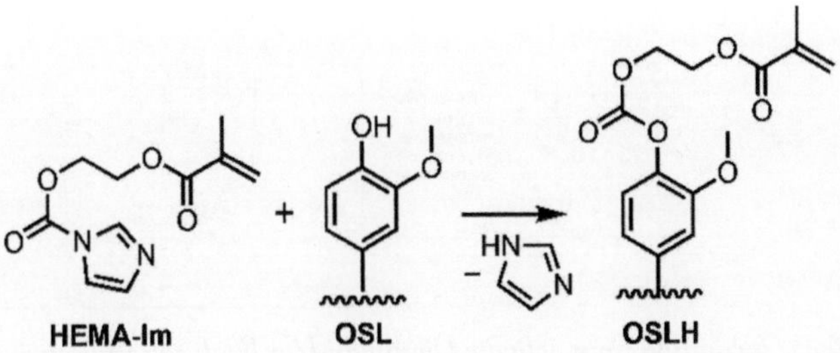

Figure 5. Modification of lignin (OSL) using HEMA-Im to produce OSLH. Reprinted with permission from (46). Copyright 2018 American Chemical Society.

Crosslinking by Crosslinkers

Adding a crosslinker is a novel and simple strategies to prepare chemical hydrogels. There are many types of crosslinkers used in hydrogels preparation while epichlorohydrin (ECH) is widely used due to its strong reactivity. Lignin hydrogels were established by dissolving two lignin (IL-lignin isolated from poplar wood and sulfate lignin) in an alkaline solution and then crosslinked with epichlorohydrin (*47*). Results showed that the hydrogel can be formed only when the lignin content was higher than 15wt%. The author next prepared PEG (Polyethylene glycol)-lignin hydrogels because of the weak mechanical strength of lignin hydrogels. The most obvious finding to emerge from this study is that it is easier to form a hydrogel (the lignin content was 10wt%) owing to the introduction of PEG. Furthermore, the author first condensed PEG with ECH to created ETPEG to prevent ECH from being crosslinked with PEG but not lignin, and then ETPEG crosslinked with lignin. This study has shown that the hydrogels formed by this way had higher water absorption and tensile strength.

With regard to the research methods of lignin hydrogels (only use lignin as raw materials) preparation, some limitations were acknowledged according to previous reports. Therefore, much of the literature on lignin-based hydrogels seems to have been based on two or more polymers. A lignin-cellulose hydrogel was prepared by crosslinking alkali lignin and microcrystalline cellulose (*48*). Rheological tests proved that alkali lignin significantly enhanced the mechanical strength of the hydrogel and the swelling rate of lignin-cellulose hydrogels was 1.8-3.2 times larger than that of microcrystalline cellulose hydrogel. Wu et al. (*49*) used epichlorohydrin as a crosslinker to prepare lignin-PVA hydrogels by one pot method. The dense chemical crosslinking network and the hydrogen bonds formed between the hydroxyl groups of PVA and lignin increased the mechanical properties. Diana Ciolacu et al. (*50*) also used epichlorohydrin to prepare lignin-PVA hydrogels. In contrast to one pot method, the author first combined lignin and epichlorohydrin to prepare a lignin-based epoxy resin, and then crosslinked with PVA.

Carbodiimide (EDC) is another widely used crosslinker because of its non-toxicity and good biocompatibility. Water-soluble lignin which was highly compatible with hyaluronic acid was obtained by modifying the kraft pulp lignin through Mannich reaction in the presence of formaldehyde (*51*). EDC was added as a crosslinker to the hyaluronic acid-lignin solution, and then gently stirred at 25°C for 1 h to form hydrogels (Figure 6). The hydrogels had strong stability, swelling ability, and viscoelasticity.

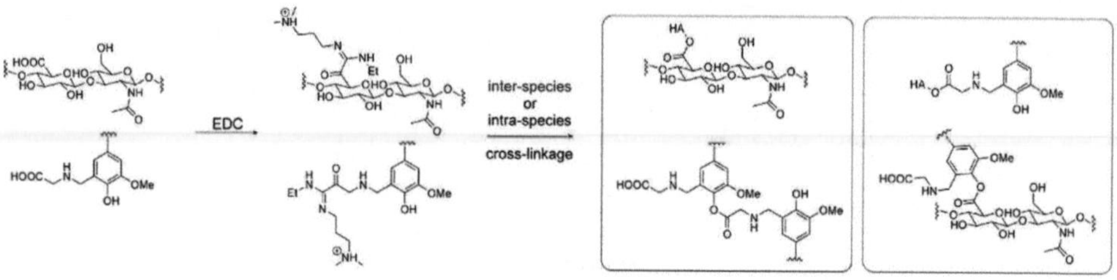

Figure 6. Cross-linkage formation from activated lignin and/or hyaluronan. Reprinted with permission from (51). Copyright 2019 Elsevier Sci Ltd.

Crosslinking by Functional Groups of Monomers

Since chemical crosslinkers are usually toxic and produce by-products, using the functional groups of lignin to directly react with polymers is another way to form a crosslinked network.

Poly(ethyleneglycol)diglycidyl ether (PEGDGE) is widely used to prepare lignin-based hydrogels because the epoxy group can open the ring under alkaline conditions and react with the hydroxyl group of lignin. Lin Dai et al. (52) synthesized lignin-based hydrogels by reacting sulfate lignin (high phenolic hydroxyl content) with PEGDGE in alkaline solutions and polymerized at 40°C for 12 hours. The findings clearly indicate that the hydrogels had excellent mechanical properties, and the second major finding was that the hydrogel has flexibility and rapid pH response. It can achieve softening/enhancing and straight/bending switching by immersing in KOH and HCl solutions alternately. Another lignin-PEGDGE hydrogel was formed at 50°C and stirred for 15 minutes (53). It was biodegradable, had no phytotoxicity, and can be used as a soil additive. A key requirement for crosslinking by functional groups of monomers is that the monomers should have high reactivity. XiaoXu Teng et al. (54) improve the reactivity of lignin by modifying sodium lignosulfonate with diethylenetriamine (DETA) to introduced amine groups on lignin. These modified lignins were then used for the preparation of lignin-PEGDGE hydrogels.

Another polymer that has been extensively studied by researchers is poly (methyl vinyl ether co-maleic acid) because the carboxyl group of it can be crosslinked with the hydroxyl group of lignin through an esterification reaction. Juan Domínguez-Robles et al. (55) mixed Poly (methyl vinyl ether co-maleic acid) and different technical lignins in ammonium hydroxide solution, then cured in an oven at 80 °C for 24 h to prepare hydrogels with high swelling rate (130 g/g). What is innovating in this article is that toxic reagent was not used in the preparation process and the main by-product was water(Figure 7). Eneko Larrañeta et al. (56) combined lignin with poly (ethylene glycol) and Poly (methyl vinyl ether co-maleic acid) through esterification to prepare hydrogels. In this work, microwave radiation can accelerate the preparation of hydrogels and the crosslinking time was reduced from 24 hours to 1 hour, thereby reducing energy consumption.

Some recent attention has focused on other types of chemical reactions. Kuan-Ting Lin et al. (57) used lignin extracted from a eutectic solvent of choline chloride and lactic acid instead of resorcinol to prepare a lignin-formaldehyde hydrogels. The hydroxyl and carbonyl groups of lignin provided reaction sites for crosslinking with formaldehyde through electrophilic addition and aldol condensation reactions. Hyoe Hatakeyama et al. (58) used water as a solvent to prepare sodium lignosulfonate-based polyurethane hydrogels with sodium lignosulfonate and hexamethylene diisocyanate (HDI) through urethane linkage.

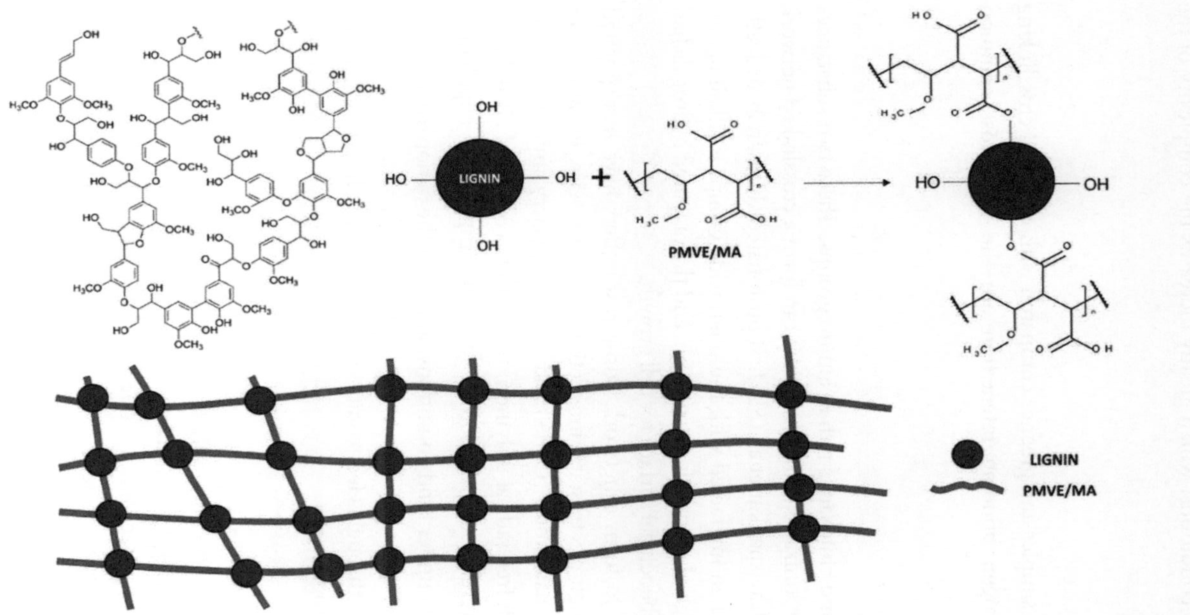

Figure 7. Chemical structures and proposed chemical reactions taking place during the crosslinking process. Reprinted with permission from (55). Copyright 2018 Elsevier Science BV.

Crosslinking by crosslinkers is a common way to prepare chemically cross-linked hydrogels because most of them are prepared by the one-pot reaction system, which is simple and fast. While crosslinking by functional groups of monomers is a greener way to prepare hydrogels due to most crosslinkers are toxic, but it has higher requirements on the purity of lignin (in order to ensure the activity of the reaction). Crosslinking through free radical polymerization usually needs to initiate free radicals under light, heat and other conditions, which greatly increases the complexity of the experiment.

Physical Crosslinking

The physical crosslinking process is simpler and greener compared with chemical crosslinking and electrostatic interaction and hydrogen bonding interaction are the two most common crosslinking methods.

Crosslinking by Electrostatic Interaction

Lignin is rich in phenoxide groups and chitosan is rich in amino groups. These two substances can form oppositely charged groups after ionization, and then they can form a crosslinked network through electrostatic interaction. Kartik Ravishankar et al. (*59*) first protonated chitosan in an acetic acid solution to form ammonium, and then interacted with the phenoxide groups in alkali lignin, forming a chitosan-lignin hydrogel through electrostatic interaction, and the prepared hydrogels had good biocompatibility and can provide a beneficial surface for cell growth.

In addition to directly using lignin to form a network, some researchers have incorporated lignin nanoparticles into hydrogels through electrostatic interaction in recent years. Jinshun Lu et al. (*60*) used TEMPO-oxidized cellulose nanofibers (CNFs) and N-isopropylacrylamide (NIPAM) to prepare a crosslinked network through free radical polymerization. Then combined alkali lignin and NH_2-PEG-COOH through an esterification reaction to form an amphiphilic copolymer. Nanoparticles were formed autonomously under mild conditions, and then the nanoparticles were added to the above-mentioned crosslinked network to prepare hydrogels. Lignin nanoparticles were incorporated into the polymer network through electrostatic interaction, which can adjust the elasticity and viscous modulus of the hydrogel.

Crosslinking by Hydrogen-Bonding Interactions

Polyvinyl alcohol (PVA) and cellulose are commonly used to prepare lignin-based hydrogels through hydrogen-bonding interactions owing to abundant hydroxyl groups. In these hydrogels, lignin plays an essential role in strengthening mechanical properties. Lili Zhang et al. (*61*) prepared lignocellulose hydrogels with different lignin content by dissolving lignocellulose in the N-methylmorpholine N-oxide solvent system. The hydrogel had the best viscoelasticity and compressibility when the lignin content was 6.5%. Amaia Morales et al. (*62*) designed 6 synthetic pathways to synthesize green lignin-PVA hydrogels through repeated freezing and vacuum. The influence of lignin and PVA concentration on the swelling rate and mechanical properties is clearly supported by the findings.

Hydrogen-bonding interactions provide a simpler strategy to prepare composite hydrogel with three polymers. Ryan M et al. (*63*) used 1-ethyl-3-methylimidazolium acetate ([C2C1im][OAc]) as a solvent to prepare physically crosslinked hydrogels from a mixture of cellulose, xylan, and lignin. The evidence from this study suggests that adding lignin and xylan to the cellulose hydrogels can improve its mechanical strength and stiffness. Yiwen Zhang et al. (64) prepared a novel lignin–chitosan–PVA composite hydrogel through simple physical crosslinking(Figure 8).Chitosan interacted with PVA through hydrogen bonding to form a crosslinked network and the sulfate groups of lignin and amine groups of chitosan formed ionic bonds and increased the mechanical strength of the hydrogel.

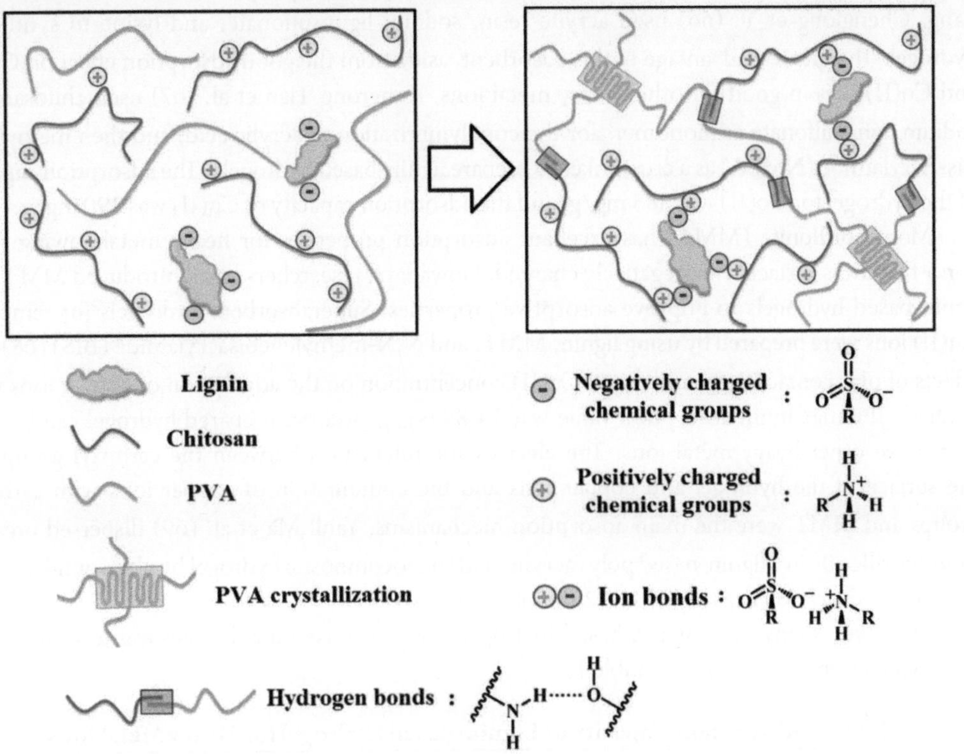

Figure 8. The crosslinking mechanism of LCPHs. Reprinted with permission from (64). Copyright 2019 Elsevier.

Applications

Lignin-based hydrogels are widely used in water purification, biomedical, electronic device, and other fields due to their excellent performance. In this section, we summarize the studies which highlight the beneficial effects of lignin-based hydrogels.

Water Treatment

Much of the serious environmental pollution can be attributed to heavy metal wastewater. Absorption is a convenient method to remove heavy metal ions in wastewater. The aromatic rings and abundant surface groups of lignin are beneficial to the application of lignin-based hydrogels in this field.

Lignosulfonates hydrogels have great potential to serve as an efficient absorbent in water treatment. It not only can interact with metal ions electrostatically due to many negatively charged groups, but the unshared electron pairs on oxygen atoms can form coordination bonds with metal ions to produce chelation. Lignosulfonate-graphene hydrogels for Pb(II) adsorption was synthesized through a one-step process (65). The three-dimensional (3D) porous structure of the hydrogel provided multiple channels for Pb(II) transportation and can be achieved by using lignosulfonate with the different oxygen-containing groups. A negatively charged surface was formed owing to the carboxyl groups of the reduced graphene oxide sheets and the sulfonic acid groups of the lignosulfonate. Another important adsorption factor is that many oxygen-containing groups on the lignosulfonate chain can form complexes with Pb(II). The hydrogel exhibited ultra-high Pb(II) adsorption capacity (1210 mg/g) compared with most previously reported adsorbents. Likewise, Jiang, Chenglong et al. (*66*) used acrylic resin, sodium lignosulfonate, and lysine to synthesize hydrogel. The greatest advantage of this adsorbent, aside from the good adsorption effect on Cu(II) and Co(II), lies in good for other heavy metal ions. Rongrong Tian et al. (*67*) used chitosan and sodium lignosulfonate as monomers for the copolymerization of acrylic acid, and then methylene-Bis-Acrylamide (NMBA) as a crosslinker to prepare lignin-based hydrogels. The adsorption capacity of the hydrogel for Co(II) was 385 mg/g, and the adsorption capacity of Cu(II) was 290 mg/g.

Montmorillonite (MMT) has excellent adsorption properties for heavy metals owing to its nano-thickness surface with negatively charged. Nowadays, researchers have introduced MMT into lignin-based hydrogels to improve adsorptive properties. Superabsorbent hydrogels for removing Cu(II) ions were prepared by using lignin, MMT, and N,N-methylenebisacrylamide (BIS) (*68*).The effects of pH, contact time, and initial Cu(II) concentration on the adsorption of copper ions were studied. The maximum adsorption value was 74.88 mg/g, and the prepared hydrogel can be used to remove other heavy metal ions. The electrostatic interaction between the carboxyl groups on the surface of the hydrogel and copper ions and the combination of copper ions with carboxyl groups and MMT were the main adsorption mechanisms. Yanli Ma et al. (*69*) dispersed organo-montmorillonite in a lignin-based poly (acrylic acid) nanocomposite hydrogel by ultrasound, and the gel had an adsorption capacity of 1.192 mmol/g for Pb(II).

Moreover, many other lignin-based hydrogels were also used for the removal of heavy metal ions, which were summarized in Table 1.

Table 1. Absorption Capacity of Lignin-Based Hydrogel for Heavy Metal Ions

Hydrogels	*Source of Lignin*	*Heavy metal ions*	*Ref.*
Lignocellulose hydrogels	Poplar wood chips	Pb (II), Cu (II)	(*4*)
Alkali lignin grafted PAA hydrogels	Alkali lignin	Pb (II), Cu (II), Cd (II)	(*5*)
Lignin-Chitosan-PAM hydrogels	Air-dried rice husk	Pb (II), Cu (II), Cd (II)	(*6*)
Lignin-cellulose hydrogels	Alkali lignin	Pb (II), Hg (II), Ni (II)	(*7*)
Starch/acryl amide/lignin hydrogels	From black liquor	Cu (II), Ni (II)	(*8*)
Hemicelluloses based hydrogels	Hemicelluloses	As (V), Ni (VI)	(*9*)

In addition to heavy metal ions, pollution caused by printing and dyeing wastewater has been one of the serious environmental problems. Lignin-based hydrogels play an essential role in removing dyes due to its excellent adsorption capacity and swelling rate. Lili Zhang et al. (*75*) has prepared mesoporous lignocellulose hydrogels composed of cellulose, lignin, and hemicellulose. The hydrogels had a large specific surface area and can expose many active sites of functional groups which were beneficial to the adsorption of organic dyes. The results showed that the adsorption rate of lignin-containing samples (LCGA) was faster than that of samples without lignin (CGA), and the maximum adsorption capacity was also higher than that of samples without lignin (Figure 9). Jiarong Li et al. (*76*) prepared a ligninsulfonate hydrogel for the adsorption of cationic dye methylene blue

(MB) by a simple and effective method. The adsorption capacity of the hydrogel for MB reached 495 mg/g at 30°C. Electrostatic interaction and π-π interaction were the main adsorption mechanisms, and sulfonic acid groups of sulfonate lignin enhanced π-π interaction. A new type of hybrid hydrogel which MMT was incorporated into the crosslinked polymer was successfully synthesized (77). The hydrogels were used to remove methylene blue in aqueous solution (the best adsorption value was 9646.92 mg/g) and had excellent regeneration ability.

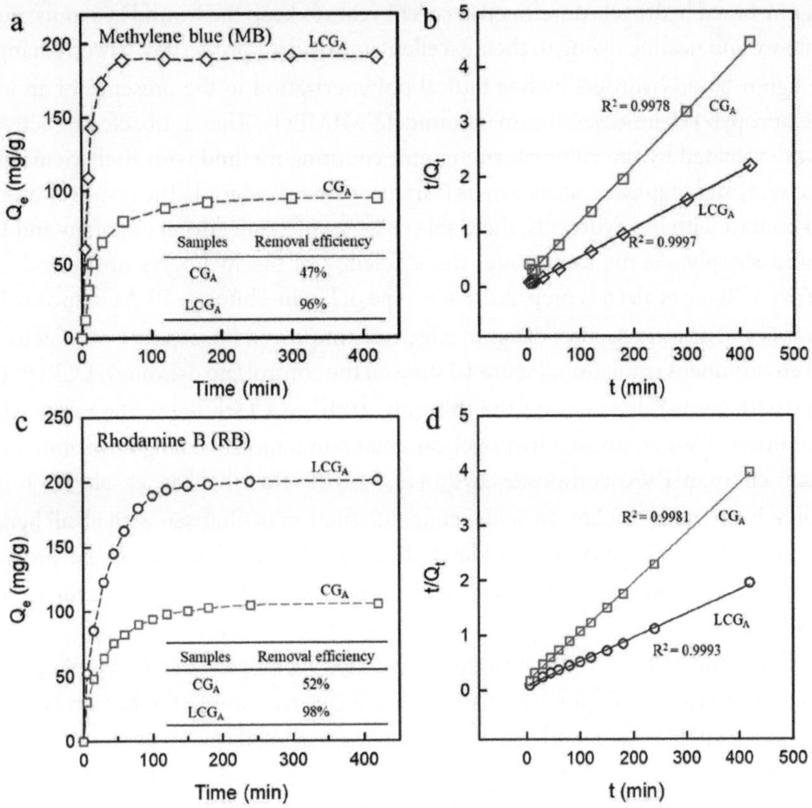

Figure 9. Dye adsorption capacity of lignin-free hydrogel (CGA) and lignin-containing hydrogel (LCGA) for methylene blue (MB) (a) and rhodamine B (RB) (c). The pseudo-second-order model of CGA and LCGA for MB (b) and RB (d) (organic dye: 50 mg L^{-1} (100 mL), adsorbent (o.d.) 25 mg). Reprinted with permission from (75). Copyright 2019 American Chemical Society.

Super absorption capacity is also an important factor for dye absorbers. hence, super-swelling hydrogels are also widely used for dyes adsorption. Linjun Wu et al. (49) used alkali lignin as raw material to prepare Lignin-PVA superabsorbent hydrogel. The swelling rate of the hydrogel increased from 92 g/g to 456 g/g when the lignin concentration increased from 0% to 5%. The adsorption capacity of lignin-PVA hydrogel for rhodamine 6G, crystal violet, and methylene blue dyes were as high as 196,169 and 179 mg/g, respectively. The research group also used long-chain hydroxyethyl cellulose and short-chain PVA to prepare lignin-based hydrogels with a high swelling rate of up to 1220 g/g (78). Juan Domínguez-Robles et al. (55) used different industrial lignins (organosolv lignin, wheat straw soda lignin, and softwood kraft lignin) in ammonium and sodium hydroxide solutions to prepare lignin-based hydrogels with high swelling capacity. The resulting hydrogels showed a very wide water absorption capacity, ranging from 13 to 130 g/g. Studies have shown that the hydrogel can remove cationic dyes in wastewater and the hydrogel made of wheat straw soda lignin had stronger MB retention capacity.

Biomedical Field

Skin injury is one of the most common injuries in the human body. Unreasonable treatment programs often cause bacterial infections, tissue dehydration, and even more serious secondary trauma. Lignin-based hydrogels developed in recent years to keep the wound in a moist environment and promote wound healing owing to their excellent antioxidant properties. Yiwen Zhang et al. (*79*) prepared a lignin-based hydrogel by free radical polymerization in the presence of an ionic liquid, 3-butyl-1-isopropyl-1H-imidazol-3-ium bromide(VMIMBr). The antibacterial activity of the hydrogel was evaluated by the colony forming unit counting method with Escherichia coli (Gram-negative bacteria) and Staphylococcus aureus (Gram-positive bacteria). The results showed that after 24 hours of contact with the hydrogels, the viable colonies of Staphylococcus aureus and Escherichia coli decreased sharply. At the same time, the efficiency of the hydrogels on wound healing was explored. Yiwen Zhang et al. (*64*) prepared a new type of lignin-chitosan-PVA composite hydrogel as a wound dressing. The introduction of lignin effectively improved the capacity of protein adsorption and wound environment regulation. Figure 10 showed the control (no dressing), LCPH0 (no lignin), and LCPH1(with lignin) dressings on the 0th, 5th, 10th, and 15th days. The results showed that the lignin-chitosan-PVA composite hydrogel can maintain a moist healing environment and faster healing than chitosan-PVA composite hydrogel. Kartik Ravishankar et al. (*59*) prepared a biocompatible hydrogel by mixing an acidic aqueous solution of chitosan with alkali lignin. Studies have shown that NIH 3T3 mouse fibroblasts show good cell migration in the presence of the hydrogel. The characteristics indicated that the hydrogel may be the best choice for wound healing applications.

An excellent drug carrier has the advantages of good biocompatibility and biodegradability and requires a moist surface and affinity with tissues to reduces irritation to the human body. Xiaohong Wang et al. (*80*) prepared a sodium lignosulphonate-grafted poly(acrylic acid-co-poly(vinyl pyrrolidone)) hydrogel which showed high water swelling, good pH responsiveness, and excellent sustained release properties in the intestine. Studies have shown that the hydrogel swelled in a pH 7.4 solution and then shrank in a pH 1.2 solution. Therefore, the release behavior can be controlled by adjusting the pH of the medium. Amoxicillin was used as a model drug to study its release characteristics in the non-enzyme simulated gastrointestinal fluid in two pH conditions. Results indicated that the drug release characteristics of the hydrogel were sensitive to pH, and the cumulative release percentage increased rapidly in the first hour, so the drug can be delivered in a short time.

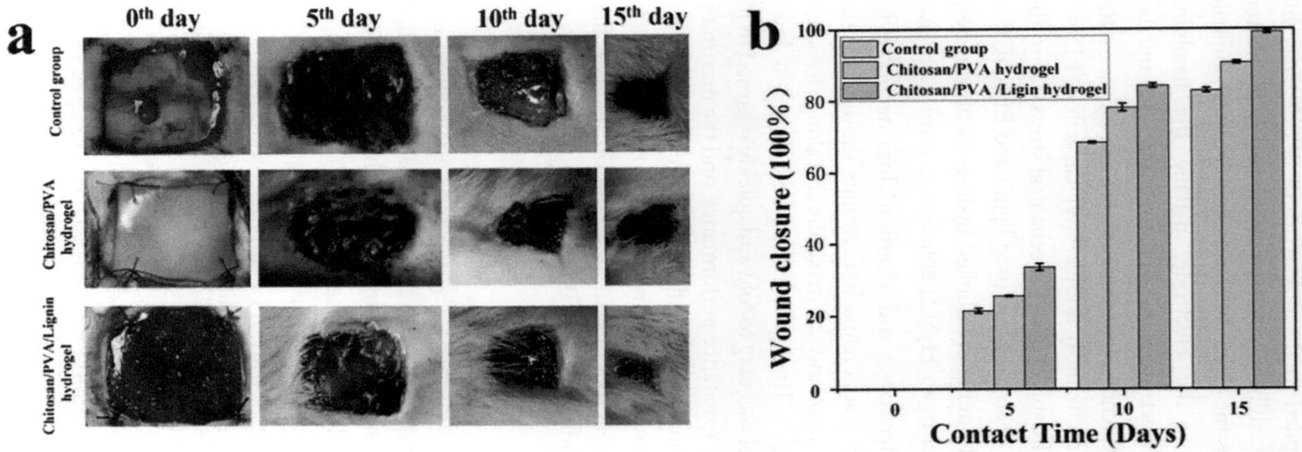

Figure 10. (a) Photographs of wounds with control group (no dressing), LCPH0 dressing, and LCPH1 dressing on the 0th, 5th, 10th and 15th day; (b) the wound recovery rate calculated from the wound recovery area treated with control group (no dressing), LCPH0, and LCPH1 dressing. Reprinted with permission from (64). Copyright 2019 Elsevier.

Supercapacitors

Materials of the traditional flexible wearable electronic devices have the problems of insufficient biocompatibility and mechanical properties that do not match human skin. Long-term wear causes discomfort and skin inflammation, which greatly affected the practical application of it. In recent years, lignin-based hydrogels have gradually been applied to the researches of flexible wearable electronics due to their excellent pore structure, good flexibility, and biocompatibility. Lignin-based hydrogel prepared by lignin and PEGDGE was served as the electrolyte and lignin/polyacrylonitrile nanofiber prepared by electrospinning was used as the electrode, making the flexibility supercapacitor (*81*). Results showed that the electrode/electrolyte interface was continuous and had good adhesion, resulting in lower contact resistance. Cyclic voltammetry (CV) and galvanostatic charge–discharge (GCD) curves indicated ideal electrochemical double-layer capacitor (EDLC) characteristics. The device had excellent elasticity and durability under mechanical stress due to the flexibility and the strong continuous interface. Tao Liu et al. (*82*) also used lignin and PEGDGE to prepare lignin-based hydrogel. In this system, a hybrid double crosslinked hydrogel with excellent compressibility was created by immersing the hydrogel in an H_2SO_4 solution to form a physical crosslinking network. The hydrogel was used as an electrolyte and a carbon cloth deposited by polyaniline (PANI) was used as an electrode to prepare a supercapacitor. Results showed that the double-crosslinked hydrogel had higher mechanical strength and the maximum breaking stress was 4.74 Mpa (40 times of the chemical crosslinked hydrogel). The lignin hydrogel did not breakdown after 100 consecutive cycles, indicating that the hydrogel was easy to fold and quickly returned to its original state. Results showed that the device has good capacitance performance, and the maximum specific capacitance reached 190 F/g (better than previously reported flexible supercapacitors based on hydrogel electrolyte). Junqi Cai et al. (*83*) used lignin and polyvinyl alcohol as raw materials to prepare a conductive hydrogel with high strength and super toughness. The composite hydrogel had a strong tensile strength of up to 13.3 MPa and good conductivity of 1.0 S/m.

Other Applications

Some lignin-based hydrogels have pH or temperature responsiveness, and the volume can change with pH or temperature. This ability makes the hydrogels smart and expands the application of lignin-based hydrogels. Liu, Wu et al. (*41*) prepared smart lignin hydrogels with pH responsive and magnetic properties. The hydrogel has a different swelling rate under different PH, the optimum swelling condition is PH=6.8. Jin, Can et al. (*84*) prepared lignosulfonate/N-isopropylacrylamide(NIPAM)/itaconic acid (IA) hydrogels with pH and temperature responsive properties. All the hydrogels showed temperature sensitive behaviors and exhibited pH sensitivity in the range of pH 3.0 to 9.1.

Aerogels can be obtained after drying the hydrogels which can expand the application of hydrogels. Aerogels with strong mechanical performance were prepared through cellulose and lignin (*85*). It would have promising implications for sound-adsorption materials which developed a path for lignin-based hydrogels to an advanced utilization. Jingxian Jiang et al. (*86*) freeze-dried the lignin-based hydrogel to obtain a hydrophobic, compressible aerogel with an absorption capacity of 5 g/g~22 g/g for different oils and make it possible for water oil separation field.

Conclusion and Future Perspective

In conclusion, researchers have extensively introduced lignin into hydrogels to prepare lignin-based hydrogels with excellent performance due to the abundance of lignin. The preparation methods include chemical crosslinking and physical crosslinking. Chemical crosslinking methods mainly include free radical polymerization, adding a crosslinker or chemical reaction of lignin functional groups. Physical crosslinking methods are usually formed by electrostatic or hydrogen bond interaction. These lignin-based hydrogels often include other polymers to impart additional properties to the hydrogels, such as PEG, PVA, chitosan, cellulose, graphene, and more. Lignin-based hydrogels are widely used in the field of water purification due to their rich active groups and excellent swelling ability, which can effectively remove heavy metal ions and dyes from water. They are also used in the biomedical field because of their excellent biocompatibility and antioxidant properties. Furthermore, electronic devices and other filed are also covered.

Although researchers have developed many excellent lignin-based hydrogels, there are still some problems:

(1) The chemical crosslinking process usually requires the addition of many toxic reagents, while the hydrogel prepared by physical crosslinking usually has poor mechanical properties.
(2) Most lignin-based hydrogels are formed by crosslinking with other monomers or polymers. The possible reason is that industrial lignin contains a lot of impurities, which cannot be completely purified to expose more active groups. Furthermore, the complexity of the molecular structure of lignin increases the difficulty of gel formation (compared to cellulose).
(3) There is a lack of in-depth research on the toxicity of lignin-based hydrogels, and there are still safety problems in applying them to the fields of medicine and food.

Therefore, how to use a greener method to prepare lignin-based hydrogels with excellent performance, how to increase the content of lignin-based hydrogels, and how to ensure the safety of lignin-based hydrogels are the issues that the researchers need to disclose in the future.

References

1. Wen, J. L.; Sun, S. L.; Xue, B. L.; Sun, R. C. Recent Advances in Characterization of Lignin Polymer by Solution-State Nuclear Magnetic Resonance (NMR) Methodology. *Mater.* **2013,** *6* (1), 359–391.
2. Yang, S.; Yuan, T. Q.; Sun, R. C. Structural Elucidation of Whole Lignin in Cell Walls of Triploid of Populus tomentosa Carr. *ACS Sustain. Chem. Eng.* **2016,** *4* (3), 1006–1015.
3. Liitia, T. M.; Maunu, S. L.; Hortling, B.; Toikka, M.; Kilpelainen, I. Analysis of technical lignins by two- and three-dimensional NMR spectroscopy. *J. Agric. Food Chem.* **2003,** *51* (8), 2136–2143.
4. Liu, C. J.; Lei, F. H.; Li, P. F.; Jiang, J. X.; Wang, K. Borax crosslinked fenugreek galactomannan hydrogel as potential water-retaining agent in agriculture. *Carbohydr. Polym.* **2020,** *236*, 9.
5. Li, S. M.; Chen, G. Agricultural waste-derived superabsorbent hydrogels: Preparation, performance, and socioeconomic impacts. *J. Clean. Prod.* **2020,** *251*, 11.

6. Song, B.; Liang, H. X.; Sun, R. R.; Peng, P.; Jiang, Y.; She, D. Hydrogel synthesis based on lignin/sodium alginate and application in agriculture. *Int. J. Biol. Macromol.* **2020**, *144*, 219–230.
7. Dreiss, C. A. Hydrogel design strategies for drug delivery. *Curr. Opin. Colloid Interface Sci.* **2020**, *48*, 1–17.
8. Han, Z. L.; Wang, P.; Mao, G. Y.; Yin, T. H.; Zhong, D. M.; Yiming, B. R. B.; Hu, X. C.; Jia, Z.; Nian, G. D.; Qu, S. X.; Yang, W. Dual pH-Responsive Hydrogel Actuator for Lipophilic Drug Delivery. *ACS Appl. Mater. Interfaces* **2020**, *12* (10), 12010–12017.
9. Hamedi, H.; Moradi, S.; Hudson, S. M.; Tonelli, A. E. Chitosan based hydrogels and their applications for drug delivery in wound dressings: A review. *Carbohydr. Polym.* **2018**, *199*, 445–460.
10. Sun, W. X.; Jiang, H. T.; Wu, X.; Xu, Z. Y.; Yao, C.; Wang, J.; Qin, M.; Jiang, Q.; Wang, W.; Shi, D. Q.; Cao, Y. Strong dual-crosslinked hydrogels for ultrasound-triggered drug delivery. *Nano Res.* **2019**, *12* (1), 115–119.
11. Ghawanmeh, A. A.; Ali, G. A. M.; Algarni, H.; Sarkar, S. M.; Chong, K. F. Graphene oxide-based hydrogels as a nanocarrier for anticancer drug delivery. *Nano Res.* **2019**, *12* (5), 973–990.
12. Peers, S.; Montembault, A.; Ladaviere, C. Chitosan hydrogels for sustained drug delivery. *J. Controlled Release* **2020**, *326*, 150–163.
13. Rong, Q. F.; Lei, W. W.; Liu, M. J. Conductive Hydrogels as Smart Materials for Flexible Electronic Devices. *Chem.-Eur. J.* **2018**, *24* (64), 16930–16943.
14. Ryplida, B.; Lee, K. D.; In, I.; Park, S. Y. Light-Induced Swelling-Responsive Conductive, Adhesive, and Stretchable Wireless Film Hydrogel as Electronic Artificial Skin. *Adv. Funct. Mater.* **2019**, *29* (32), 9.
15. Zhao, S. W.; Tseng, P.; Grasman, J.; Wang, Y.; Li, W. Y.; Napier, B.; Yavuz, B.; Chen, Y.; Howell, L.; Rincon, J.; Omenetto, F. G.; Kaplan, D. L. Programmable Hydrogel Ionic Circuits for Biologically Matched Electronic Interfaces. *Adv. Mater.* **2018**, *30* (25), 10.
16. Chang, Q.; Darabi, M. A.; Liu, Y. Q.; He, Y. F.; Zhong, W.; Mequanin, K.; Li, B. Y.; Lu, F.; Xing, M. M. Q. Hydrogels from natural egg white with extraordinary stretchability, direct-writing 3D printability and self-healing for fabrication of electronic sensors and actuators. *J. Mater. Chem. A* **2019**, *7* (42), 24626–24640.
17. Batista, R. A.; Espitia, P. J. P.; Quintans, J. D. S.; Freitas, M. M.; Cerqueira, M. A.; Teixeira, J. A.; Cardoso, J. C. Hydrogel as an alternative structure for food packaging systems. *Carbohydr. Polym.* **2019**, *205*, 106–116.
18. Rodriguez, L. M. D. L.; Hemar, Y. Prospecting the applications and discovery of peptide hydrogels in food. *Trends Food Sci. Tech.* **2020**, *104*, 37–48.
19. Lim, H. P.; Ho, K. W.; Singh, C. K. S.; Ooi, C. W.; Tey, B. T.; Chan, E. S. Pickering emulsion hydrogel as a promising food delivery system: Synergistic effects of chitosan Pickering emulsifier and alginate matrix on hydrogel stability and emulsion delivery. *Food Hydrocoll.* **2020**, *103*, 10.
20. Li, Q. F.; Du, X. D.; Jin, L.; Hou, M. M.; Wang, Z. L.; Hao, J. H. Highly luminescent hydrogels synthesized by covalent grafting of lanthanide complexes onto PNIPAM via one-pot free radical polymerization. *J. Mater. Chem. C* **2016**, *4* (15), 3195–3201.

21. Sadeghi, I.; Yi, H.; Asatekin, A. A Method for Manufacturing Membranes with Ultrathin Hydrogel Selective Layers for Protein Purification: Interfacially Initiated Free Radical Polymerization (IIFRP). *Chem. Mater.* **2018**, *30* (4), 1265–1276.
22. Ji, H. F.; Song, X.; Cheng, H. T.; Luo, L. B.; Huang, J. B.; He, C.; Yin, J. R.; Zhao, W. F.; Qiu, L.; Zhao, C. S. Biocompatible In Situ Polymerization of Multipurpose Polyacrylamide-Based Hydrogels on Skin via Silver Ion Catalyzation. *ACS Appl. Mater. Interfaces* **2020**, *12* (28), 31079–31089.
23. Hu, D.-n.; Sun, Y.-f.; Tao, L.; Yuan, J.-y.; Sui, X.-f.; Wei, Y. Environmentally Responsive Hydrogels Based on Cellulose. *Acta Polym. Sin.* **2020**, *51* (8), 880–889.
24. Zhang, H.; Luan, Q.; Huang, Q. D.; Tang, H.; Huang, F. H.; Li, W. L.; Wan, C. Y.; Liu, C. S.; Xu, J. Q.; Guo, P. M.; Zhou, Q. A facile and efficient strategy for the fabrication of porous linseed gum/cellulose superabsorbent hydrogels for water conservation. *Carbohydr. Polym.* **2017**, *157*, 1830–1836.
25. Liu, M.; Zhang, Y.; Li, J.; Zhou, C. Chitin-natural clay nanotubes hybrid hydrogel. *Int. J. Biol. Macromol.* **2013**, *58*, 23–30.
26. Yu, J.; Chen, X. X.; Yang, Y. X.; Zhao, X. D.; Chen, X. J.; Jing, T. H.; Zhou, Y.; Xu, J. P.; Zhang, Y. F.; Cheng, Y. L. Construction of supramolecular hydrogels using imidazolidinyl urea as hydrogen bonding reinforced factor. *J. Mater. Chem. B* **2020**, *8* (15), 3058–3063.
27. Song, G. S.; Zhao, Z. Y.; Peng, X.; He, C. C.; Weiss, R. A.; Wang, H. L. Rheological Behavior of Tough PVP-in Situ-PAAm Hydrogels Physically Cross-Linked by Cooperative Hydrogen Bonding. *Macromolecules* **2016**, *49* (21), 8265–8273.
28. Huang, S. C.; Xia, X. X.; Fan, R. X.; Qan, Z. G. Programmable Electrostatic Interactions Expand the Landscape of Dynamic Functional Hydrogels. *Chem. Mater.* **2020**, *32* (5), 1937–1945.
29. Criado-Gonzalez, M.; Wagner, D.; Fores, J. R.; Blanck, C.; Schmutz, M.; Chaumont, A.; Rabineau, M.; Schlenoff, J. B.; Fleith, G.; Combet, J.; Schaaf, P.; Jierry, L.; Boulmedais, F. Supramolecular Hydrogel Induced by Electrostatic Interactions between Polycation and Phosphorylated-Fmoc-Tripeptide. *Chem. Mater.* **2020**, *32* (5), 1946–1956.
30. Lima-Tenorio, M. K.; Tenorio-Neto, E. T.; Guilherme, M. R.; Garcia, F. P.; Nakamura, C. V.; Pineda, E. A. G.; Rubira, A. F. Water transport properties through starch-based hydrogel nanocomposites responding to both pH and a remote magnetic field. *Chem. Eng.J.* **2015**, *259*, 620–629.
31. Zhang, Q.; Chen, B.; Tao, L.; Yan, M.; Chen, L.; Wei, Y. Microorganism inspired hydrogels: hierarchical super/macro-porous structure, rapid swelling rate and high adsorption. *RSC Adv.* **2014**, *4* (61), 32475–32481.
32. Liang, R. X.; Yu, H. J.; Wang, L.; Lin, L.; Wang, N.; Naveed, K. U. R. Highly Tough Hydrogels with the Body Temperature-Responsive Shape Memory Effect. *ACS Appl. Mater. Interfaces* **2019**, *11* (46), 43563–43572.
33. Rogina, A.; Ressler, A.; Matie, I.; Gallego Ferrer, G.; Marijanovic, I.; Ivankovic, M.; Ivankovic, H. Cellular hydrogels based on pH-responsive chitosan-hydroxyapatite system. *Carbohydr. Polym.* **2017**, *166*, 173–182.
34. Green, R. Elastic and conductive hydrogel electrodes. *Nat. Biomed. Eng.* **2019**, *3* (1), 9–10.

35. Qian, C.; Higashigaki, T.; Asoh, T. A.; Uyama, H. Anisotropic Conductive Hydrogels with High Water Content. *ACS Appl. Mater. Interfaces* **2020**, *12* (24), 27518–27525.
36. Zhang, X. F.; Ma, X. F.; Hou, T.; Guo, K. C.; Yin, J. Y.; Wang, Z. G.; Shu, L.; He, M.; Yao, J. F. Inorganic Salts Induce Thermally Reversible and Anti-Freezing Cellulose Hydrogels. *Angew. Chem.-Int. Edit.* **2019**, *58* (22), 7366–7370.
37. Huang, B.; Liu, M. X.; Zhou, C. R. Chitosan composite hydrogels reinforced with natural clay nanotubes. *Carbohydr. Polym.* **2017**, *175*, 689–698.
38. Thakur, S.; Chaudhary, J.; Kumar, V.; Thakur, V. K. Progress in pectin based hydrogels for water purification: Trends and challenges. *J. Environ. Manage.* **2019**, *238*, 210–223.
39. Zerpa, A.; Pakzad, L.; Fatehi, P. Hardwood Kraft Lignin-Based Hydrogels: Production and Performance. *Acs Omega* **2018**, *3* (7), 8233–8242.
40. Ji, X.; Zhang, Z.; Chen, J.; Yang, G.; Chen, H.; Lucia, L. A. Synthesis and Characterization of Alkali Lignin-based Hydrogels from Ionic Liquids. *Bioresources* **2017**, *12* (3), 5395–5406.
41. Liu, W.; Ye, Z. H.; Liu, D. X.; Wu, Z. P. Hydrogels Derived from Lignin with pH Responsive and Magnetic Properties. *Bioresources* **2018**, *13* (4), 7281–7293.
42. Meng, Y.; Liu, X. Q.; Li, C. X.; Liu, H.; Cheng, Y.; Lu, J.; Zhang, K.; Wang, H. S. Super-swelling lignin-based biopolymer hydrogels for soil water retention from paper industry waste. *Int. J. Biol. Macromol.* **2019**, *135*, 815–820.
43. Banerjee, S. L.; Swift, T.; Hoskins, R.; Rimmer, S.; Singha, N. K. A muscle mimetic polyelectrolyte-nanoclay organic-inorganic hybrid hydrogel: its self-healing, shape-memory and actuation properties. *J. Mater. Chem. B* **2019**, *7* (9), 1475–1493.
44. Tong, R. P.; Chen, G. X.; Pan, D. H.; Tian, J. F.; Qi, H. S.; Li, R. A.; Lu, F. C.; He, M. H. Ultrastretchable and Antifreezing Double-Cross-Linked Cellulose Ionic Hydrogels with High Strain Sensitivity under a Broad Range of Temperature. *ACS Sustain. Chem. Eng.* **2019**, *7* (16), 14256–14265.
45. Jiang, P.; Sheng, X. R.; Yu, S.; Li, H. M.; Lu, J.; Zhou, J. H.; Wang, H. S. Preparation and characterization of thermo-sensitive gel with phenolated alkali lignin. *Sci. Rep.* **2018**, *8*, 10.
46. Rajan, K.; Mann, J. K.; English, E.; Harper, D. P.; Carrier, D. J.; Rials, T. G.; Labbe, N.; Chmely, S. C. Sustainable Hydrogels Based on Lignin-Methacrylate Copolymers with Enhanced Water Retention and Tunable Material Properties. *Biomacromolecules* **2018**, *19* (7), 2665–2672.
47. Shen, X.; Berton, P.; Shamshina, J. L.; Rogers, R. D. Preparation and comparison of bulk and membrane hydrogels based on Kraft- and ionic-liquid-isolated lignins. *Green Chem.* **2016**, *18* (20), 5607–5620.
48. Dai, L.; Zhu, W. Y.; Lu, J. S.; Kong, F. G.; Si, C. L.; Ni, Y. H. A lignin-containing cellulose hydrogel for lignin fractionation. *Green Chem.* **2019**, *21* (19), 5222–5230.
49. Wu, L. J.; Huang, S. Q.; Zheng, J.; Qiu, Z. J.; Lin, X. L.; Qin, Y. L. Synthesis and characterization of biomass lignin-based PVA super-absorbent hydrogel. *Int. J. Biol. Macromol.* **2019**, *140*, 538–545.
50. Ciolacu, D.; Cazacu, G. New Green Hydrogels Based on Lignin. *J. Nanosci. Nanotechnol.* **2018**, *18* (4), 2811–2822.

51. Musilova, L.; Mracek, A.; Kovalcik, A.; Smolka, P.; Minarik, A.; Humpolicek, P.; Vicha, R.; Ponizil, P. Hyaluronan hydrogels modified by glycinated Kraft lignin: Morphology, swelling, viscoelastic properties and biocompatibility. *Carbohydr. Polym.* **2018**, *181*, 394–403.

52. Dai, L.; Ma, M. S.; Xu, J. K.; Si, C. L.; Wang, X. H.; Liu, Z.; Ni, Y. H. All-Lignin-Based Hydrogel with Fast pH-Stimuli Responsiveness for Mechanical Switching and Actuation. *Chem. Mater.* **2020**, *32* (10), 4324–4330.

53. Mazloom, N.; Khorassani, R.; Zohury, G. H.; Emami, H.; Whalen, J. Lignin-based hydrogel alleviates drought stress in maize. *Environ. Exp. Bot.* **2020**, *175*, 8.

54. Teng, X. X.; Xu, H.; Song, W. J.; Shi, J. W.; Xin, J. N.; Hiscox, W. C.; Zhang, J. W. Preparation and Properties of Hydrogels Based on PEGylated Lignosulfonate Amine. *Acs Omega* **2017**, 2 (1), 251–259.

55. Dominguez-Robles, J.; Peresin, M. S.; Tamminen, T.; Rodriguez, A.; Larraneta, E.; Jaaskelainen, A. S. Lignin-based hydrogels with "super-swelling" capacities for dye removal. *Int. J. Biol. Macromol.* **2018**, *115*, 1249–1259.

56. Larraneta, E.; Imizcoz, M.; Toh, J. X.; Irwin, N. J.; Ripolin, A.; Perminova, A.; Dominguez-Robles, J.; Rodriguez, A.; Donnelly, R. F. Synthesis and Characterization of Lignin Hydrogels for Potential Applications as Drug Eluting Antimicrobial Coatings for Medical Materials. *ACS Sustain. Chem. Eng.* **2018**, *6* (7), 9037–9046.

57. Lin, K. T.; Ma, R. S.; Wang, P. P.; Xin, J. N.; Zhang, J. W.; Wolcott, M. P.; Zhang, X. Deep Eutectic Solvent Assisted Facile Synthesis of Lignin-Based Cryogel. *Macromolecules* **2019**, *52* (1), 227–235.

58. Hatakeyama, H.; Hatakeyama, T. Thermal properties of freezing bound water restrained by sodium lignosulfonate-based polyurethane hydrogels. *J. Therm. Anal. Calorim.* **2019**, *135* (4), 2039–2048.

59. Ravishankar, K.; Venkatesan, M.; Desingh, R. P.; Mahalingam, A.; Sadhasivam, B.; Subramaniyam, R.; Dhamodharan, R. Biocompatible hydrogels of chitosan-alkali lignin for potential wound healing applications. Mater. Sci. Eng. C. **2019**, *102*, 447–457.

60. Lu, J. S.; Zhu, W. Y.; Dai, L.; Si, C. L.; Ni, Y. H. Fabrication of thermo- and pH-sensitive cellulose nanofibrils-reinforced hydrogel with biomass nanoparticles. *Carbohydr. Polym.* **2019**, *215*, 289–295.

61. Zhang, L. L.; Lu, H. L.; Yu, J.; Fan, Y. M.; Ma, J. X.; Wang, Z. G. Contribution of lignin to the microstructure and physical performance of three-dimensional lignocellulose hydrogels. *Cellulose* **2019**, *26* (4), 2375–2388.

62. Morales, A.; Labidi, J.; Gullon, P. Assessment of green approaches for the synthesis of physically crosslinked lignin hydrogels. *J. Ind. Eng. Chem.* **2020**, *81*, 475–487.

63. Kalinoski, R. M.; Shi, J. Hydrogels derived from lignocellulosic compounds: Evaluation of the compositional, structural, mechanical and antimicrobial properties. *Ind. Crop Prod.* **2019**, *128*, 323–330.

64. Zhang, Y. W.; Jiang, M. M.; Zhang, Y. Q.; Cao, Q. P.; Wang, X.; Han, Y.; Sun, G. W.; Li, Y.; Zhou, J. H. Novel lignin-chitosan-PVA composite hydrogel for wound dressing. *Mater. Sci. Eng. C* **2019**, *104*, 8.

65. Li, F. F.; Wang, X. L.; Yuan, T. Q.; Sun, R. C. A lignosulfonate-modified graphene hydrogel with ultrahigh adsorption capacity for Pb(II) removal. *J. Mater. Chem. A* **2016**, *4* (30), 11888–11896.
66. Jiang, C. L.; Wang, X. H.; Hou, B. X.; Hao, C.; Li, X.; Wu, J. B. Construction of a Lignosulfonate-Lysine Hydrogel for the Adsorption of Heavy Metal Ions. *J. Agric. Food. Chem.* **2020**, *68* (10), 3050–3060.
67. Tian, R. R.; Liu, Q.; Zhang, W.; Zhang, Y. Y. Preparation of Lignin-Based Hydrogel and Its Adsorption on Cu2+ Ions and Co2+ Ions in Wastewaters. *J. Inorg. Organomet. Polym. Mater.* **2018**, *28* (6), 2545–2553.
68. Sun, X. F.; Hao, Y. W.; Cao, Y. Y.; Zeng, Q. H. Superadsorbent hydrogel based on lignin and montmorillonite for Cu(II) ions removal from aqueous solution. *Int. J. Biol. Macromol.* **2019**, *127*, 511–519.
69. Ma, Y. L.; Lv, L.; Guo, Y. R.; Fu, Y. J.; Shao, Q.; Wu, T. T.; Guo, S. J.; Sun, K.; Guo, X. K.; Wujcik, E. K.; Guo, Z. H. Porous lignin based poly (acrylic acid)/organo-montmorillonite nanocomposites: Swelling behaviors and rapid removal of Pb (II) ions. *Polymer* **2017**, *128*, 12–23.
70. Liu, M. Y.; Liu, Y.; Shen, J. J.; Zhang, S. Y.; Liu, X. Y.; Chen, X. X.; Ma, Y. L.; Ren, S. X.; Fang, G. Z.; Li, S. J.; Li, C. T.; Sun, T. Simultaneous removal of Pb2+, Cu2+ and Cd2+ ions from wastewater using hierarchical porous polyacrylic acid grafted with lignin. *J. Hazard. Mater.* **2020**, *392*, 10.
71. Ma, J. H.; Li, T.; Liu, Y. T.; Cai, T.; Wei, Y. F.; Dong, W. Y.; Chen, H. Rice husk derived double network hydrogel as efficient adsorbent for Pb(II), Cu(II) and Cd(II) removal in individual and multicomponent systems. *Bioresour. Technol.* **2019**, *290*, 7.
72. Meng, Y.; Li, C. X.; Liu, X. Q.; Lu, J.; Cheng, Y.; Xiao, L. P.; Wang, H. S. Preparation of magnetic hydrogel microspheres of lignin derivate for application in water. *Sci. Total Environ.* **2019**, *685*, 847–855.
73. Penaranda, J. E.; Sabino, M. A. Effect of the presence of lignin or peat in IPN hydrogels on the sorption of heavy metals. *Polym. Bull.* **2010**, *65* (5), 495–508.
74. Dax, D.; Chavez, M. S.; Xu, C. L.; Willfor, S.; Mendonca, R. T.; Sanchez, J. Cationic hemicellulose-based hydrogels for arsenic and chromium removal from aqueous solutions. *Carbohydr. Polym.* **2014**, *111*, 797–805.
75. Zhang, L. L.; Lu, H. L.; Yu, J.; McSporran, E.; Khan, A.; Fan, Y. M.; Yang, Y. Q.; Wang, Z. G.; Ni, Y. H. Preparation of High-Strength Sustainable Lignocellulose Gels and Their Applications for Antiultraviolet Weathering and Dye Removal. ACS Sustain. *Chem. Eng.* **2019**, 7 (3), 2998–3009.
76. Li, J. R.; Li, H.; Yuan, Z.; Fang, J.; Chang, L. F.; Zhang, H. Y.; Li, C. L. Role of sulfonation in lignin-based material for adsorption removal of cationic dyes. *Int. J. Biol. Macromol.* **2019**, *135*, 1171–1181.
77. Wang, Y.; Xiong, Y.; Wang, J. Y.; Zhang, X. D. Ultrasonic-assisted fabrication of montmorillonite-lignin hybrid hydrogel: Highly efficient swelling behaviors and super-sorbent for dye removal from wastewater. *Colloid Surf. A-Physicochem. Eng. Asp.* **2017**, *520*, 903–913.

78. Huang, S. Q.; Wu, L. J.; Li, T. Z.; Xu, D. Y.; Lin, X. L.; Wu, C. D. Facile preparation of biomass lignin-based hydroxyethyl cellulose super-absorbent hydrogel for dye pollutant removal. *Int. J. Biol. Macromol.* **2019**, *137*, 939–947.
79. Zhang, Y.; Yuan, B.; Zhang, Y.; Cao, Q.; Yang, C.; Li, Y.; Zhou, J. Biomimetic lignin/poly(ionic liquids) composite hydrogel dressing with excellent mechanical strength, self-healing properties, and reusability. *Chem. Eng.J.* **2020**, *400*.
80. Wang, X. H.; Zhou, Z. L.; Guo, X. W.; He, Q.; Hao, C.; Ge, C. W. Ultrasonic-assisted synthesis of sodium lignosulfonate-grafted poly(acrylic acid-co-poly(vinyl pyrrolidone)) hydrogel for drug delivery. *RSC Adv.* **2016**, *6* (42), 35550–35558.
81. Park, J. H.; Rana, H. H.; Lee, J. Y.; Park, H. S. Renewable flexible supercapacitors based on all-lignin-based hydrogel electrolytes and nanofiber electrodes. *J. Mater. Chem. A* **2019**, *7* (28), 16962–16968.
82. Liu, T.; Ren, X. L.; Zhang, J. M.; Liu, J.; Ou, R. X.; Guo, C. G.; Yu, X. Y.; Wang, Q. W.; Liu, Z. Z. Highly compressible lignin hydrogel electrolytes via double-crosslinked strategy for superior foldable supercapacitors. *J. Power Sources* **2020**, *449*, 8.
83. Cai, J.; Zhang, X.; Liu, W.; Huang, J.; Qiu, X. Synthesis of highly conductive hydrogel with high strength and super toughness. *Polymer* **2020**, *202*.
84. Jin, C.; Song, W. J.; Liu, T.; Xin, J. N.; Hiscox, W. C.; Zhang, J. W.; Liu, G. F.; Kong, Z. W. Temperature and pH Responsive Hydrogels Using Methacrylated Lignosulfonate Cross-Linker: Synthesis, Characterization, and Properties. *ACS Sustain. Chem. Eng.* **2018**, *6* (2), 1763–1771.
85. Wang, C.; Xiong, Y.; Fan, B. T.; Yao, Q. F.; Wang, H. W.; Jin, C. D.; Sun, Q. F. Cellulose as an adhesion agent for the synthesis of lignin aerogel with strong mechanical performance, Sound-absorption and thermal Insulation. *Sci. Rep.* **2016**, *6*, 9.
86. Jiang, J. X.; Zhang, Q. H.; Zhan, X. L.; Chen, F. Q. Renewable, Biomass-Derived, Honeycomblike Aerogel As a Robust Oil Absorbent with Two-Way Reusability. *ACS Sustain. Chem. Eng.* **2017**, *5* (11), 10307–10316.

Chapter 10

Prospects and Challenges of Using Lignin for Thermoplastic Materials

Aditi Nagardeolekar,[1] Mathew Ovadias,[1] Prajakta Dongre,[1] and Biljana Bujanovic[1,2,*]

[1]Department of Chemical Engineering, State University of New York College of Environmental Science and Forestry, Syracuse, New York 13210, United States

[2]USDA-FS-Forest Products Laboratory, Madison, Wisconsin 53726, United States

[*]Email: biljana.bujanovic@usda.gov.

Lignin is a multifunctional biopolymer with high application potential in the synthesis of thermoplastic materials. Industrial and investigational lignins have been incorporated into various commercial polymers, including biobased polymers, to improve their biodegradability, cost, and thermomechanical stability. Lignin is also expected to bring additional benefits to the formulations, such as flame retardancy, UV-protecting ability, and antioxidant and anti-weathering capacity. Strategies focused on lignin size reduction, fractionation, and modification via esterification, etherification, and graft copolymerization, and plasticization and compatibilization have been employed to improve the physicochemical properties of lignin-based thermoplastic materials as promising alternatives to petroleum-based plastics.

Introduction

The high carbon footprint of petroleum-derived plastics, and the dwindling petroleum supplies have brought bioplastics in sharp focus as sustainable alternatives to synthetic plastics (*1*). However, only 1 % of the 350 million tons of plastic produced annually are sourced from biomass (*1*, *2*). Hence, research and industrial efforts have been urgently directed towards increasing their use for a more sustainable world. Bioplastics is an umbrella term that includes the plastics that are derived from bio-based/renewable sources, as well as those that are biodegradable (*1*). Examples of the biodegradable bioplastics are polyhydroxyalkanoates (PHAs) and polylactic acid (PLA), which contribute to 12 % and 15 % of the worldwide bioplastics production, respectively (*3*). There are examples of bioplastics derived from biorenewable platform chemicals, which themselves are non-biodegradable, such as bio-polyethylene (PE). The most readily available biodegradable bioplastics may be derived from biopolymers such as cellulose, chitin, starch, and lignin produced in biological systems (*3*). However, in contrast to their availability, their exploitation in the production of plastics is only an emerging field.

Thermoplastic materials are a subset of plastics. These are polymeric materials that can be thermally softened to a flowable state as the macromolecular chains slide past each other, molded into desired shapes, cooled, and then reprocessed by passing through the same cycle while having a limited effect on their inherent properties, allowing for multiple re-use cycles. As a result, desirable properties of thermoplastics include low melt viscosity, good processability, and high thermal and mechanical stability (*4*). Thus, research efforts are focused on endowing biorenewable thermoplastics, including those produced from biopolymers such as lignin-based thermoplastics, with performances similar to their synthetic counterparts to produce sustainable alternatives.

Lignin is a main structural constituent of lignocellulosic biomass, and the most abundant aromatic polymer on earth. It is readily available as a byproduct of the pulp and paper industry. As of today, approximately 50-70 million tons of lignin are produced worldwide annually (*5, 6*). Kraft lignin (KL) is the largest source of lignin currently available, which accounts for 85 % of total lignin produced, and approximates 50 million ton/year (*7*). Kraft lignin is the product of sulfate (kraft) pulping process. Kraft pulping uses an aqueous solution of sodium hydroxide and sodium sulfide to solubilize lignin, and inevitably some hemicelluloses, to produce the so-called black liquor (*8, 9*). During this process, lignin undergoes degradation and condensation reactions, resulting in its structural modification (*10*). Conventionally, concentrated black liquor is burnt in the recovery boiler to regenerate chemicals and produce heat and power, whereas acidification of black liquor to recover lignin is rare. LignoBoost is one such patented process, which involves acidifying a portion of the black liquor, followed by filtration, and resuspension of lignin in conjunction with an elaborate washing procedure. It yields a low-ash kraft lignin of high quality (*11*). A second patented process, LignoForce, includes oxidation of the black liquor to control volatile sulfur-containing compound emissions in the acidification step that follows (*12*). Organosolv lignins (OSL) are recovered from different organosolv delignification processes conducted mostly on angiosperms. The delignification processes include those using organic solvents, such as the ethanol-based Alcell process (*13*), and organic acids, such as acetic acid-based acetosolv and formic acid-based formasolv. The Alcell process was commercialized in the 1990s (Repap Enterprise Inc.) and later discontinued. A total of 3,500 tons of Alcell lignin was sold as technical lignin by 2005, and widely used in the studies of lignin valorization, including in the production of thermoplastic materials (*14*). Soda lignins (e.g. commercially available Protobind) are recovered from spent liquors of soda pulping, which is conducted mainly on herbaceous species (grasses, and agricultural residues; AGR) and thereafter, soda lignins are mostly non-wood lignins (*13*). Kraft and sulfite pulping processes yield sulfur-containing lignins; KL and lignosulfonates (LS). Sulfur-free pulping processes, such as soda and organosolv produce sulfur-free lignins (Table 1).

Although the majority of the lignin is currently produced during kraft delignification (*19*), emerging lignocellulosic-based biorefineries are expected to produce large amounts of lignin as a byproduct. Hydrolytic lignin (HL) recovered from autohydrolysis-based pretreatment liquors, (e.g. steam explosion- or hot-water extracted lignin), and lignin recovered as an insoluble residue at the end of enzymatic hydrolysis (enzymatic hydrolysis lignin; EHL) are typical representatives of biorefinery lignins (*8*). It has been estimated that by 2022, 62 million tons of lignin per year may be generated from the cellulosic biofuel industry in the US alone, with about 60 % more lignin generated than needed to meet internal biorefinery energy needs by its combustion (*20*). Currently, the majority of lignin is used as a low-cost fuel due to its high heating value of 23.3-25.6 MJ/kg (*21*). However, the combustion of lignin is a thermodynamically and economically unfavorable utilization pathway. Hence, alternate commercial uses of lignin that preserve its aromatic structure must be further explored. Lignin is exceptionally valuable for the production of versatile polymeric materials, due to its polyphenolic nature, ease of derivatization and ability to participate in cross-linking reactions, intense UV-absorption and radical quenching capacity, biocompatibility, and antimicrobial properties (*22, 23*). Valorization of lignin through the production of polymeric materials, including thermoplastics would generate additional revenues for lignocellulosic biorefineries, and would help cellulosic biofuels become more cost-effective and competitive in price with fossil fuels. Moreover, exploiting lignin in polymer form would increase overall sustainability by preserving precious biochemical energy invested by plants into lignin biosynthesis (*5, 20, 21, 24*).

Table 1. Global Production of Technical Lignins

Lignin	*Pulping/Recovery method*	*Main producer*	*Global Production (t/year)*
Indulin	Kraft	Ingevity[a]	265,000 (*15*)
BioChoice	Kraft/LignoBoost	Domtar, Plymouth, NC	25,000 (*16*)–27,300 (*14*)
LignoBoost	Kraft/LignoBoost	Stora Enso, Sunila, Finland	50,000 (*14*)
LignoForce	Kraft/LignoForce	West Fraser, Canada	10,000 (*15*)
Lignosulfonates	Sulfite	Borregard	1 million (*15, 17*)
Protobind 1000	Soda	GreenValue, Granit, Switzerland	5,000–10,000 (*17*)
CIMV	Organosolv [b]	CIMV, DECHEMA/ Fraunhofer, Dedini	3,000 (*17*)

[a] Formerly part of WestRock, (Mead)Westvaco. [b] Acetic acid:formic acid:water (50:30:20)18.

Lignin-Based Thermoplastic Materials

Benefits of Incorporation of Lignin in Thermoplastic Materials

Lignin is a natural thermoplastic material up to certain temperatures, with a relatively high glass transition temperature (T_g) as compared to synthetic thermoplastic polymers. In its native form, lignin from hardwoods (HW) has a lower T_g (~65-85 °C) than from softwoods (SW) (~90-105 °C), mostly due to the less condensed nature of HW lignins (*19*). Lignins after isolation are characterized by a higher T_g. For example, softwood KL, LignoBoost has been reported to contain different fractions characterized by T_gs ranging from 74 °C to 190 °C (*25*). Hence, a thermoplastic containing lignin is expected to benefit from the relatively superior thermal properties of lignin.

Synthetic polymers are known for their excellent mechanical and chemical stability, as well as processability. However, they exhibit a poor thermal stability and high flammability. Therefore, their use in applications requiring flame-retardant grades such as in electronics, automotive, and aerospace industries is limited. Flame-retardant (FR) agents are used to confer fire-resistant properties with effects strongly dependent on their ability to form char during the thermal degradation. The char coats the polymeric materials and provides a good barrier against heat and oxygen diffusion; thus, reducing the combustion rate of the polymeric materials. Halogenated agents, which are commonly used in this area are toxic and may cause severe health issues. Therefore, alternative FR agents are being actively sought. In this context, lignin as a polyphenolic compound, which results in high char yields (50-60 % (*26*)) upon thermal decomposition in an inert atmosphere, presents a strong opportunity to improve the flame retardancy of thermoplastic polymers. Various lignin types have been studied as the next-generation FR agents (*26–32*). Beneficial effects of lignin on the thermal properties of the thermoplastics include an increase in the T_{max} and the residue/char yield (Table 2) for polymers such as polyamide (PA11), polyethylene terephthalate (PET), low-density polyethylene (LDPE), PLA, polypropylene (PP), and polyhydroxybutyrate (PHB) (*31, 33–40*).

Table 2. Thermal Properties of Select Polymers and Their Blends with Lignin

Lignin Thermoplastic Material			*Pure Polymer Degradation*			*Lignin Thermoplastic Material Degradation*			*Atm*
Lignin		*Polymer*							
Type	*Content (w/w, %)*		*$T_{5\%}$ (°C)*	*T_{max} (°C)*	*Residue (%)*	*$T_{5\%}$ (°C)*	*T_{max} (°C)*	*Residue (%)*	
LS (Um) (*33*)	20	PA11	396	423	0.7	285	468	13.5	N_2
KL (Um) (*33*)						341	435	12.4	N_2
HL (Um) (*34*)	5	PET	403	438	14.8	396	439	15.9	N_2
	20					365	436	19.5	N_2
	5		396	434	17.9	394	440	19.4	Air
	20					365	431	26.7	Air
KL (Um) (*35*)	5	LDPE	340[a]	349	N/A	390[a]	446	N/A	Air
KL (Dp) (*35*)	5				N/A	404[a]	450	N/A	Air
HL (Um) (*35*)	5				N/A	343[a]	405	N/A	Air
HL (Dp) (*35*)	5				N/A	393[a]	446	N/A	Air
KL (Um) (*36*)	5	PLA	264	314	7.5	337	367	8.6	N_2
KL (Ac) (*36*)	5					325	360	9.5	N_2
OSL (Um) (*36*)	5					335	365	9.7	N_2
OSL (Ac) (*36*)	5					318	365	7.3	N_2
SWOSL (Dc) (*37*)	1		309.7	331.6	9.7	292.2	346.1	5.9	N_2
	25					270.7	357.1	16.1	N_2
HWOSL (Dc) (*37*)	1					309.8	336.3	8.4	N_2
	25					266.7	339.1	12.5	N_2

Table 2. (Continued). Thermal Properties of Select Polymers and Their Blends with Lignin

Lignin Thermoplastic Material			*Pure Polymer Degradation*			*Lignin Thermoplastic Material Degradation*			*Atm*
Lignin									
Type	*Content (w/w, %)*	*Polymer*	$T_{5\%}$ (°C)	T_{max} (°C)	*Residue (%)*	$T_{5\%}$ (°C)	T_{max} (°C)	*Residue (%)*	
Soda (Um) (*38*)	20	PP	409	450	0.07	319	452	5.7	N_2
	30					297	453	9.3	N_2
Soda (Um)[b] (*38*)	20					308	461	6.13	N_2
	30					286	461	9.5	N_2
KL (Um) (*39*)	5		272	331[c]	0	280	365[c]	2.1	Air
	20					251	343[c]	11	Air
Alkali (Um)[b] (*31*)	30		324[d]	418	0	316[d]	472	12.5	N_2
Alkali (Gr)[b, e] (*31*)	30					361[d]	483	21.0	N_2
Acidolysis (Um) (*40*)	5	PHB	261	293	0	279	306	0	Air
	15					283	310	3	Air

Um: Unmodified; Dp: Depolymerized; Ac: Acetylated; Dc: Dodecanoated; Gr: Grafted; SWOSL: Softwood organosolv lignin; HWOSL: hardwood organosolv lignin. [a] $T_{10\%}$. [b] With maleated PP compatibilizer (5 % w/w). [c] $T_{50\%}$. [d] Onset of the DTG peak. [e] Lignin by grafting phosphorous-nitrogen.

Lignin, as a structural constituent of lignocellulosic biomass evolved to provide strength and rigidity to the cell wall, can improve the mechanical properties of the thermoplastic materials. The Young's modulus for pine lignin (*Pinus radiata*) has been found to be moisture-content and isolation-method dependent, and in the ~3-6 GPa range (*41*). KL has been reported to have the Young's modulus in a similar range (~3.9 GPa), with the tensile strength of ~23 MPa, and elongation at break of ~0.6 % (*42*). Alkylated (ethylated methylated) KL has shown a slightly lower Young's modulus at ~1.9 GPa, a higher tensile strength of ~37 MPa, and elongation at break of ~2 %, which were found to be quite similar to the values observed for polystyrene (PS) (*43*). There are reports where lignin has been employed as a toughening agent in PP formulations (at 5 % lignin), where PP-lignin formulation exhibited an increase in the impact energy and a decrease of 50 % in elongation at break (*39*). The addition of lignin has been shown to assist β-crystallization of PP, where lignin serves as a β-nucleation agent (*39*). Additionally, lignin can assist as a plasticizer and compatibilizer in the thermoplastic formulations. It can also improve their performance as a UV absorber and optical modifier, and serve as an alternative for or act synergistically with synthetic antioxidants due to its strong radical quenching capacity (*44*). This property in concert with its other intrinsic properties, including antimicrobial, antifungal, and cytotoxic activities, and potential to reduce oxygen permeability makes lignin a promising candidate in the production of food and active packaging (*45*). Lignin is well positioned in the development of strong light-weight materials due to its low density; e.g. Indulin softwood kraft lignin (SWKL) and hardwood organosolv lignin (HWOSL) are characterized by $\rho = 1.3$ gcm^{-3} and $\rho = 1.24$ gcm^{-3}, respectively (*46–49*).

Additional benefits offered by lignin as a component of thermoplastic materials include its low cost (KL; USD 70–1,300 per ton, depending on the application, versus synthetic thermoplastic precursors USD ~1,700 per ton (*15*)) and smaller carbon footprint, as compared to the petroleum-based alternatives (*1, 50*). This would also create additional revenue streams for biorefineries and the pulp and paper industries by providing an opportunity for utilization of lignins (*4*). Furthermore, there are reports of lignin-based thermoplastic materials benefitting from the antibacterial (*45*), antioxidizing (*51, 52*), UV-protecting (*53, 54*), anti-weathering (*55*), anti-aging (*56*), and biodegradable (*52, 57, 58*) properties of lignin.

Challenges in Incorporation of Lignin in Thermoplastic Materials

Despite the proposed advantages, utilization of lignin for value-added applications such as in thermoplastic materials, presents considerable challenges due to the inherently heterogenous nature of lignin. Lignin structure and properties are dependent on numerous factors, including the species, morphological tissue, and age of the lignocellulosic source from which lignin is isolated. Further complications are caused by the changes that it undergoes by different processing conditions during separation and recovery depending on the isolation process (*5*).

Due to strong intra- and inter-molecular interactions between the structural units of lignin, it shows immiscibility with a large number of polymers and displays a poor dispersion in the polymer matrix. This is especially true in the case of non-polar, hydrophobic polymers (*59*). Lignin self-interactions, including hydrogen bonds based on the hydroxyl (OH) groups (primarily free phenolic hydroxyl - PhOH groups), ether oxygens, and carboxylic acid groups, van der Waals attraction forces, and π-π stacking of aromatic rings (*60*) give rise to a brittle material of globular structure (*61*). The OH and carboxylic acid groups also make lignin a relatively polar molecule (*61, 62*), limiting its compatibility with hydrophobic polymers. Lignin's mutual insolubility and incompatibility with the polymers lead to its agglomeration, resulting in two-phase systems with hindered transfer of loads at the interface of the immiscible polymers; hence, a reduced strength (*48*). However, evidence of lignin's miscibility with various polymers, including polyethylene oxide (PEO) (*42*), polyvinyl pyrrolidone (PVP) (*63, 64*), and PET (*61*) has been driving research and industry efforts toward future applications of these miscible lignin-based thermoplastic blends. In this context, solubility parameters for polymers developed by Hildebrand and Hansen have been instrumental in predicting the miscibility of lignin with the other polymers. Hildebrand is a single solubility parameter, δ, while Hansen describes three separate solubility parameters, each representing different forces that impact miscibility; δ_d, δ_p, and δ_H (*65, 66*).

The molecular interactions are also responsible for the high T_g of lignin, which is advantageous in terms of thermal stability (*67*), but results in poor thermal processibility (*68*). Lignin also shows poor film-forming properties (*68, 69*). Furthermore, during processing in laboratory or industrial conditions, lignin easily changes color due to the abundance of chromophores formed during the separation/recovery process (*70*). Dark color has been seen as an obstacle for using technical lignins in polymer applications (*71, 72*), limiting lignin's suitability for applications where color is not a critical feature (*48*). Additional challenges associated with lignin as compared to synthetic polymers are its relatively lower molecular weight, higher polydispersity, uncertain reactivity, variable content of free phenolic, hydroxyl, and carboxylic groups, degree of cross-linking, and amount of residual carbohydrates (*4, 48*). To circumvent these challenges, different strategies have been developed to incorporate lignin in the thermoplastic formulations (Figure 1). Lignin can be mixed/blended with thermoplastic polymers in specific proportions to produce a material of desired properties and lower cost (73). Various strategies such as lignin size reduction and fractionation to produce more uniform lignin have been explored. Meanwhile, plasticization, compatibilization, and lignin derivatization have been employed to improve its miscibility in thermoplastic blends. Lignin graft copolymers (in both, one and multi-component blends) and elastomers have also been produced as a relatively new approach in producing lignin-based thermoplastic materials.

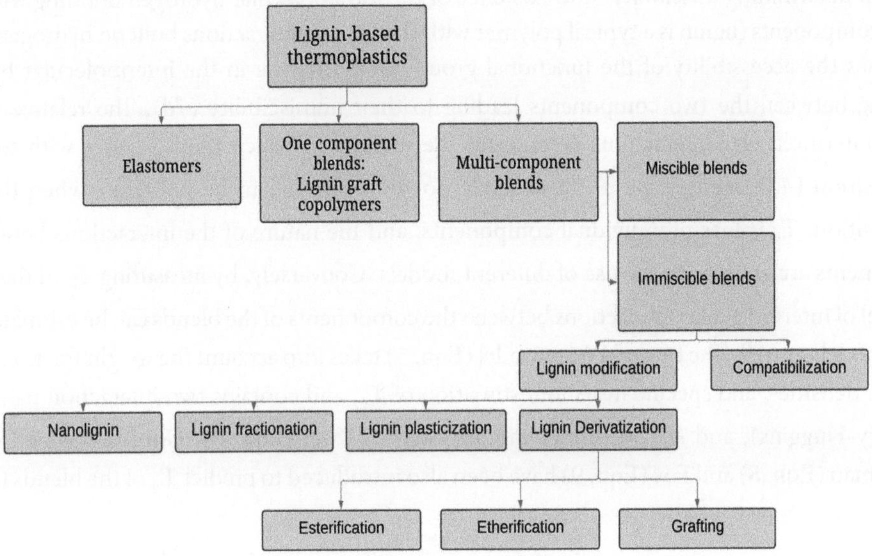

Figure 1. Lignin-thermoplastic materials and major strategies employed in their improvement.

Miscible Blends

A negative free energy of mixing ($\Delta Gm < 0$) is a prerequisite for spontaneous mixing of components to produce a single-phase system. ΔGm is determined by the enthalpy of mixing (ΔHm) and the entropy of mixing (ΔSm) as (73, 74):

$$\Delta Gm = \Delta Hm - T\Delta Sm \quad \text{Eqn. 1}$$

ΔHm is further related to the Flory-Huggins interaction parameter, χ_{AB} and the Hildebrand solubility parameter (δ) through χ_{AB} as (42):

$$\Delta Hm = \chi_{AB}\, RT\phi_A\phi_B \quad \text{Eqn. 2}$$

$$\chi_{AB} = \frac{V_r(\delta_A - \delta_B)^2}{RT} \quad \text{Eqn. 3}$$

$$\text{Hence, } \Delta Hm = (\delta_A - \delta_B)2\ \phi_A\phi_B \quad \text{Eqn. 4}$$

where φ_A, φ_B are volume fractions of the polymeric components, T is the absolute temperature, δ_A and δ_B are the solubility parameters of the components, and V_r is a reference volume taken as 100 cm^3 (42). Therefore, ΔHm is critically dependent on the solubility parameters of the components of the blend and decreases as the difference between these values decreases.

Generally, ΔGm values of the polymer blends are largely dependent on ΔHm, and to a lesser extent on ΔSm, which is quite small for large polymers, and hence, ΔHm must be negative (*73*). Consequently, exothermic interactions between the polymeric components result in the formation of miscible single-phase binary blends. Therefore, miscibility of polymers is defined by the ability of two different polymers to create various polymer-polymer interactions such as hydrogen bonding, acid-base interactions (*4*, *61*), and π-electron interactions (*75*). In this context, the extent of the formation of intermolecular hydrogen bonds that can occur between the components is a critical factor in determining miscibility. The existence of the intramolecular hydrogen bonding within one of the components (lignin is a typical polymer with strong self-interactions built on hydrogen bonds) decreases the accessibility of the functional groups to participate in the intermolecular hydrogen bonding between the two components leading to their immiscibility (*76*). The relative strength of the intermolecular interactions determines the manner in which the T_g varies with the blend composition (*42*). Hence, the T_g in miscible polymer blends can be predicted when the blend composition, T_g values of individual components, and the nature of the interactions between the components are known, by the use of different models. Conversely, by measuring T_g of the blends, the level of intermolecular interactions between the components of the blends can be estimated using these models as well. The Lu and Weiss model (Eqn. 5) takes into account the weight fractions, molar masses, densities, and specific heats for estimation of T_g, and contains two interaction parameters, χ (Flory-Huggins), and k (*77*). Other models such as Kwei (Eqn. 6), Gordon-Taylor (Eqn. 7), Couchman (Eqn. 8) and Fox (Eqn. 9) have been also introduced to predict T_g of the blends (*42*).

$$T_g = \frac{w_1 T_{g1} + k w_2 T_{g2}}{w_1 + k w_2} + \frac{A w_1 w_2}{(w_1 + k w_2)(w_1 + b w_2)(w_1 + c w_2)^2} \quad \text{Eqn. 5}$$

$$where, A = -\frac{\chi R(T_{g1} - T_{g2})c}{M_1 \Delta c_{p1}} \text{ and } k = \frac{(\Delta c_{p2} - w_1 \delta\, c_p^1)}{(\Delta c_{p1} - w_2 \delta\, c_p^g)}$$

$$T_g = \frac{w_1 T_{g1} + k w_2 T_{g2}}{w_1 + k w_2} + A w_1 w_2 \quad \text{Eqn. 6}$$

$$T_g = \frac{w_1 T_{g1} + k w_2 T_{g2}}{w_1 + k w_2} \quad \text{Eqn. 7}$$

$$\ln T_g = \frac{w_1 \Delta c_{p1} \ln T_{g1} + w_2 \Delta c_{p2} \ln T_{g2}}{w_1 \Delta c_{p1} + w_2 \Delta c_{p2}} \quad \text{Eqn. 8}$$

$$\frac{1}{T_g} = \frac{w_1}{T_{g1}} + \frac{w_2}{T_{g2}} \quad \text{Eqn. 9}$$

Thermal analysis (DSC, DMTA) is the most common method used to evaluate miscibility; other methods include microscopy, solid-state NMR, and FT-IR spectroscopy (*42*, *78*). It has been postulated that a single composition-dependent T_g indicates the formation of a homogeneous one-phase miscible blend through intermolecular interactions. Conversely, two T_gs signify a heterogeneous two-phase immiscible blend. In addition, a comparison between melting temperatures of a neat crystalline polymer and of that after blending with an amorphous polymer provides evidence regarding the miscibility of these polymers, i.e. level of their interactions.

Table 3. Hansen Solubility Parameters of Reference Lignins

Lignin	$\delta_d\,(MPa^{1/2})^a$	$\delta_p\,(MPa^{1/2})^a$	$\delta_H\,(MPa^{1/2})^a$
SWKL	16.7 (*84*)	13.7	11.7
HWKL	16.7 (*84*)	13.5	11.3
Björkman Spruce Lignin	21.9 (*62*)	14.1	16.9

[a] Hildebrand solubility parameter, δ based on the energy of vaporization relates to Hansen solubility parameters as: $\delta^2 = \delta_d^2 + \delta_p^2 + \delta_H^2$ (*66*) where, δ_d = Hansen solubility parameter based on dispersion interaction forces; δ_p = Hansen solubility parameter based on dipolar interaction forces; δ_H = Hansen solubility parameter based on hydrogen bond interaction forces.

Table 4. Hansen Solubility Parameters of Thermoplastic Polymers Miscible with Lignin

Polymer	*Hansen Solubility Parameters, ($MPa^{1/2}$)*			*Lignin Type and Content (%)*	*Type of interaction*
	δ_d	δ_p	δ_H		
PEO	17 (*85*)	10 (*85*)	5 (*85*)	HWKL (12.5-87.5 %) (*42*)	Hydrogen bonds between PhOH from lignin and ether groups in PEO
				HW Alcell (25-95 %) (*61*)	
PANI	17.4 (*86*)	8.1 (*86*)	10.7 (*86*)	KL (7-36 %) (*87*)	Hydrogen bonds between imine-nitrogen and OH groups from lignin
PET	18.2 (*88*)	7.3 (*88*)	7.9 (*88*)	HWKL (25-95 %) (*61, 67*)	π-π interactions between aromatic rings of lignin and PET
PBAT	19.0 (*89*)	4.8 (*89*)	9.1 (*89*)	SWKL Indulin (30 %) (*90*)	Hydrogen bonds between OH group from lignin and the aromatic ring of PBAT. π-π bonds between the lignin aromatic rings of lignin and PBAT
PHB	16.5 (*65*)	9.0 (*65*)	8.6 (*65*)	AGR Soda Lignin (10-40 %) (*91, 92*)	Hydrogen bonds between carbonyl groups in PHB and OH groups in lignin
PVP	21.4 (*85*)	11.6 (*85*)	21.6 (*85*)	Grass Alkali Lignin (4-20 %) (*63, 64*)	Hydrogen bonds between OH groups of lignin and pyridine ring of PVP or VP monomers in P(VP-co-VAc)
P(VP-co-VAc) [a]	20.9 (*88*)	11.3 (*88*)	9.7 (*88*)	Organosolv (*68*) (20-80 %)	
PA6	17.0 (*93*)	10.6 (*93*)	3.4 (*93*)	AGR Soda Lignin (5-30 %) (*94*)	Hydrogen bonds between OH groups from lignin and amine groups from PA6 and PA11
PA11	18.2 (*93*)	9.1 (*93*)	8.1 (*93*)	AGR Soda Lignin (12.5-25 %) (*95*)	

[a] P(VP-co-VAc) with at least 30 mol % vinyl pyrrolidone (VP); Hansen solubility parameters are for PVAc homopolymer.

Lignin has been shown to impact the degree and rate of crystallization, as well as crystal morphology in thermoplastic blends. It has been hypothesized that lignin with low to no miscibility (immiscible) may act as a nucleating agent at lower concentrations, promoting crystallinity in the blend (*39, 79*). If there are strong interactions (high miscibility) between the components of the blend, lignin may disrupt or interfere with the crystal structure. This could result in decreased crystallinity, altering crystal morphology, and reducing the rate of crystallization (*40, 80, 81*).

A few polymers are spontaneously miscible with lignin (Table 3), leading to the formation of miscible thermoplastic blends (Table 4). Miscibility results in improved mechanical properties of the blends, as the load transfer in the single-phase systems is not interrupted at the interface between phase-separated components (*75, 82*). Based on eqn. 1-4, miscibility of components is influenced by their inherent solubility characteristics, generally defined in terms of δ (*83*) (Tables 3 and 4).

Immiscible Blends

Low compatibility between polymers of dissimilar solubility parameters leads to interfacial tension, poor adhesion, poor dispersion of the minor component in the mixture, and ultimately phase separation. The diameter of the particles (d) of the dispersed phase may be predicted using the following simplified equation:

$$d = \frac{8\alpha\gamma_{AB} f(\eta_{rel})}{\pi\eta_m}\phi \approx \frac{8\alpha\gamma_{AB}}{\pi\eta_m}\phi \qquad \text{Eqn. 10}$$

where γ_{AB} is the interfacial tension, $f(\eta_{rel})$ is a function of the relative viscosity of the two components and $f(\eta_{rel}) \rightarrow 1$, α is the coalescence probability or probability that collision between two particles would lead to their fusion, φ is the volume fraction of the particles, and η_m is the viscosity of the matrix polymer. Smaller values for d (~5-15nm) indicate miscibility (*42*). In addition, the interfacial tension can be expressed as a function of the Flory-Huggins interaction parameter, χ_{AB} (Eqn. 3), as defined below:

$$\gamma_{AB} = \frac{bRT\chi_{AB}^{\frac{1}{2}}}{V_r} \qquad \text{Eqn. 11}$$

where b is the effective length of a monomer unit, V_r is a reference volume taken as 100 cm^3, and T is the absolute temperature. Therefore, the diameter of the particles is correlated to the Flory-Huggins interaction parameter as follows:

$$d = k_1 \frac{bRT\chi_{AB}^{\frac{1}{2}}}{V_r} \qquad \text{Eqn. 12}$$

where $k_1 = 8\alpha/\pi\eta_m$. Hence, greater miscibility is achieved between components that have δ-values closer to each other, which leads to a decrease in the Flory-Huggins interaction parameter, and results in a decrease in the diameter of the particles, *d*. The difference between solubility parameters ($\Delta\delta$) of two polymers has been suggested as a criteria to predict their miscibility; good miscibility if $\Delta\delta \leq 2$ $MPa^{1/2}$ ($\Delta\delta \leq 1.0$ $cal^{1/2}$ $cm^{-3/2}$) (*96*), partial miscibility if 2 MPa $< \Delta\delta <$ 10 $MPa^{1/2}$, and immiscibility if $\Delta\delta >$ 10 $MPa^{1/2}$ (*97*). This theory is supported by the well-documented reports showing lignin ($\delta =$ ~22 $MPa^{1/2}$) as largely immiscible with non-polar synthetic polymers such as polypropylene (PP; δ = 16 $MPa^{1/2}$), polystyrene (PS; δ =18.6 $MPa^{1/2}$) (*65*, *71*) and LDPE (δ = 17.6 $MPa^{1/2}$) (*83*). Its immiscibility with biorenewable polymers such as PLA (δ = 19.7 $MPa^{1/2}$) has also been confirmed (*71*).

Miscibility between lignin and commercial polymers may be influenced by employing different strategies to enhance the quality of lignin-based thermoplastic blends, including the use of plasticizers, compatibilizers, and lignin derivatization.

Compatibilizers

Tailoring lignin structure by derivatization to enhance its miscibility with immiscible polymers further contributes to the cost of producing thermoplastic materials, reducing economic benefits of this approach. Thus, compatibilization is an alternative approach, which may increase the competitiveness of lignin-based thermoplastic blends by increasing the mutual miscibility of lignin with polymers of contrasting solubilities through physical and chemical means (*3*, *78*, *83*).

Compatibilizers can be *block or graft polymers*. Their multifunctional structure provides a dual action through chemical bonding and/or formation of hydrogen-bonds, and good blending based on similar solubility at the interface between polymers. They assist blending by decreasing the interfacial tension, promoting adhesion of the components, and reducing the coalescence of polymer particles. Therefore, the dispersion of particles in the polymer matrix is stabilized. In addition, compatibilizers may be *coupling agents*. These are lower-molecular weight compounds such as cross-linking agents and free radical initiators, which promote formation of bonds between polymers during blending; *in situ* compatibilization (*98–100*).

The lignins used in the production of compatibilized blends include unmodified technical lignins, such as KL (*46*, *98*, *99*, *101*, *102*) and LS (*103*), and lignin derivatives, such as lignin phthalate (*104*), maleated lignin (*99*), and hydroxypropyl lignin (*105*).

Polymeric Compatibilizers in Lignin-Based Thermoplastic Blends

Various copolymers containing two or more constituents have been studied, with maleic anhydride and glycidyl methacrylate grafted (co)polymers as the most commonly attempted; polymer-g-MA and polymer-g-GMA copolymers, respectively (Figure 2). Maleic anhydride grafted PP (PP-g-MA) has been used in the production of PP-lignin blends, by itself and more recently, in concert with ethylene-butyl acrylate-glycidyl methacrylate (EBGMA) (*38*, *46*, *103*, *106*, *107*). The compatibilizing effect of a hyperbranched polymer lubricant (HBPL), synthesized from oleic acid and amino-terminated hyperbranch polymer, has also been studied on PP-lignin blends, and found to be superior to that of a commercial PP-g-MA compatibilizer (*107*). Other maleic anhydride copolymers, maleic anhydride grafted PE (PE-g-MA) and maleic anhydride grafted styrene-ethylene-butylene copolymer (SEBS-g-MA) have been explored in the production of PE-lignin (*104*) and PS-lignin blends (*101*, *102*), respectively. Hydroxymethylated wheat straw alkali lignin grafted by flame-retardant elements, phosphorus and nitrogen, has been blended with PP in the presence of PP-g-MA as a compatibilizer, and provided improvements in flame retardancy of PP (Table 2) (*31*). Similarly, glycidyl methacrylate grafted PE (PE-g-GMA) and PLA (PLA-g-GMA) polymers have been explored in the production of PE-lignin blends (*108*) and PLA-lignin blends (*55*, *109*), respectively. The PLA-lignin blends in these studies employed a biorefinery lignin produced as a residue from the conversion of giant reed grass (*Arundo donax*) to bioethanol. The grass lignin was transformed to nanosized particles, as an additional strategy to increase its available surface area, along with the use of PLA-g-GMA as a compatibilizer. The compatibilizing effect of MA- and GMA-based compatibilizers have been reported to take place through the formation of ester (*38*, *101*) and ether (*109*) bonds with lignin OH groups (Figure 2), and hydrogen bonding between OH groups of lignin and various groups in compatibilizing agents, such as imino groups of HBPL mentioned above (*107*).

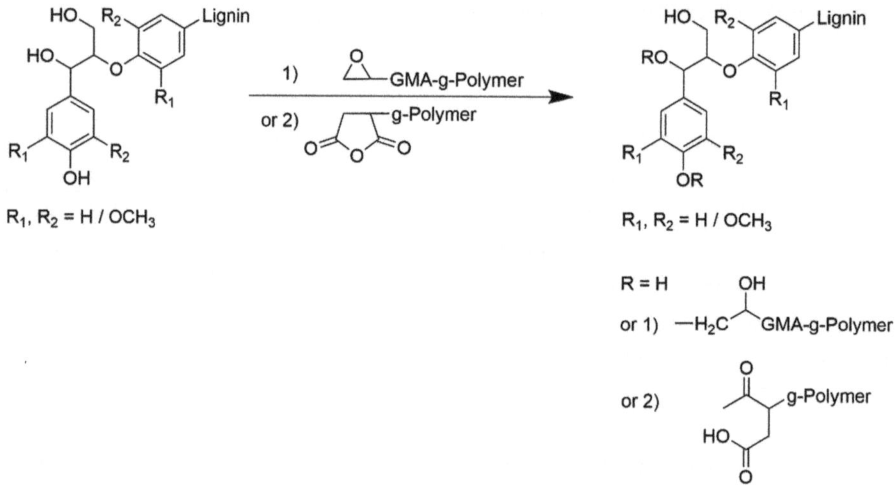

Figure 2. Reactions proposed to occur between lignin and glycidyl methacrylate (1) and maleic acid anhydride (2) copolymers as compatibilizing agents.

Other copolymers that have also been used to improve interactions between lignin and synthetic polymers include ethylene-vinylacetate (EVA) in PE-lignin blends (EVA; 28 % vinyl acetate; 10 % addition) (*110*) and ethylene acrylic acid copolymer (EAA; 5 % addition), which was added together with a titanate coupling agent (0.5 % addition) to LPDE-, high-density PE- (HDPE-), and PP-lignin blends (*72*). Lignin-based copolymers have also been suggested for use as compatibilizers. A star-like copolymer between lignin and PS (LPS) was explored for its compatibilizing effect in the production of PS-lignin blends. LPS was synthesized via grafting of NCO-capped PS into hydroxypropyl lignin, (HPL): PS-NC(=O) + HO-CH(CH_3)-CH_2-O-Lignin→PS-NH-C(=O)-O-CH(CH_3)-CH_2-O-Lignin (*105*).

This led to an enhanced compatibility between PS and HPL, documented by SEM indicating an improved anchoring between HPL particles and PS-matrix in the presence of LPS; at 5 and 10 % lignin content (*105*). In more recent studies on improving PS-lignin blends, random copolymers of styrene and vinyl phenol; PS-ran-4VPh have been investigated. Various levels of 4VPh have been introduced into copolymers to provide a range of concentrations of hydrogen bonding sites through the available PhOH groups. Neutron refractometry results revealed that the 30 % 4VPh copolymer creates the thickest interface layer between lignin and the copolymer indicating the highest degree of interpenetration, i.e. closest interaction. Additionally, the 20 % 4VPh copolymer provides the greatest level of miscibility, i.e. lowest level of phase separation. The compatibilizing effect was reported to be induced by formation of hydrogen bonds through the OH groups of PS-ran-4VPh (*82*). These studies demonstrated a need for a fine tuning of the level of PhOH groups introduced into PS-ran-4VPh to maximize their accessibility for the hydrogen bond formation with lignin, i.e. minimize steric shielding and formation of hydrogen bonds inside the copolymer (*82*).

Polymeric methylene diphenyl diisocyanate (PMDI) has also been used for compatibilization of lignin-containing Arboform or liquid wood, (a thermoplastic composite consisting of cellulosic fibers, lignin as a bonding matrix, and additives) with polybutylene succinate (PBS) (*111*). Inclusion of 1 % PMDI in the composite formulation was reported to enhance the tensile, flexural, and impact strength of the material, along with the heat deflection temperature (*111*). Arboform composites are fully degradable and recyclable and are applicable in various areas, including automotive interiors, furniture industry, and electronics. Different types of lignin and lignin derivatives can be used as a bonding matrix to create a strong composite material resistant to impact, bending, and compression (*106, 112*).

Coupling Agents as Compatibilizers in Lignin-Based Thermoplastic Blends

Studies have been performed to explore benefits of radical coupling reactions in increasing compatibility between lignin and polymers. For example, in efforts to improve PLA-lignin blending, lignin was exposed to electron-beam irradiation (EBI; 30-90 kGy) prior to blending with PLA in reactive extrusion in the presence of triallyl isocyanurate (TAIC, 3 %) as a cross-linking agent. Peroxy and poly-conjugated radicals in lignin formed during EBI are proposed to generate radical sites in TAIC; $\cdot CH_2$-CH-CH_2-N- and $\cdot$CH(CH_2)-CH_2-N-, and in PLA; R-C(=O)-O$\cdot$ and -$\cdot$CH(CH_3)-C(=O)-O). The TAIC and PLA couple creating a PLA-TAIC-lignin cross-linked structure. The formation of this three-component structure, which serves as a compatibilizer at the interface between lignin and PLA was confirmed by FT-IR studies of blends. These studies included optimization of the EBI dosage to suppress lignin-self cross-linking as an excessive formation of lignin free radicals may cause this undesirable effect at greater irradiation levels (60-90 kGy) (*98*).

Effect of Compatibilizers on Polymer Properties

The effect of compatibilizers is evaluated by measuring T_g of the blends made with and without compatibilizers. For this purpose, DSC (*98*, *102*, *109*) and DMTA (*38*, *102*, *103*, *107*) are commonly used methods. The increase in T_g of blends with compatibilizers has been attributed to the formation of lignin-compatibilizer-matrix cross-linked structure, which restricts the motion of the matrix chains as in the case of using HBPL as a compatibilizer in lignin-PS blends (*107*). Increase in T_g was also noticed for PLA-lignin blends produced in the presence of TAIC as a cross-linking agent with lignin exposed to EBI prior to blending. Processing the data through the Gordon-Taylor equation (Eqn. 7) to estimate the quality of lignin-PLA interactions revealed that the negative values of the k interaction parameter determined for PLA-lignin blends (lignin content 5 and 20 %) increased to positive values for the blends made from EB-irradiated lignin (30 kGy) in the presence of TAIC (3 phr). This clearly indicates a compatibilizing effect from the combination of EBI of lignin and TAIC (*98*). In addition, it was shown that the elongation at break and impact strength increased even compared to those measured for the neat polymer with increases of ~21 % and ~9 %, respectively. These results highlighted the importance of the EBI of lignin as without EBI of lignin these properties increased for 125 % and 630 %, respectively, but the properties of the neat PLA were not restored (*98*).

Compatibilization has been observed to improve tensile strength, impact strength, and elongation at break. In studies of LDPE-lignin blends (30 % lignin) reported earlier, the addition of EVA was found to produce a two-fold increase in tensile strength and a thirteen-fold increase in elongation at break demonstrating a strong compatibilizing effect of EVA (*110*). The improvement in the mechanical properties was also reported for lignin phthalate-LDPE (40 % lignin) blends compatibilized with LDPE-g-MA. In studies of PP-lignin blends, the impact strength of blends increased by 52.3 % with the addition of HBPL and for 33.6 % with the addition of PP-g-MA compatibilizer (3-5 % addition level) showing superior effects of HBPL (*107*). In studies of PLA-lignin blends which have explored the use of nanosized lignin (1 % lignin content) and PLA-g-GMA as compatibilizers, mechanical properties were similar to those measured for PLA. The elongation at break increased more than five times compared to that of both PLA and PLA-g-GMA in the formulation containing PLA-g-GMA and lignin (99:1) (*55*). Use of SEBS-g-MA in compatibilization of PS-lignin blends with 60 % of lignin improved tensile modulus (increase of 1.8 %), tensile strength (increase of 21 %) and elongation at break (increase of 33 %), but did not restore the properties of the neat PS polymer (*101*).

The effects of compatibilizing agents have also been evaluated for the quality of dispersion of lignin through morphological studies conducted by particle size analysis using microscopy (*101, 102*) and SEM (*38, 55, 98, 99, 103–108*). They have provided evidence that the addition of effective compatibilizers leads to improvements in the appearance of blends, which convert from heterogeneous ruptured structures to smoother more continuous surfaces with reduced number of cavities indicating better networking between dispersed particles and the matrix, better dispersion of the particles, and improved interfacial adhesion. The particle size of lignin decreases during processing of blends and it is reduced compared to the initial size even in the blends produced without compatibilizers. However, the addition of compatibilizers facilitates stress transfer from the matrix to the particles, and interferes with fusion of the particles during collision; thus producing smaller particles. Morphological studies have proved instrumental in selecting effective compatibilizers and adjusting their concentration in lignin-based thermoplastic blends.

Compatibilization is a simple strategy to circumvent the challenges posed by lignin's immiscibility with common polymers for the synthesis of thermoplastic blends. The multifunctional structure of compatibilizing agents facilitates the interactions between lignin and the immiscible polymers at the interface improving their mutual miscibility and providing benefits to blends properties, including mechanical properties.

Nanolignin

Thermoplastic materials where lignin plays a role of a filler has long been investigated (*22*). However, challenges such as large particle size and limited dispersion in the matrix hinder implementation of these materials. Reduction of lignin particle size with an associated increase in specific surface area is one of the emerging and relatively simple methods in this area. Ultrasonication followed by freezing at -196 °C and then freeze-drying has been shown to produce lignin sheet-like particles with thickness of ~200 nm. With 2 % addition of these sheet-like particles in PP, improvements in mechanical properties and decrease in UV-induced and thermo-oxidative degradation was observed. An increase in elongation at break compared to the neat PP (95.5 % versus 54.0 %) with comparable yield stress and elastic modulus, clearly demonstrates benefits of freeze-drying (*113*). However, freezing at such low temperatures is an energy-consuming method (estimated four-to-ten times higher energy consumption than in hot-air drying), and alternative methods to reduce the size of lignin particles to produce nanosized-lignin have been explored. Some recent attempts include methods such as ultrasonication only, acidolysis, and homogenization (*109, 114–116*). Nanolignin has shown to promote strong interfacial bonding with the matrix (*109, 114*);

thus, improving the mechanical properties, thermal stability, and barrier properties of the composite (*115*).

Nanolignin was prepared by ultrasonicating KL (Indulin) and remained suspended in water even after five months. The increased polarity in nanolignin was attributed to the chemical modification during sonication; mostly oxidation, as observed via FT-IR. Polyurethane- (PU)-KL and PU-nanolignin composites (lignin content 5–20 % in both cases) were fabricated, where the former showed large agglomerates, while an even dispersion was seen in the latter. Improvements in tensile strength and Young's modulus, but a compromised elongation at break were observed with PU-nanolignin. However only a slight improvement in thermal stability was observed (*114*). Similar results regarding thermal stability, tensile strength, Young's modulus, and elongation at break were obtained from PU-nanolignin (1–7 %), where the nanolignin was prepared by acidolysis (hydrochloric, sulfuric, and phosphoric acids) of biorefinery corn stover lignin. However, this composite also exhibited a significant improvement in hydrophobicity and resistance to UV-light, as well as demonstrating a frequency-dependent behavior for electrical conductivity (*116*).

Lignin nanoparticles were produced by hydrochloric acidolysis and incorporated into PLA in varying amounts of 1–3 % by either extrusion or solvent casting. Incorporating lignin had little to no effect on the T_g, and only a slight improvement in thermal stability was observed. For extruded polymers, 1 % lignin-PLA blends showed better tensile strength than pure PLA, and 3 % lignin-PLA blends. However, adding lignin decreased tensile strength for solvent-cast polymers. Elongation at break of a 3 % lignin-PLA extruded blend was radically higher than of all the other polymers, which is noteworthy. This may have been due to the flexibility of the lignin structure, which enabled it to act as a plasticizer. Overall, extruded blends demonstrated better mechanical properties than solvent-cast. This was attributed to the uneven distribution of lignin in the PLA matrix for solvent-cast polymers (*117*). The effect of nanolignin on the crystallinity of lignin-PLA blends has also been studied. Blending of PLA with underivatized lignin nanoparticles resulted in more than a two-fold increase in crystallinity at 5 % lignin content, with crystallinity continuing to increase at 10 % lignin content (*79*). While the small particle size and a more even distribution of lignin aided in increasing crystallinity, the immiscibility of underivatized lignin in PLA allowed lignin to act as a potent nucleating agent, as mentioned in the earlier section on miscibility.

Nanolignin prepared by acidolysis was used in the production of various thermoplastic blends with PLA and PLA-g-GMA by extrusion, employing a masterbatch approach (1 % lignin in all blends). Nanolignin in combination with the proposed masterbatch approach and with the assistance of PLA-g-GMA as a compatibilizer produced blends with tailored mechanical, optical, and thermal properties (*55*).

SWKL was homogenized (15,000 rpm for 1–4 h) to yield nanolignin (*115*). This homogenization method did not result in noticeable chemical changes in the lignin, as no cleavage of β-aryl ether bonds was recorded, and the aliphatic, phenolic and carboxylic OH content remained the same. The nanolignin did not exhibit change in molecular weight distribution either. Solvent-cast polyvinyl acetate (PVA)-SWKL and PVA-nanolignin films were produced. The PVA-nanolignin films demonstrated superior thermal stability compared to PVA-SWKL films. The SEM revealed an even distribution of nanolignin in the PVA matrix, whereas agglomerates were present with SWKL.

Although some work on nanolignin has been done, there are several opportunities that are yet to be explored. The studies so far are promising, and show that nanolignin has the ability to improve properties and performance of the thermoplastic materials.

Solvent Fractionation of Lignin

Fractionation of lignin using single or sequential organic solvents is another approach in modifying lignin by its partition into homogenous fragments with distinct physicochemical properties, depending on their differential solubilities in the solvent(s) (*118*). Thus, fractionation offers an opportunity to influence miscibility/compatibility of lignin with thermoplastic polymers by tailoring its molecular weight and functional group profile, especially the PhOH and aliphatic OH group content, which is responsible for lignin's extensive internal hydrogen bonding (*119–129*). Higher molecular weight (HMW)-fractions have been found to induce higher thermostability owing to their relatively rigid structure and higher T_g (*128*). Furthermore, HMW-fractions contain a relatively lower amount of PhOH groups than lower molecular weight (LMW)-fractions because of the differential partitioning of PhOH groups in organic solvents (*119, 121–124, 126, 129*). Therefore, HMW-fractions can be advantageous in thermoplastic formulations as they are expected to avoid issues related to excessive cross-linking and incompatibility with hydrophobic polymers and solvents (*24*). On the other hand, LMW-fractions can be good plasticizers in thermoplastics due to their relatively lower T_g and flexible structure (*130, 131*). For example, an LMW-fraction obtained from sequential fractionation of softwood soda lignin using isopropanol/ethanol (4:1 v/v) and methanol was found to result in better dispersion of lignin in lignin-polyvinyl chloride (PVC) blends (2.5–10 % lignin), a faster plasticization effect as compared to the HMW-fraction (*131*). This blend also showed a higher tensile strength as compared to the HMW-fraction-PVC blend and the pristine lignin-PVC blend. Similar plasticizing effect was also observed for lignin-PE blends (5-25 % lignin), containing LMW-fractions of SWKL fractionated with acetone (*130*). LMW-fractions have also been reported to result in improved compatibility between lignin and synthetic polymers, and hence in improved blend morphology (*59*). Fractionation has additionally been used in concert with lignin graft copolymerization, as described in subsequent sections (*24, 132*).

Plasticizers

Plasticizers are used to modify polymers to improve their processability, as they are able to interrupt the internal bonds and associations originally present in the polymers. As the interactions between the polymer chains weaken, and the chains are spaced apart, their free volume and mobility increase, leading to a decrease in T_g and improved thermal processability. The melt viscosity and elastic modulus also decrease (*4*). Numerous agents have been explored for lignin plasticization, including polyethers, esters of various organic acids, lignin-like aromatic compounds, and commercial polymer plasticizers (*3, 133–137*). The plasticization effect increases with the concentration of plasticizers, as estimated by the decrease in T_g of lignin in their presence (Table 5).

Sakata and Senju (1975) performed a thorough initial study of more than 30 organic plasticizers, including phthalates, organophosphates, and aliphatic acid esters on their effects on two types of lignin; thiolignin (SWKL; T_g = 174 °C) and dioxane lignin (laboratory isolated lignin; T_g = 138 °C) (*138*). Plasticizers were characterized by different δ-values, ranging from 15.4 to 26.2 $MPa^{1/2}$, and with low hydrogen-bonding capacity (mainly esters). The δ-value of lignin was assumed to be 22.5 $MPa^{1/2}$ based on the results of lignin solubility studies by Schuerch (*139*); this value was also close to later estimates for softwood lignin (*78*) $(24\ MPa)^{1/2}$. It was shown that agents with solubility lower than 18.4 $MPa^{1/2}$ were not able to decrease the T_g of lignin to any extent. As expected, the plasticization effect increased with greater compatibility between lignin and plasticizer, i.e. with the decrease in difference between the δ-values of lignin and plasticizer; Δδ. Later reports have supported these results and established that a polymer and a plasticizer are compatible, if $\Delta\delta < 3.1$ MPa$^{1/2}$ for the pair. Furthermore, a linear decrease in the T_g of lignin was noticed with the increase in plasticizer concentration up to 20 %. A nonlinear negative relationship between T_g and water content in lignin was noted, indicating water as a comparatively good lignin plasticizer, despite the lignin-water incompatibility ($\delta_{H2O} = 47.8$ MPa$^{1/2}$). This effect was attributed to the hydrogen-bonding capacity of water and its ability to swell lignin (*139*). Moreover, superior plasticization was revealed in the case of combined addition of water and plasticizer, as it was greater than the corresponding individual plasticization effects (*138*). The plasticizing effect of water was also reported in a later comprehensive study of ~20 plasticizers with wide variations in compatibility with SWKL (Indulin, pine) (*135*). In this study, the solubility differences between lignin and plasticizers were expressed by their distance in the Hansen space, $\Delta\delta_{3D}$. A negative two-step pseudo-linear correlation between the water content and the T_g of lignin was determined with the initial fast decrease in T_g. This effect diminishes when the water content reaches beyond the lignin saturation level. This change in the rate of lignin plasticization by water was attributed to their intrinsic incompatibility, with $\Delta\delta_{3D} = 26.7$ (J cm^{-3})$^{-1/2}$ (*135*). It was concluded that unsaturated/dry lignin is more readily plasticized by molecules that can interfere with hydrogen bonds in lignin as demonstrated by the superior plasticization effect of water compared to compounds more compatible with lignin in terms of $\Delta\delta_{3D}$. Conversely, saturated lignin is more readily plasticized by more compatible molecules, i.e. those closer to lignin in the Hansen space; for example, lignin-like molecules, such as vanillin, $\Delta\delta_{3D} = 3.9$ (J cm^{-3})$^{-1/2}$. This study suggested that lignin plasticization results from the cleavage of hydrogen bonds and some other disentanglement mechanism, as most of the predicted plasticizing values were underestimates of the experimentally determined values. In contrast, for some agents plasticizing results were lower than expected and were attributed to the agent self-interactions (ethylene glycol, for example) or steric hindrance effects. Acetylated lignin samples showed T_g at ~100 °C which is close to the T_g of water-saturated lignin confirming that water-based lignin plasticization proceeds through the cleavage of the hydrogen bonds up to lignin saturation point (*135*).

The first high-lignin content thermoplastics were made using plasticizers. The combined plasticization action of diethylene glycol dibenzoate and indene (combined plasticizers accounted for < 2.5 % mass of blends) resulted in successful blending of underivatized SWKL (pine, *Pinus banksiana*) and PVA with lignin accounting for 85 % of the mass of these blends (*134*). More generally, these results may be taken as an illustration of the importance of Hildebrand solubility parameter of plasticizers on their efficiency (diethylene glycol dibenzoate, $\delta = 20.7$ MPa$^{1/2}$ and SWKL, $\delta = 24.6$ MPa$^{1/2}$) (*78*).

In the development of lignin-based thermoplastic blends, various plasticizers may be used to facilitate lignin dispersion in thermoplastic polymers by weakening the strong self-interactions of lignin. They reduce the T_g and enhance the thermal processability of lignin, and their efficiency has been shown to depend on their δ-value.

Table 5. Select Plasticizers Studied for Lignin; Hildebrand Solubility Parameters, δ

Plasticizer (Content)	*Solubility δ (MPa$^{1/2}$)*	*Lignin Type; (Solubility δ) (MPa$^{1/2}$)*	*Lignin T_g (°C)*	*Plasticized Lignin T_g (°C)*
Butylbenzyl phthalate (15 %) (*136*)	20.2	Alcell (28.0)	97	66
Alkylbenzyl phthalate (15 %) (*136*)	19.5			66
Diethylene glycol dibenzoate (15 %) (*136*)	20.7			65
Tricresyl phosphate (15 %) (*136*)	20.2			69
Diethylhexyl phthalate (35 %) (*133*)	16.8	Indulin (20.5)	142	126
Diethylene glycol dibenzoate (35 %) (77)	20.7			80
Tricresyl phosphate (35 %) (*133*)	20.2			28
Dimethyl phthalate (20 %) (*138*)	21.5	Dioxane (22.5)	137	62
Diethyl phthalate (20 %) (*138*)	20.3			72
Dibutyl phthalate (20 %) (*138*)	19.2			110
Dihexyl phthalate (20 %) (*138*)	18.6			117
Diphenyl phthalate (20 %) (*138*)	26.2			98
Triethyl phosphate (20 %) (*138*)	19.7			67
Tributyl phosphate (20 %) (*138*)	18.4			71
Triphenyl phosphate (20 %) (*138*)	21.3			73
Tricresyl phosphate (20 %) (*138*)	19.8			101
Cresyldiphenyl phosphate (20 %) (*138*)	21.7			80
Diphenyl-2-ethylahexyl phosphate (20 %) (*138*)	19.6			94
Trichloroethyl phosphate (20 %) (*138*)	22.3			80
Dibutyl maleate (20 %) (*138*)	18.4			97
o-nitrodiphenyl (20 %) (*138*)	22.5			80

Esterification

It has been documented that technical lignins can resemble thermoset-like materials at elevated temperatures. At temperatures around or above the T_g, condensation polymerization is induced via free radicals captured by lignin's free PhOH groups, increasing molecular weight and T_g of lignin (*140, 141*). The abundance of OH groups instigates polar character of lignin limiting its miscibility, and therefore compatibility, with typically hydrophobic polymers, as postulated through the Hansen solubility theory (*53, 84*). Chemical derivatization of lignin, such as esterification has been explored to improve its thermoplasticity and compatibility (i.e. miscibility) with most commercial polymers, among other purposes (*142, 143*). It is important to note that this is particularly the case with more hydrophobic polymers, as it has been demonstrated that lignin following acetylation is actually less compatible with PEO than the unmodified lignin (*42*).

Esterification results in significant changes in lignin properties while it is relatively easy to carry out regarding reagents and reaction parameters. Typically, esterification of lignin is conducted in a reaction with carboxyl acids, their acid halides (often chlorides) or anhydrides (*144*). Acetylation, perhaps the most widely studied type of lignin esterification, is commonly conducted by using one-to-one mixtures of pyridine and acetic anhydride in laboratory conditions (*145*). At industrial scale, use of a mixture of alkanoic acid, the corresponding anhydride and metal (sodium or potassium) alkanoate is considered more practical. Studies on acetylation of numerous technical lignins using either pyridine or sodium alkanoate as a catalyst found the same degree of acetylation achieved by the two methods (*145*). Similar reaction conditions are often employed for esterification reactions that utilize reactants with longer chain lengths. In some cases an additional lignin solvent may be utilized either to reduce or completely replace the need for pyridine, a well-known toxin (*36, 146, 147*). In one scenario, the esterification reaction was performed under elevated temperatures (60 °C) and in the presence of an argon catalyst to ensure the most favorable reaction kinetics (*148*). The use of 1-methylimidazole as a catalyst for esterification reactions using acetic, propionic, n-butyric, maleic, and methacrylic anhydrides has led to complete esterification of SWKL and HWKL within 2 h in contrast to 48 h required in the presence of pyridine (*84*). Subsequently, 1-methylimidazole has been used on a number of different technical lignins (*149*), also with the goal of optimizing reaction conditions (*81*). In addition to traditional synthetic chemical reactions to produce esterified lignin, enzymatic approaches have been introduced as well. Lipase-catalyzed transesterification (lipases from *Candida antarctica, Pseudomonas cepacia* and *Mucor miehei*) of commercial KL with ethyl oleate ($C_{18:1}$) was conducted in ionic liquid media resulting in lignin oleate and transesterification yield of up to ~30 % (*150*). An additional method of producing lignin esters has been explored without the use of any solvents, through microwave assisted acylation with various acid anhydrides. This work produced lignin esters with the degree of substitution varying from as low as 39 % with maleic anhydride, up to 99 % with acetic anhydride (*151*).

Lignin esterification has been explored as a method to create a more dynamic and valuable material since at least the 1940s. US patents were issued to Lewis and Brauns on "Esters of lignin material" in the 1940s and to Pruett and coworkers on "Lignin and lignin derivatives as copolymerizable colorants for polyesters" in the 1980s. Lewis and Brauns work focused on the production of lignin esters with the reactant chain-length varying from C_2-C_{18}. Lignin esters synthesized by Lewis and Brauns were used in molded plastics and served as plasticizers and internal mold lubricants. On the other hand, lignin esters covered in Pruett's patent were C_2-C_4 lignin esters utilized in the production of plastic containers with low light transmittance in the UV-range (*145, 148*). In 2015, a patent was granted to Pietarinen and coworkers for their invention on using tall oil fatty acids (TOFAs; a byproduct of kraft pulping) to replace fatty acid chlorides to produce lignin esterified with at least one fatty acid. The reaction was conducted on acetylated lignin at high temperature to enable distillation of the acetic acid formed during the reaction (*152*). Lignin esters with TOFA have been examined in barrier surface coating of paper to produce sustainable packaging material and replace synthetic/nonrenewable barrier materials (*69, 153, 154*).

Increased hydrophobicity and thermal stability, as well as decreased hydrogen bonding and T_g of lignin have been reported following esterification (Figure 3) (*146–148, 155*). However, a recent TGA study of acetylated lignin and a lignin laurate has suggested that there may be a thermal sensitivity associated with the ester linkage around 200 °C (*53*). The decrease in T_g (ΔT_g) may be controlled by the extent of the acylation reaction and the size of the moiety added through the ester linkage; short chain acylation (acetylation) results in ΔT_g of ~20 °C versus long chain fatty acid acylation which results in ΔT_g in the range from 20 to 100 °C (*148, 156*). Acetylation leading to greater compatibility was also reported to affect crystallization behavior of the blends. In experiments with acetylated KL, a decrease in crystallinity of PLA was noticed when compared to the pure polymer or to that of the blend containing underivatized alkaline lignin (*36*). DSC analysis in another study suggested that the formation of α′ crystals was favored over that of α crystals, the tighter and more orderly crystal polymorph, when blends containing acetylated lignin were compared to the pure polymer. Furthermore, underivatized lignin showed the opposite trend, favoring the formation of α crystals over α′ crystals (*53*).

Increased hydrophobicity following esterification has been supported by a higher water contact angle (*157*), reduced water vapor transmission rate (WVTR) (*69, 153*), as well as decrease in the δ_H and δ_P parameters (*84*). Predicted greater decrease in the δ_H and δ_P parameters than in the δ_D parameter following different types of esterification (with exception to maleation) was confirmed through the use of the Hoy model (*84*).

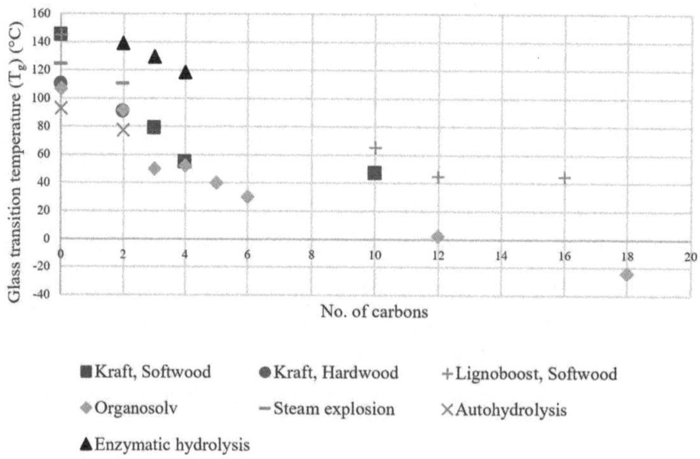

Figure 3. Effect of number of carbons of esterifying agents on the glass transition temperature of different technical lignins. Number of carbons = 0 represents the underivatized lignin (4, 145, 148, 159).

While properties of lignin-based thermoplastic blends change by lignin esterification performed prior to blending, the extent of change has been found to directly correlate with the length of the added chain. Therefore, lignin esters with properties tailored for the given application may be proposed. For example, esterification of SWKL and HWKL utilizing TOFA chlorides (which included saturated C_{16} – C_{20} and more than 85 % unsaturated $C_{18:1}$ – $C_{18:3}$ fatty acids) has been shown to exhibit a favorable reduction in oxygen transmission rate compared to lignin esterified with shorter saturated fatty acids (C_{12} and C_{16}) (*69, 153, 154*). Lignin esterified with TOFA also exhibited a low T_g, even below 0 °C (*69, 153*). Due to the more noticeable increase in hydrophobicity with increasing chain length, lignin esterified with C_{17} and $C_{18:1}$ acyl chains have been used as compatibilizers for recycling plastics and producing blends of LDPE either with PP or PS (*145*). Lignin esters produced with stearoyl chloride (C_{18}) have also been proposed as plasticizers with PS (*144*). Additionally, miscibility of esterified lignin depends strongly on the chain length. For example, lignin acetate (C_2) was shown to be completely immiscible with poly(ε-caprolactone)

(PCL), whereas butyryl and valeryl substitutions were especially effective for the enhancement of miscibility between alkyl-esterified OSL and PCL (*158*).

Esterification is an effective and well-studied technique utilized to modify polar phenolic and aliphatic OH groups on lignin. Its positive effects include increasing miscibility and compatibility with more hydrophobic commercial polymers (Table 6), which can lead to improved physicochemical properties, such as processability by having a plasticizing effect on lignin, oxygen and water vapor transmittance behavior, and mechanical properties of the resulting thermoplastic.

Etherification

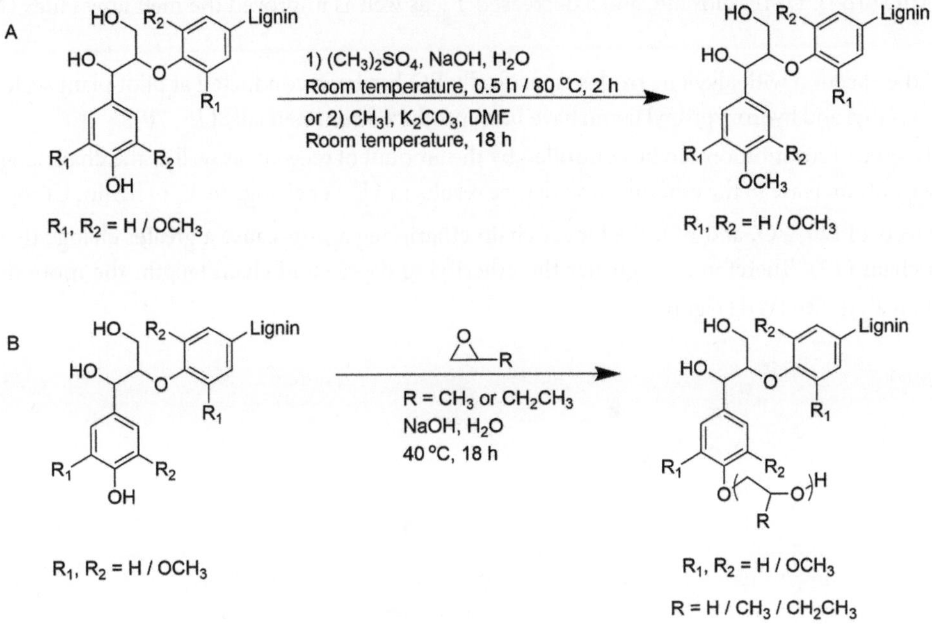

Figure 4. Mechanism for methylation (A) and hydroxyalkylation (B) for lignin. Adapted with permission from Reference (143). Copyright 2012 American Chemical Society.

Etherification is another approach for lignin derivatization and improving its thermal processibility. Etherification includes alkylation and alkoxylation, which are conducted via the mechanism of nucleophilic aromatic substitution (S_NAr) to produce lignin ethers and hydroxyalkyl lignins, respectively (*19*); thus masking the PhOH groups on lignin and reducing the degree of intra-molecular hydrogen binding; thus, the T_g. In fact, alkoxylation converts the PhOH groups to less acidic secondary aliphatic OH (*19, 160*), causing the same effects. These groups can further serve as reaction sites providing uniform functionality for additional lignin modification, such as in synthesis of lignin copolymers, as described in the later section on lignin copolymers (*161*). Lignins etherified by various etherifying agents have been blended with numerous polymers, such as PP, PEO, PLA, PVA, and polypropylene carbonate (PPC). Though most studies employ kraft lignin, more recent attempts of using biorefinery lignins have been described.

Alkylation (Figure 4 A), most frequently as methylation can be performed by various reagents, including dimethyl sulfate, methyl iodide, or dimethyl carbonate in aqueous alkali solutions or in organic solvents such as dimethylformamide/dimethyl sulfoxide (DMF/DMSO). Dimethyl sulfate has been found to be a more effective methylation agent than methyl iodide (*143*). Other alkylating agents such as 1-iodododecane (C_{10}) and 1-iodohexadecane (C_{16}) have been used (*162*), but in a limited capacity. Moreover, studies on blending these alkylated lignins with other polymers have not been commonly performed.

Hydroxyalkylation (Figure 4 B) is the reaction with alkylene oxides such as ethylene oxide (EO), propylene oxide (PO) or butylene oxide (BO) performed in aqueous alkali solution or organic solvents, with hydroxypropylation being the most common method (*13*).

Both methods of etherification selectively substitute PhOH groups. The reagents act as internal plasticizers for lignin altering its physical and chemical properties. The main effects include increased solubility in organic solvents such as tetrahydrofuran (THF) (used in SEC) (*163*), chloroform (used in NMR) (*164*), DMF, dioxane, and a decreased T_g, as well as improved the melt flow index (MFI) (*13*).

Etherification with alkylene oxides, specifically EO has been conducted at pilot plant scale, and hydroxyethyl and hydroxypropyl lignin have been produced commercially (*13, 70*).

Degree of substitution can be controlled by the amount of reagent, as well as the chain length of the reagent. Increase in the etherification degree results in a larger change in T_g of lignin. Comparing the effects of EO, PO, and BO, the longer chain etherifying agents cause a greater change than the short chain (*13*). Therefore, the greater the etherifying degree and chain length, the more drastic effect on T_g is observed (Figure 5).

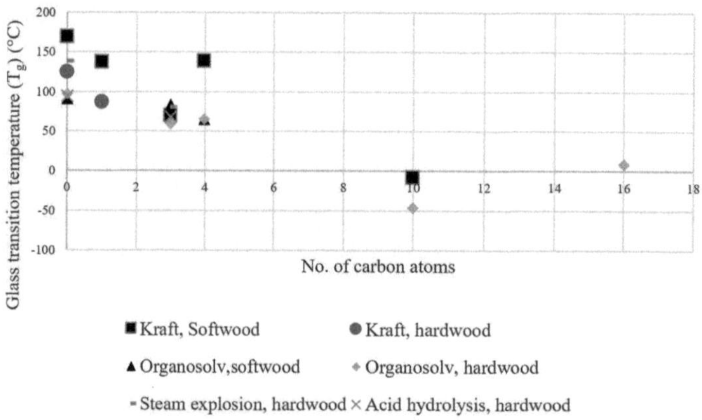

Figure 5. Effect of number of carbons of etherification agents on the glass transition temperature of different technical lignins. Number of carbons = 0 represents the underivatized lignin (4, 162).

Comparative studies of methylation and hydroxypropylation have been performed. The average chain length of the grafted polypropylene glycol in hydroxypropylation reaction was found to affect the molecular weight, T_g, and functionality of the final lignin derivatives. Comparison of methylated and hydroxypropylated SWKL (HP-SWKL) demonstrated a decrease in T_g, but the former was more thermally stable (*141*). However, as expected both methods successfully replace PhOH groups of lignin, but no change was observed in the content of lignin aliphatic OH groups.

Methylating agents dimethyl sulfate and methyl iodide are highly toxic and hazardous. In contrast, dimethyl carbonate is of low toxicity, noncorrosive, and has been suggested as a biodegradable alternative for lignin methylation. Dimethyl carbonate is produced by catalytic synthesis from carbon dioxide and methanol (*4, 19, 165, 166*). Similarly, propylene oxide is risky to handle owing to its high vapor pressure (588 hPa at 20 °C), flammability, toxicity, and carcinogenicity. In contrast, five-member cyclic organic carbonates with low toxicity and vapor pressure (0.04 hPa at 20 °C), high boiling points and flash points have been proposed as alternatives. They also have the added advantage of being biodegradable (*167*). Moreover, reactions with these reagents can take place at high temperatures at atmospheric pressure; thus, they do not require pressurized reactors. Their high boiling points also afford them the ability to act as solvents and reagents (*168*).

Varying ratios of unmodified and methylated HWKL and PEO were extruded to make PEO-lignin blends. Based on the presence of a single T_g for all compositions, it was concluded that lignin and PEO were miscible. A strong hydrogen bond between the aromatic hydroxyl proton of the unmodified lignin and the ether oxygen of PEO was observed (Table 4). Since the PhOH groups are replaced by methoxy groups, the formation of hydrogen bonds between methylated lignin and PEO is less likely, as confirmed by the FT-IR results. Moreover, blends with the lignin-to-PEO ratio > 1 demonstrated superior mechanical properties to pure lignin and PEO polymers. This was attributed to PEO disrupting the intermolecular bonds of lignin and the supramolecular lignin complexes (*42*).

Dichloromethane and chlorobenzene were used as alkylating and arylating reagents, respectively, to modify sugarcane bagasse biorefinery lignin to produce PP-lignin blends via melt extrusion (lignin content 5–25 %). Improvements in thermal stability with increasing lignin content for both alkylation and arylation were observed. Crystallinity was improved with the addition of alkylated lignin (up to 15 % lignin content) compared to pure PP. In addition, MFI of the lignin blends remained stable with five extrusion cycles, while pure PP showed a continuous increase. However, tensile strength was compromised with only a slight improvement in Young's modulus. The authors concluded that both alkylated and arylated lignins are viable processing stabilizers for PP; however, lignin content above 15 % in PP-lignin blends was not beneficial (*169*).

Alkylation of HWKL with bromododecane was performed prior to producing lignin-PP blends. The compatibility of alkylated lignin and PP was superior to that of unmodified lignin-PP blends. The SEM micrographs of alkylated lignin-PP blends showed a homogenous blend with a single T_g and better thermal stability. Although, with increasing lignin content, tensile, flexure, and impact strength decreased, a higher impact strength at 10 % lignin content was observed. Blends with higher lignin content were also more brittle, as indicated by the high elastic modulus, but had better damping effect (absorption of energy) demonstrated by the wide tan δ peaks (*170*).

HP-SWKL was blended with PVA (the degree of hydrolysis from 0 to 96 %; hence, various MW (3,500–96,000), T_gs (32–77 °C), and δ (19.6–25.8 $MPa^{1/2}$)) by solvent casting. HP-SWKL had an assigned δ value of 22.7 $MPa^{1/2}$. All blends displayed similarly homogenous morphology, indicating that hydrogen bonding between the polymers can occur over a wider range of solubility parameters than previously expected. The authors also observed changes in crystalline structure indicated by the changes apparent in the melt endotherm in the DSC. A decrease in melt endotherm peak area indicates polymer-polymer interactions resulting in a disorder in the crystal, thereby reducing crystallinity. Increasing lignin content resulted in an increase of T_g of the blends. This was attributed to the hydrogen bonding between OH groups of HP-SWKL and PVA, yielding quasi cross-links and increasing the T_g (*96*).

Up to 40 % HP-KL was incorporated in to PPC via melt extrusion (*171*). As in most cases, PPC-lignin blends showed superior thermal properties to neat PPC. In addition, a significant improvement was seen in tensile strength, Young's modulus, storage modulus, and elongation at break with increasing lignin content. However, water contact angle measurements showed that the wettability of the blends increased with HP-lignin content. This was viewed positively by the authors, as it would aid in biodegradability of the material, as corroborated by the results of soil biodegradation tests.

HP-KL has been blended (up to 30 % HP-KL) with biodegradable copolyester from 1,4-butanediol, adipic acid, and terephthalic acid to produce films by melt-blowing on an industrial scale (*172*). It has been demonstrated that lignin provides resistance against photodegradation and compostability to these films, which have been already used in the semi-commercial production of waste bags (Xylobags). Other injection-molded products have also been produced from HP-KL (Figure 6).

Figure 6. Prototype and semi-commercial products made from HP-KL-polymer blends. Reprinted with permission from Reference (172). Copyright 2017 W.G. Glasser.

Four cyclic carbonates - ethylene, propylene, vinyl ethylene, and glycerol were studied as low cost, non-toxic, and green alternative etherifying agents for lignin (*168*). A complete conversion of PhOH was within short reactions times. Cyclic carbonates were reported to react with aliphatic and carboxylic OH groups, unlike conventional etherifying agents, which selectively replace PhOH groups.

In conclusion, etherification can successfully mask PhOH groups of lignin, improve solubility in organic solvents (Table 6), and decrease T_g to facilitate lignin use in thermoplastic applications. Methylation and hydroxypropylation are the most studied methods, and in addition to PhOH, the latter also has the potential to replace aliphatic OH. However, methylated lignin has demonstrated better thermal stability. The reagents used in methylation and hydroxypropylation are hazardous. Therefore, dimethyl and five-member cyclic carbonates generated from CO_2 have been proposed as non-toxic, environmentally friendly alternatives. In almost all cases, addition of lignin improves thermal properties. However, mechanical properties are often compromised with higher lignin loadings, which is a challenge yet to be overcome.

Table 6. Effects of Lignin Esterification and Etherification on Solubility

Lignin	*Hansen Solubility Parameters*			*Hildebrand or Total Hansen Solubility Parameter* ($MPa^{1/2}$)
	δ_d ($MPa^{1/2}$)	δ_p ($MPa^{1/2}$)	δ_h ($MPa^{1/2}$)	
Unmodified SWKL (*84*)	16.7	13.7	11.7	24.6
Acetylated SWKL (*84*)	16.1	12.0	10.2	22.5
Propionated SWKL (*84*)	16.3	11.5	9.7	22.1
Butyrated SWKL (*84*)	16.4	11.1	9.3	21.9
Maleated SWKL (*84*)	15.6	13.2	10.7	23.0
Methacrylated SWKL (*84*)	15.9	11.7	9.2	21.8
Unmodified HWKL (*84*)	16.7	13.5	11.3	24.3
Acetylated HWKL (*84*)	16.1	12.1	10.3	22.6
Propionated HWKL (*84*)	16.3	11.5	9.7	22.2
Butyrated HWKL (*84*)	16.4	11.1	9.3	21.9
Maleated HWKL (*84*)	15.6	13.1	10.7	23.1
Methacrylated HWKL (*84*)	15.9	11.7	9.2	21.8
Indulin Kraft (*173*)	N/A	N/A	N/A	22.5
Hydroxypropyl Indulin Kraft (*96*)	N/A	N/A	N/A	22.7

Lignin Copolymers

The typical multifunctional structure of lignin enables an alternative approach to lignin modification by covalently linking lignin (or its derivatives) with immiscible polymers (graft copolymerization) to synthesize a thermoplastic copolymer, or to facilitate blending of lignin with other thermoplastic polymers (*56*). Most notably, the OH groups on the lignin (PhOH groups on unmodified lignin, or artificially created secondary aliphatic OH groups on derivatized lignin) are utilized for lignin copolymerization. Lignin graft copolymers can be synthesized via two approaches: "grafting onto", and "grafting from" (Figure 7). In the first approach, polymers are coupled to the OH or other functionalities by techniques such as diisocyanate and boron mediated coupling and azide alkyne Huisgen cycloaddition (click chemistry). In the second approach, monomers are polymerized directly from lignin cores via free radical polymerization (FRP), ring opening polymerization (ROP), chemo-enzymatic polymerization, and atom transfer radical polymerization (ATRP) (*174*). Lignin graft copolymers have been employed in synthesis of thermoplastic materials as a strategy for lignin derivatization, and facilitating eventual blend formation between the graft copolymer and previously incompatible polymers (*"binary/multi-component blends"*) or as thermoplastic materials by themselves (*"one component blends"*).

Lignin Graft Copolymers in Binary Blends

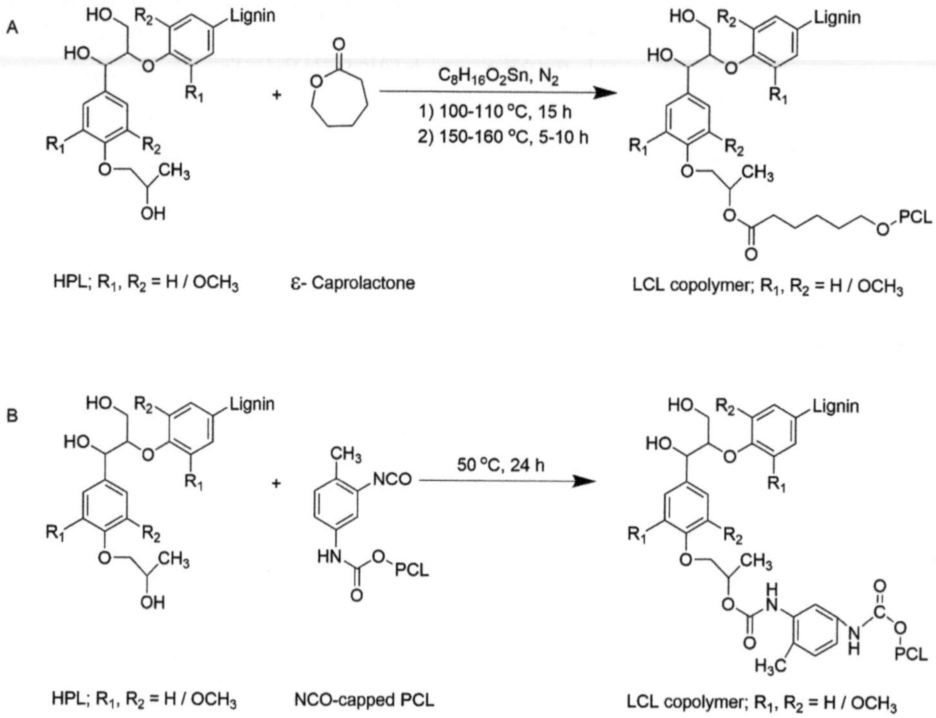

Figure 7. Copolymerization of hydroxypropylated lignin with ε-caprolactone monomer (grafting from) (A) and NCO-capped poly(ε-caprolactone) prepolymer (grafting onto) (B) to form lignin-caprolactone copolymer (LCL). Adapted with permission from Reference (161). Copyright 1994 American Chemical Society.

This method consists of creating lignin graft copolymers, which are then blended with other polymers. In one study, multiphase block co-polymer was formed between HP-SWKL (Indulin) and PCL, which was then blended with PVC. HP-SWKL served as the core, while PCL served as the flexible arm segments (2-7 per core) of star-like lignin-polycaprolactone (LCL) co-polymer. Both, anionic copolymerization between HP-SWKL and CL (Figure 7 A), and grafting with monoisocyanate (NCO)-terminated PCL were reported (Figure 7 B). The resulting copolymer architecture was found to be dependent on the average molecular weight, molecular weight distribution, lignin content, and T_g of the starting lignin and its derivatives, presenting an opportunity to customize the copolymer properties per specific requirements. The copolymer was characterized by FT-IR, GPC, DSC, and optical microscopy. During eventual blending of LCL copolymer and PVC, the blend was found to benefit significantly from the innate compatibility between PCL and PVC (56). An improved compatibility between the components was evidenced by a single T_g of the blend, as measured by DSC, and supported by DMTA. Copolymerization of lignin was also found to help with the mechanical properties of the blend, where inclusion of LCL in the formulation (up to 50 %) improved the elongation from 200 to 600 %, as compared to pure PCL. LCL was also found to slow down the characteristic crystallization of PCL in PCL-PVC blends, which leads to their embrittlement (56). Hence, LCL was beneficial to reduce the effects of aging on the thermoplastic blend. 'Grafting onto' approach via ROP was employed to synthesize lignin-g-poly-L-lactide (lignin-g-PLLA) copolymers from soda lignin which were then blended with

PLA (5-15 % copolymer) (*132*). The lignin-g-PLLA-PLA blends showed a tensile strength that was higher than that observed for pristine lignin-PLA blends, and comparable to that observed for neat PLA. Most importantly, lignin-g-PLLA-PLA blends showed UV-blocking properties that increased with increasing content of the copolymer in the blend. In a similar fashion, synthesis of dodecylated (C_{12} esterified) alkaline lignin-g-PLA-PLA blends (50 % copolymer) was reported, with effective UV-blocking properties, and a high elongation, up to 40 times elongation at break of the neat PLA (*175*). 'Grafting from' approach was used in synthesis of OSL- and biorefinery-lignin-based copolymers (lignin-g-polyacrylate) by FRP, initiated by calcium chloride and hydrogen peroxide. The resulting lignin-g-polyacrylate copolymers were blended with PLA, providing a UV-blocking effect and enhancing mechanical properties (*54*). Apart from the UV-blocking and anti-aging properties, lignin copolymers have also been reported to possess antioxidant, biodegradable, and biocompatible properties (*52*). Specifically, a lignin-PCL-co-lactide copolymer (lignin-PCLLA) from alkali lignin via solvent-free ROP, was further blended with PCL or PLA to produce electrospun composite fibers for medical applications (*52*).

Lignin Graft Copolymers as One Component Blends

Examples of one-component blend are lignin-PS (up to 19.6 % lignin) and lignin-polymethyl methacrylate (PMMA) (up to 22.1 % lignin) copolymers prepared by ATRP between KL and the corresponding synthetic polymers ('grafting from') (*176*). Both copolymers were reported to exhibit 10 times increase in toughness, while the lignin-PMMA copolymer showed twice the elongation as compared to the non-grafted, binary blend system. Lignin-PMMA copolymers were reported to have higher elongation and toughness values than lignin-PS copolymers. The mechanical properties of the blends were found to be mostly dictated by lignin than by the synthetic polymers. A single T_g, which was higher than T_gs of individual homopolymers, was reported for the copolymers indicating presence of strong interactions between lignin and the polymers. Transmission electron microscopy (TEM) showed a uniform dispersion of the components in the blend. In another example of one component blend between lignin and PS, styrene was 'grafted onto' HWKL (average T_g ~65-85 °C (*19*)) via ATRP to synthesize a lignin-g-PS copolymer (*177*). A single T_g was observed at ~107 °C. Thermal decomposition behavior of the copolymer showed a similar trend as of pure PS. 'Grafting from' via ATRP was employed in synthesis of another one component blend between HWKL and *N*-isopropylacrylamide (NIPAM) (*178*). The thermal decomposition temperature of the lignin-g-poly(NIPAM) copolymer was reported to be significantly higher than poly(NIPAM). The copolymer showed a thermally activated phase transition from hydrophilic to hydrophobic above 32 °C, similar to poly(NIPAM) (lower critical solution temperature, LCST ~32 °C). The use of ATRP (activator regenerated electron transfer – ARGET ATRP) has also been employed in combination with the convenience and cost-effectiveness of photo cross-linking, to synthesize UV-cross-linkable lignin-g-PS(co-acryloyl benzophenone) copolymers (*174*). The photo-cross-linking ability was provided by inclusion of acryloyl benzophenone (ABP), a UV-sensitive compound as a comonomer. The resulting copolymer exhibited enhanced thermoplasticity as compared to the unmodified kraft lignin, and higher thermal stability than PS homopolymer. Lignin copolymerization has also been reported to be employed along with fractionation approach to enhance lignin compatibility (*24*, *132*). Sequentially fractionated and cross-linked HWKL was reacted with dicarboxy-terminated polybutadiene (PBD) to form lignin-PBD copolymers (12-38 % lignin) (*24*). The resulting copolymer films from HMW-fractions oflignin showed a two-phase viscoelastic behavior. Increasing lignin content resulted in increasing storage modulus at rubbery plateau. Lignin-PCL one component blends have also been reported, in a solvent-free bulk polymerization between KL and CL, via ROP (*179*). Formation of the copolymer was confirmed by change in the solubility of the copolymer in organic solvents. The resulting copolymer showed an enhanced thermal processability and extrusion behavior compared to pristine lignin. Biorefinery lignin-g-PCL copolymers have also been reported to possess UV-blocking effect (*180*).

Lignin copolymerization has also been used to achieve improvement in the biodegradability of the blends. Graft copolymers between SWKL and 1-ethenylbenzene (10–51 % lignin), synthesized by FRP was reported to be biodegradable by white rot (*Pleuroteus ostreatus*, *Phanerochaete chrysosporium*, and *Trametes versicolor*), and brown rot fungi (*Gloeophyllum trabeum*). The rate of biodegradation was found to increase with increasing lignin content (*58*).

Lignin grafts offer an effective approach for creating thermoplastic blends between lignin and commercial off-the-shelf polymers, through the formation of covalent linkages that facilitate the blend formation, and improving the thermomechanical properties of the resulting blends. This is a unique approach, especially for creating single component blends from polymers that are otherwise immiscible with each other.

Lignin Elastomers

Thermoplastic elastomers are a class of materials combining both thermoplastic and elastomeric properties (*181*). Stable, high-functioning lignin-based thermoplastic materials can offer an alternative to the synthetic elastomers (*182*). Lignin-based thermoplastic elastomers can be synthesized by blending lignin with rubbers/polymers, or by chemically linking them together. An example of the former is an elastomer formulation (50 % lignin) prepared by blending acrylonitrile butadiene rubber (NBR-33; $\delta = 19.2–20.3\ MPa^{1/2}$) and various types of lignin (HWOSL, HWSL, SWKL, and methanol-fractionated SWKL) (*182*). Among these, HWOSL lignin provided characteristics similar to an elastomer containing organic fillers (5-7 MPa tensile failure stress, 500-600 % elongation). Methanol-fractionated SWKL-NBR blend also showed good mechanical properties.

Chemical synthesis was used to produce lignin-based elastomers as multi-armed star shaped lignin-g-PMMA-co-BA copolymers via ATRP (< 1 % lignin) (*181*). It was reported that the mechanical properties of these copolymers were significantly improved as compared to the linear PMMA-co-BA polymers. The T_g of lignin-g-copolymers was found to be similar to that of linear PMMA-co-BA polymer. The copolymers showed a higher thermostability than unmodified lignin, possibly due to their star-shaped architecture, which restricts the chain mobility. They also showed a higher storage modulus and stress at break as compared to the linear PMMA-co-BA polymer. Creep-recovery tests confirmed the property of a rubber-like elasticity. Inclusion of lignin was also shown to impart UV-absorbing capacity to the formulation. In another innovative study, EHL was linked with MA-grafted saturated polyolefin elastomer (POE-MA) rubber matrix (30 % lignin), by constructing coordination-based energy sacrificial bonds at the interface between lignin and POE (*183*). The coordination bonds were reported to facilitate lignin dispersion into the rubber matrix, and help orient the chains during stretching. The resulting formulation showed 100 % increase in the toughness, and 35 % increase in the stress at break, as compared to the formulation without lignin. The material also showed excellent shape memory performance.

Conclusions

Lignin shows tremendous potential as a biopolymeric component in the formulation of thermoplastic materials. Although it can form miscible blends with a few polymers, it largely shows immiscibility/incompatibility with most polymers, due to its multifunctional, polar structure characterized by strong self-interactions (such as hydrogen bonding). Several strategies dealing with this incompatibility exist as discussed within this chapter. Inclusion of lignin in the thermoplastic formulation not only helps in lowering the cost and the carbon footprint of plastics, but also provides additional benefits such as thermomechanical stability, flame retardancy, biodegradability, UV-protecting, antioxidant and anti-weathering capacity. These benefits make lignin-based thermoplastics attractive alternatives for petroleum-based thermoplastics, in numerous applications from packaging to the biomedical field.

References

1. Andreeßen, C.; Steinbüchel, A. Recent Developments in Non-Biodegradable Biopolymers: Precursors, Production Processes, and Future Perspectives. *Appl. Microbiol. Biotechnol.* **2019**, *103* (1), 143–157. https://doi.org/10.1007/s00253-018-9483-6.
2. Gruter, G.-J. Confusion about Terminology and Definitions for Bio-Based and Biodegradable Plastics. *Chim. Oggi.* **2019**, *34* (4), 54–55.
3. Imre, B.; Pukánszky, B. Compatibilization in Bio-Based and Biodegradable Polymer Blends. *Eur. Polym. J.* **2013**, *49* (6), 1215–1233. https://doi.org/10.1016/j.eurpolymj.2013.01.019.
4. Wang, C.; Kelley, S. S.; Venditti, R. A. Lignin-Based Thermoplastic Materials. *ChemSusChem.* **2016**, *9* (8), 770–783. https://doi.org/10.1002/cssc.201501531.
5. Grossman, A.; Vermerris, W. Lignin-Based Polymers and Nanomaterials. *Curr. Opin. Biotechnol.* **2019**, *56*, 112–120. https://doi.org/10.1016/j.copbio.2018.10.009.
6. Kai, D.; Tan, M. J.; Chee, P. L.; Chua, Y. K.; Yap, Y. L.; Loh, X. J. Towards Lignin-Based Functional Materials in a Sustainable World. *Green Chem.* **2016**, *18* (5), 1175–1200. https://doi.org/10.1039/C5GC02616D.
7. Thébault, M.; Müller, U.; Kandelbauer, A.; Zikulnig-Rusch, E.; Lammer, H. Review on Impregnation Issues in Laminates Manufacture: Opportunities and Risks of Phenol Substitution by Lignins or Other Natural Phenols in Resins. *Eur. J. Wood Prod.* **2017**, 75 (6), 853–876. https://doi.org/10.1007/s00107-017-1206-7.
8. Vishtal, A. G.; Kraslawski, A. Challenges in Industrial Applications of Technical Lignins. *BioResources.* **2011**, *6* (3), 3547–3568.
9. Duval, A.; Lawoko, M. A Review on Lignin-Based Polymeric, Micro- and Nano-Structured Materials. *React. Funct. Polym.* **2014**, *85*, 78–96. https://doi.org/10.1016/j.reactfunctpolym.2014.09.017.
10. Crestini, C.; Lange, H.; Sette, M.; Argyropoulos, D. S. On the Structure of Softwood Kraft Lignin. *Green Chem.* **2017**, *19* (17), 4104–4121. https://doi.org/10.1039/C7GC01812F.
11. Ohman, F.; Theliander, H.; Tomani, P.; Axegard, P. *Method for Separating Lignin from Black Liquor, a Lignin Product, and Use of a Lignin Product for the Production of Fuels or Materials.* U.S. Patent 20,100,325,947,A1, December 30, 2010.

12. Ohman, F.; Theliander, H.; Tomani, P.; Axegard, P. *Method for Separating Lignin from Black Liquor, a Lignin Product, and Use of a Lignin Product for the Production of Fuels or Materials.* U.S. Patent 20,100,325,947,A1, December 30, 2010.
13. Lora, J. H.; Glasser, W. G. Recent Industrial Applications of Lignin: A Sustainable Alternative to Nonrenewable Materials. *Journal of Polymers and the Environment* **2002**, *10* (1–2), 39–48. https://doi.org/10.1023/A:1021070006895.
14. Li, T.; Takkellapati, S. The Current and Emerging Sources of Technical Lignins and Their Applications. *Biofuels, Bioprod. Biorefin.* **2018**, *12* (5), 756–787. https://doi.org/10.1002/bbb.1913.
15. Dessbesell, L.; Paleologou, M.; Leitch, M.; Pulkki, R.; Xu, C. (Charles). Global Lignin Supply Overview and Kraft Lignin Potential as an Alternative for Petroleum-Based Polymers. *Renewable and Sustainable Energy Reviews* **2020**, *123*, 109768. https://doi.org/10.1016/j.rser.2020.109768.
16. *LignoBoost plant at Domtar's Plymouth mill in North Carolina* https://www.valmet.com/more-industries/bio/biofuels-and-biomaterials/lignoboost-plant-at-domtars-plymouth-mill-in-north-carolina/ (accessed Oct 11, 2020).
17. Tribot, A.; Amer, G.; Abdou Alio, M.; de Baynast, H.; Delattre, C.; Pons, A.; Mathias, J.-D.; Callois, J.-M.; Vial, C.; Michaud, P.; Dussap, C.-G. Wood-Lignin: Supply, Extraction Processes and Use as Bio-Based Material. *Eur. Polym. J.* **2019**, *112*, 228–240. https://doi.org/10.1016/j.eurpolymj.2019.01.007.
18. Abdelkafi, F.; Ammar, H.; Rousseau, B.; Tessier, M.; El Gharbi, R.; Fradet, A. Structural Analysis of Alfa Grass (Stipa Tenacissima L.) Lignin Obtained by Acetic Acid/Formic Acid Delignification. *Biomacromolecules.* **2011**, *12* (11), 3895–3902. https://doi.org/10.1021/bm2008179.
19. Sen, S.; Patil, S.; Argyropolulos, D. Thermal Properties of Lignin in Copolymers, Blends, and Composites: A Review. *Green Chem.* **2015**, *17*, 4862–4887.
20. Ragauskas, A. J.; Beckham, G. T.; Biddy, M. J.; Chandra, R.; Chen, F.; Davis, M. F.; Davison, B. H.; Dixon, R. A.; Gilna, P.; Keller, M.; Langan, P.; Naskar, A. K.; Saddler, J. N.; Tschaplinski, T. J.; Tuskan, G. A.; Wyman, C. E. Lignin Valorization: Improving Lignin Processing in the Biorefinery. *Science.* **2014**, *344* (6185), 1246843–1246843. https://doi.org/10.1126/science.1246843.
21. Bujanovic, B. M.; Goundalkar, M. J.; Amidon, T. E. Increasing the Value of a Biorefinery Based on Hot-Water Extraction: Lignin Products. *Tappi J.* **2012**, *11* (1), 19–26. https://doi.org/10.32964/10.32964/TJ11.1.19.
22. Thakur, V. K.; Thakur, M. K.; Raghavan, P.; Kessler, M. R. Progress in Green Polymer Composites from Lignin for Multifunctional Applications: A Review. *ACS Sustainable Chem. Eng.* **2014**, *2* (5), 1072–1092. https://doi.org/10.1021/sc500087z.
23. Gil-Chávez, G. J.; Padhi, S. S. P.; Pereira, C. V.; Guerreiro, J. N.; Matias, A. A.; Smirnova, I. Cytotoxicity and Biological Capacity of Sulfur-Free Lignins Obtained in Novel Biorefining Process. *Int. J. Biol. Macromol.* **2019**, *136*, 697–703. https://doi.org/10.1016/j.ijbiomac.2019.06.021.
24. Saito, T.; Brown, R. H.; Hunt, M. A.; Pickel, D. L.; Pickel, J. M.; Messman, J. M.; Baker, F. S.; Keller, M.; Naskar, A. K. Turning Renewable Resources into Value-Added Polymer:

Development of Lignin-Based Thermoplastic. *Green Chem.* **2012**, *14* (12), 3295–3303. https://doi.org/10.1039/C2GC35933B.

25. Gioia, C.; Lo Re, G.; Lawoko, M.; Berglund, L. Tunable Thermosetting Epoxies Based on Fractionated and Well-Characterized Lignins. *J. Am. Chem. Soc.* **2018**, *140* (11), 4054–4061. https://doi.org/10.1021/jacs.7b13620.
26. Mandlekar, N.; Cayla, A.; Rault, F.; Giraud, S.; Salaün, F.; Guan, J. Valorization of Industrial Lignin as Biobased Carbon Source in Fire Retardant System for Polyamide 11 Blends. *Polymers.* **2019**, *11* (1), 1–18. https://doi.org/10.3390/polym11010180.
27. Canetti, M.; Bertini, F.; De Chirico, A.; Audisio, G. Thermal Degradation Behaviour of Isotactic Polypropylene Blended with Lignin. *Polym. Degrad. Stab.* **2006**, *91* (3), 494–498. https://doi.org/10.1016/j.polymdegradstab.2005.01.052.
28. De Chirico, A.; Armanini, M.; Chini, P.; Cioccolo, G.; Provasoli, F.; Audisio, G. Flame Retardants for Polypropylene Based on Lignin. *Polym. Degrad. Stab.* **2003**, *79* (1), 139–145. https://doi.org/10.1016/S0141-3910(02)00266-5.
29. Réti, C.; Casetta, M.; Duquesne, S.; Bourbigot, S.; Delobel, R. Flammability Properties of Intumescent PLA Including Starch and Lignin. *Polym. Adv. Technol.* **2008**, *19* (6), 628–635. https://doi.org/10.1002/pat.1130.
30. Zhang, R.; Xiao, X.; Tai, Q.; Huang, H.; Hu, Y. Modification of Lignin and Its Application as Char Agent in Intumescent Flame-Retardant Poly(Lactic Acid). *Polym. Eng. Sci.* **2012**, *52* (12), 2620–2626. https://doi.org/10.1002/pen.23214.
31. Yu, Y.; Fu, S.; Song, P.; Luo, X.; Jin, Y.; Lu, F.; Wu, Q.; Ye, J. Functionalized Lignin by Grafting Phosphorus-Nitrogen Improves the Thermal Stability and Flame Retardancy of Polypropylene. *Polym. Degrad. Stab.* **2012**, *97* (4), 541–546. https://doi.org/10.1016/j.polymdegradstab.2012.01.020.
32. Costes, L.; Laoutid, F.; Aguedo, M.; Richel, A.; Brohez, S.; Delvosalle, C.; Dubois, P. Phosphorus and Nitrogen Derivatization as Efficient Route for Improvement of Lignin Flame Retardant Action in PLA. *Eur. Polym. J.* **2016**, *84*, 652–667. https://doi.org/10.1016/j.eurpolymj.2016.10.003.
33. Mandlekar, N.; Cayla, A.; Rault, F.; Giraud, S.; Salaün, F.; Guan, J. Development of Novel Polyamide 11 Multifilaments and Fabric Structures Based on Industrial Lignin and Zinc Phosphinate as Flame Retardants. *Molecules.* **2020**, *25* (21), 4963–4982. https://doi.org/10.3390/molecules25214963.
34. Canetti, M.; Bertini, F. Influence of the Lignin on Thermal Degradation and Melting Behaviour of Poly(Ethylene Terephthalate) Based Composites. *e-Polym.* **2009**, *9* (1), 49–59. https://doi.org/10.1515/epoly.2009.9.1.596.
35. Kabir, A. S.; Yuan, Z.; Kuboki, T.; Xu, C. De-Polymerization of Industrial Lignins to Improve the Thermo-Oxidative Stability of Polyolefins. *Ind. Crops Prod.* **2018**, *120*, 238–249. https://doi.org/10.1016/j.indcrop.2018.04.072.
36. Gordobil, O.; Egüés, I.; Llano-Ponte, R.; Labidi, J. Physicochemical Properties of PLA Lignin Blends. *Polym. Degrad. Stab.* **2014**, *108*, 330–338. https://doi.org/10.1016/j.polymdegradstab.2014.01.002.
37. Gordobil, O.; Egüés, I.; Labidi, J. Modification of Eucalyptus and Spruce Organosolv Lignins with Fatty Acids to Use as Filler in PLA. *React. Funct. Polym.* **2016**, *104*, 45–52. https://doi.org/10.1016/j.reactfunctpolym.2016.05.002.

38. Abdelwahab, M.; Misra, M.; Mohanty, A. Injection Molded Biocomposites from Polypropylene and Lignin: Effect of Compatibilizers on Interfacial Adhesion and Performance. *Ind. Crops Prod.* **2019**, *132*, 497–510.
39. Acha, B.; Marcovich, N.; Reboredo, M. Lignin in Jute Fabric-Polypropylene Composites. *J. Appl. Polym. Sci.* **2009**, *113*, 1480–1487.
40. Bertini, F.; Canetti, M.; Cacciamani, A.; Elegir, G.; Orlandi, M.; Zoia, L. Effect of Ligno-Derivatives on Thermal Properties and Degradation Behavior of Poly(3-Hydroxybutyrate)-Based Biocomposites. *Polym. Degrad. Stab.* **2012**, *97* (10), 1979–1987. https://doi.org/10.1016/j.polymdegradstab.2012.03.009.
41. Cousins, W. J. Elastic Modulus of Lignin as Related to Moisture Content. *Wood Sci. Technol.* **1976**, *10* (1), 9–17. https://doi.org/10.1007/BF00376380.
42. Kadla, J.; Kubo, S. Miscibility and Hydrogen Bonding in Blends of Poly(Ethylene Oxide) and Kraft Lignin. *Macromolecules.* **2003**, *36*, 7803–7811.
43. Li, Y.; Sarkanen, S. Thermoplastics with Very High Lignin Contents. In *Lignin: Historical, Biological, and Materials Perspectives*; Glasser, W. G., Northey, R. A., Schultz, T. P., Eds.; ACS Symposium Series 742; American Chemical Society: Washington DC, 2000; pp 351-366.
44. Gregorová, A.; Cibulková, Z.; Košíková, B.; Šimon, P. Stabilization Effect of Lignin in Polypropylene and Recycled Polypropylene. *Polym. Degrad. Stab.* **2005**, *89* (3), 553–558. https://doi.org/10.1016/j.polymdegradstab.2005.02.007.
45. Gregorova, A.; Redik, S.; Sedlarik, V.; Stelzer, F. *Lignin-Containing Polyethylene Films with Antibacterial Activity*; Brno, Czech Republic, EU: 2011 NANOCON, 2011.
46. Toriz, G.; Denes, F.; Young, R. Lignin-Polypropylene Composites. *Polym. Compos.* **2002**, *23* (5), 806–813.
47. Rials, T. G.; Glasser, W. G. Multiphase Materials with Lignin. IV. Blends of Hydroxypropyl Cellulose with Lignin. *J. Appl. Polym. Sci.* **1989**, *37* (8), 2399–2415. https://doi.org/10.1002/app.1989.070370827.
48. Holladay, J. E.; White, J. F.; Bozell, J. J.; Johnson, D. *Top Value-Added Chemicals from Biomass - Volume II-Results of Screening for Potential Candidates from Biorefinery Lignin*; Technical Report for The U. S. Department of Energy: Richland, WA, October 2007. https://doi.org/10.2172/921839.
49. Sahoo, S.; Misra, M.; Mohanty, A. Effect of Compatibilizer and Fillers on the Properties of Injection Molded Lignin-Based Hybrid Green Composites. *J. Appl. Polym. Sci.* **2013**, *127*, 4110–4121.
50. Collins, M. N.; Nechifor, M.; Tanasă, F.; Zănoagă, M.; McLoughlin, A.; Stróżyk, M. A.; Culebras, M.; Teacă, C.-A. Valorization of Lignin in Polymer and Composite Systems for Advanced Engineering Applications - A Review. *Int. J. Biol. Macromol.* **2019**, *131*, 828–849. https://doi.org/10.1016/j.ijbiomac.2019.03.069.
51. Domenek, S.; Louaifi, A.; Guinault, A.; Baumberger, S. Potential of Lignins as Antioxidant Additive in Active Biodegradable Packaging Materials. *J. Polym. Environ.* **2013**, *21*, 692–701.
52. Kai, D.; Zhang, K.; Jiang, L.; Wong, H. Z.; Li, Z.; Zhang, Z.; Loh, X. J. Sustainable and Antioxidant Lignin–Polyester Copolymers and Nanofibers for Potential Healthcare Applications. *ACS Sustainable Chem. Eng.* **2017**, *5* (7), 6016–6025. https://doi.org/10.1021/acssuschemeng.7b00850.

53. Nagardeolekar, A.; Ovadias, M.; Wang, K.-T.; Bujanovic, B. Willow Lignin Recovered from Hot-Water Extraction for the Production of Hydrogels and Thermoplastic Blends. *ChemSusChem.* **2020**, *13* (17), 4702–4721. https://doi.org/10.1002/cssc.202001259.
54. Zong, E.; Liu, X.; Liu, L.; Wang, J.; Song, P.; Ma, Z.; Ding, J.; Fu, S. Graft Polymerization of Acrylic Monomers onto Lignin with CaCl2–H2O2 as Initiator: Preparation, Mechanism, Characterization, and Application in Poly(Lactic Acid). *ACS Sustainable Chem. Eng.* **2018**, *6* (1), 337–348. https://doi.org/10.1021/acssuschemeng.7b02599.
55. Yang, W.; Dominici, F.; Fortunati, E.; Kenny, J.; Puglia, D. Effect of Lignin Nanoparticles and Masterbatch Procedures on the Final Properties of Glycidyl Methacrylate-g-Poly (Lactic Acid) Films before and after Accelerated UV Weathering. *Ind. Crops Prod.* **2015**, 77, 833–844.
56. Oliveira, W. D.; Glasser, W. G. Multiphase Materials with Lignin. XII. Blends of Poly(Vinyl Chloride) with Lignin–Caprolactone Copolymers. *J. Appl. Polym. Sci.* **1994**, *51* (3), 563–571. https://doi.org/10.1002/app.1994.070510320.
57. Silva, T. F. da; Menezes, F.; Montagna, L. S.; Lemes, A. P.; Passador, F. R. Effect of Lignin as Accelerator of the Biodegradation Process of Poly(Lactic Acid)/Lignin Composites. *Mater. Sci. Eng., B.* **2019**, *251*, 114441–114448. https://doi.org/10.1016/j.mseb.2019.114441.
58. Chen, M.-J.; Gunnells, D. W.; Gardner, D. J.; Milstein, O.; Gersonde, R.; Feine, H. J.; Hüttermann, A.; Frund, R.; Lüdemann, H. D.; Meister, J. J. Graft Copolymers of Lignin with 1-Ethenylbenzene. 2. Properties. *Macromolecules.* **1996**, *29* (5), 1389–1398. https://doi.org/10.1021/ma951150a.
59. Pouteau, C.; Dole, P.; Cathala, B.; Averous, L.; Boquillon, N. Antioxidant Properties of Lignin in Polypropylene. *Polym. Degrad. Stab.* **2003**, *81* (1), 9–18. https://doi.org/10.1016/S0141-3910(03)00057-0.
60. Chung, Y.-L.; Olsson, J.; Li, R.; Frank, C.; Waymouth, R.; Billington, S.; Sattely, E. A Renewable Lignin-Lactide Copolymer and Application in Biobased Composites. *ACS Sustainable Chem. Eng.* **2013**, *1*, 1231–1238.
61. Kubo, S.; Kadla, J. F. Poly(Ethylene Oxide)/Organosolv Lignin Blends: Relationship between Thermal Properties, Chemical Structure, and Blend Behavior. *Macromolecules.* **2004**, *37* (18), 6904–6911. https://doi.org/10.1021/ma0490552.
62. Hansen, C. M.; Björkman, A. The Ultrastructure of Wood from a Solubility Parameter Point of View. *Holzforschung.* **1998**, *52* (4), 335–344. https://doi.org/10.1515/hfsg.1998.52.4.335.
63. Cunxiu, G.; Donghua, C.; Wanjun, T.; Changhua, L. Properties and Thermal Degradation Study of Blend Films with Poly(4-Vinylpyridine) and Lignin. *J. Appl. Polym. Sci.* **2005**, *97* (5), 1875–1879. https://doi.org/10.1002/app.21465.
64. Liu, C.; Xiao, C.; Liang, H. Properties and Structure of PVP–Lignin "Blend Films". *J. Appl. Polym. Sci.* **2005**, *95* (6), 1405–1411. https://doi.org/10.1002/app.21367.
65. Pouteau, C.; Baumberger, S.; Cathala, B.; Dole, P. Lignin–Polymer Blends: Evaluation of Compatibility by Image Analysis. *C. R. Biol.* **2004**, *327* (9), 935–943. https://doi.org/10.1016/j.crvi.2004.08.008.
66. Hansen, C. M. The Universality of the Solubility Parameter. *Product R&D.* **1969**, *8* (1), 2–11. https://doi.org/10.1021/i360029a002.
67. Kubo, S.; Kadla, J. Hydrogen Bonding in Lignin: A Fourier Transform Infrared Model Compound Staudy. *Biomacromolecules.* **2005**, *6*, 2815–2821.

68. Teramoto, Y.; Lee, S.-H.; Endo, T. Molecular Composite of Lignin: Miscibility and Complex Formation of Organosolv Lignin and Its Acetates with Synthetic Polymers Containing Vinyl Pyrrolidone and/or Vinyl Acetate Units. *J. Appl. Pol. Sci.* **2012**, *125*, 2063–2070.
69. Hult, E.; Koivu, K.; Asikkala, J.; Ropponen, J.; Wrigstedt, P.; Sipilä, J.; Poppius-Levlin, K. Esterified Lignin Coating as Water Vapor and Oxygen Barrier for Fiber-Based Packaging. *Holzforschung.* **2013**, *67* (49), 899–905.
70. Glasser, W. About Making Lignin Great Again - Some Lessons from the Past. *Front. Chem.* **2019**, *7*, 565.
71. Romhányi, V.; Kun, D.; Pukánszky, B. Correlations among Miscibility, Structure, and Properties in Thermoplastic Polymer/Lignin Blends. *ACS Sustainable Chem. Eng.* **2018**, *6*, 14323–14331.
72. Kharade, A.; Kale, D. Lignin Filled Polyolefins. *J. Appl. Polym. Sci.* **1999**, *72* (41), 1321–1326.
73. Feldman, D. Lignin and Its Polyblends - A Review. In *Chemical Modification, Properties, and Usage of Lignin*; Hu, T. Q., Ed.; Springer US: Boston, MA, 2002; pp 81-99. https://doi.org/10.1007/978-1-4615-0643-0_5.
74. Muthuraj, R.; Hajee, M.; Horrocks, A. R.; Kandola, B. K. Biopolymer Blends from Hardwood Lignin and Bio-Polyamides: Compatibility and Miscibility. *Int. J. Biol. Macromol.* **2019**, *132*, 439–450. https://doi.org/10.1016/j.ijbiomac.2019.03.142.
75. Szabó, G.; Romhányi, V.; Kun, D.; Renner, K.; Pukánszky, B. Competitive Interactions in Aromatic Polymer/Lignosulfonate Blends. *ACS Sustainable Chem. Eng.* **2017**, *5* (1), 410–419. https://doi.org/10.1021/acssuschemeng.6b01785.
76. Viswanathan, S.; Dadmun, M. D. Guidelines To Creating a True Molecular Composite: Inducing Miscibility in Blends by Optimizing Intermolecular Hydrogen Bonding. *Macromolecules.* **2002**, *35* (13), 5049–5060. https://doi.org/10.1021/ma011031x.
77. Lu, X.; Weiss, R. A. Relationship between the Glass Transition Temperature and the Interaction Parameter of Miscible Binary Polymer Blends. *Macromolecules.* **1992**, *25* (12), 3242–3246. https://doi.org/10.1021/ma00038a033.
78. Kun, D.; Pukánszky, B. Polymer/Lignin Blends: Interactions, Properties. *applications. Eur. Polym. J.* **2017**, *93*, 618–641.
79. Li, Q.; Li, M.; Lin, H.-S.; Hu, C.; Truong, P.; Zhang, T.; Sue, H.-J.; Pu, Y.; Ragauskas, A. J.; Yuan, J. S. Non-Solvent Fractionation of Lignin Enhances Carbon Fiber Performance. *ChemSusChem.* **2019**, *12* (14), 3249–3256. https://doi.org/10.1002/cssc.201901052.
80. Kubo, S.; Kadla, J. The Formation of Strong Intermolecular Interactions in Immiscible Blends with Poly(Vinyl Alcohol) (PVA) and Lignin. *Biomacromolecules.* **2003**, *4*, 561–567.
81. Luo, S.; Cao, J.; McDonald, A. G. Esterification of Industrial Lignin and Its Effect on the Resulting Poly(3-Hydroxybutyrate-Co-3-Hydroxyvalerate) or Polypropylene Blends. *Ind. Crops Prod.* **2017**, *97*, 281–291. https://doi.org/10.1016/j.indcrop.2016.12.024.
82. Henry, N.; Harper, D.; Dadmun, M. Optimizing Noncovalent Interactions Between Lignin and Synthetic Polymers to Develop Effective Compatibilizers. *Macromol. Chem. Phys.* **2012**, *213* (12), 1196–1205. https://doi.org/10.1002/macp.201100633.
83. Podolyák, B.; Kun, D.; Renner, K.; Pukánszky, B. Hydrogen Bonding Interactions in Poly(Ethylene-Co-Vinylalcohol) /Lignin Blends. *Int. J. Biol. Macromol.* **2018**, *107*, 1203–1211.

84. Thielemans, W.; Wool, R. P. Lignin Esters for Use in Unsaturated Thermosets: Lignin Modification and Solubility Modeling. *Biomacromolecules.* **2005**, *6* (4), 1895–1905. https://doi.org/10.1021/bm0500345.
85. Abbott, S. Chemical Compatibility of Poly(Lactic Acid): A Practical Framework Using Hansen Solubility Parameters. In *Poly(Lactic Acid)*; Grossman, R. F., Nwabunma, D., Auras, R., Lim, L. T., Selke, S. E. M., Tsuji, H., Eds.; John Wiley & Sons, Ltd: Hoboken, NJ, 2010; pp 83–95. https://doi.org/10.1002/9780470649848.ch7.
86. Shacklette, L. W.; Han, C. C. Solubility and Dispersion Characteristics of Polyaniline. *MRS Online Proc. Libr.* **1993**, *328*. https://doi.org/10.1557/PROC-328-157.
87. Rodrigues, P.; Munaro, M.; Garcia, C.; Souza, G.; Abbate, M.; Schreiner, W.; Gomes, M. Polyaniline/Lignin Blends: Thermal Analysis and XPS. *Eur. Polym. J.* **2001**, *37*, 2217–2223. https://doi.org/10.1016/S0014-3057(01)00104-5.
88. Hansen, C. *Hansen Solubility Parameters: A User's Handbook*, Second *Edition*, second.; Hansen, C. M., Ed.; CRC Press: Boca Raton, FL, 2007.
89. Gao, J. *Using Hansen Solubility Parameters (HSPs) to Develop Antioxidant-Packaging Film to Achieve Controlled Release*, Master's thesis, Michigan State University, East Lansing, MI, 2014.
90. Chen, R.; Abdelwahab, M. A.; Misra, M.; Mohanty, A. K. Biobased Ternary Blends of Lignin, Poly(Lactic Acid) and Poly(Butylene Adipate-Co-Terephthalate): The Effect of Lignin Heterogeneity on Blend Morphology and Compatibility. *J. Polym. Environ.* **2014**, *22* (4), 439–448. https://doi.org/10.1007/s10924-014-0704-5.
91. Mousavioun, P.; Doherty, W. O. S.; George, G. Thermal Stability and Miscibility of Poly(Hydroxybutyrate) and Soda Lignin Blends. *Ind. Crops Prod.* **2010**, *32* (3), 656–661. https://doi.org/10.1016/j.indcrop.2010.08.001.
92. Mousavioun, P.; Halley, P. J.; Doherty, W. O. S. Thermophysical Properties and Rheology of PHB/Lignin Blends. *Ind. Crops Prod.* **2013**, *50*, 270–275.
93. Jubinville, D.; Chang, B. P.; Pin, J.-M.; Mohanty, A. K.; Misra, M. Synergistic Thermo-Oxidative Maleation of PA11 as Compatibilization Strategy for PA6 and PBT Blend. *Polymer.* **2019**, *179*, 121594. https://doi.org/10.1016/j.polymer.2019.121594.
94. Sallem-Idrissi, N.; Sclavons, M.; Debecker, D. P.; Devaux, J. Miscible Raw Lignin/Nylon 6 Blends: Thermal and Mechanical Performances. *J. Appl. Polym. Sci.* **2016**, *133* (6), 42963. https://doi.org/10.1002/app.42963.
95. Sallem-Idrissi, N.; Van Velthem, P.; Sclavons, M. Fully Bio-Sourced Nylon 11/Raw Lignin Composites: Thermal and Mechanical Performances. *J. Polym. Environ.* **2018**, *26* (12), 4405–4414. https://doi.org/10.1007/s10924-018-1311-7.
96. Ciemniecki, S.; Glasser, W. Multiphase Materials with Lignin: 1. *Blends of hydroxypropyl lignin with poly(methylmethacrylate). Polymer.* **1988**, *29*, 1021–1036.
97. Snowdon, M. R.; Mohanty, A. K.; Misra, M. Miscibility and Performance Evaluation of Biocomposites Made from Polypropylene/Poly(Lactic Acid)/Poly(Hydroxybutyrate-Co-Hydroxyvalerate) with a Sustainable Biocarbon Filler. *ACS Omega.* **2017**, *2* (10), 6446–6454. https://doi.org/10.1021/acsomega.7b00983.
98. Kumar, A.; Tumu, V. R.; Chowdhury, S.; R., R. S. A Green Physical Approach to Compatibilize a Bio-Based Poly (Lactic Acid) /Lignin Blend for Better Mechanical, Thermal and Degradation Properties. *Int. J. Biol. Macromol.* **2019**, *121*, 588–600.

99. Hu, L.; Stevanovic, T.; Rodrigue, D. Unmodified and Esterified Kraft Lignin-Filled Polyethylene Composites: Compatibilization by Free-Radical Grafting. *J. Appl. Polym. Sci.* **2015**, *132* (1-18), 41484.

100. Ge, X.; Chang, M.; Jiang, W.; Zhang, B.; Xing, R.; Bulin, C. Investigation on Two Modification Strategies for the Reinforcement of Biodegradable Lignin/Poly(Lactic Acid) Blends. *J. Appl. Polym. Sci.* **2020**, *137* (44), 49354. https://doi.org/10.1002/app.49354.

101. Barzegari, M.; Alemdar, A.; Zhang, Y.; Rodrigue, D. Mechanical and Rheological Behavior of Highly Filled Polystyrene with Lignin. *Polym. Compos.* **2012**, *33*, 353–361.

102. Barzegari, M.; Alemdar, A.; Zhang, Y.; Rodrigue, D. Thermal Analysis of Highly Filled Composites of Polystyrene with Lignin. *Polym. Compos.* **2013**, *21* (6), 357–366.

103. Bozsódi, B.; Romhányi, V.; Pataki, P.; Kun, D.; D., R.; K.; Pukánszky, B. Modification of Interactions in Polypropylene/Lignosulfonate Blends. *Mater. Des.* **2016**, *103*, 32–39.

104. Sailaja, R. R. N.; Deepthi, M. V. Mechanical and Thermal Properties of Compatibilized Composites of Polyethylene and Esterified Lignin. *Mater. Des.* **2010**, *31* (9), 4369–4379. https://doi.org/10.1016/j.matdes.2010.03.046.

105. Oliveira, W. de; Glasser, W. G. Multiphase Materials with Lignin. XIV. Star-Like Copolymers with Styrene. *J. Wood Chem. Technol.* **1994**, *14* (1), 119–126. https://doi.org/10.1080/02773819408003089.

106. Blanco, I.; G., C.; Latteri, A.; Saccullo, G.; El-Sabbagh, A.; Ziegmann, G. Thermal Characterization of a Series of Lignin-Based Polypropylene Blends. *J. Therm. Anal. Calorim.* **2017**, *127*, 147–153.

107. Lu, S.; Li, S.; Yu, J.; Guo, D.; Ling, R.; Huang, B. The Effect of Hyperbranched Polymer Lubricant as a Compatibilizer on the Structure and Properties of Lignin/Polypropylene Composites. *Wood Mat. Sci. Eng.* **2013**, *8* (3), 159–165.

108. Sailaja, R. Low Density Polyethylene and Grafted Lignin Polyblends Using Epoxy-Functionalized Compatibilizer: Mechanical and Thermal Properties. *Polym. Int.* **2005**, *54*, 1589–1598.

109. Yang, W.; Fortunati, E.; Dominici, F.; Kenny, J. M.; Puglia, D. Effect of Processing Conditions and Lignin Content on Thermal, Mechanical and Degradative Behavior of Lignin Nanoparticles/Polylactic (Acid) Bionanocomposites Prepared by Melt Extrusion and Solvent Casting. *Eur. Polym. J.* **2015**, *71*, 126–139. https://doi.org/10.1016/j.eurpolymj.2015.07.051.

110. Alexy, P.; Kosikova, B.; Crkonova, G.; Gregorova, A.; Martis, P. Modification of Lignin-Polyethylene Blends with High Lignin Content Using Ethylene-Vinylacetate Copolymer as Modifier. *J. Appl. Polym. Sci.* **2004**, *94*, 1855–1860.

111. Sahoo, S.; Misra, M.; Mohanty, A. K. Enhanced Properties of Lignin-Based Biodegradable Polymer Composites Using Injection Moulding Process. *Composites, Part A.* **2011**, *42* (11), 1710–1718. https://doi.org/10.1016/j.compositesa.2011.07.025.

112. Nägele, H.; Pfitzer, J.; Nägele, E.; Inone, E. R.; Eisenreich, N.; Eckl, W.; Eyerer, P. Arboform® - A Thermoplastic, Processable Material from Lignin and Natural Fibers. In *Chemical Modification, Properties, and Usage of Lignin*; Hu, T. Q. , Ed.; Springer US: Boston, MA, 2002; pp 101–119. https://doi.org/10.1007/978-1-4615-0643-0_6.

113. Chen, F.; Liu, W.; Seyed Shahabadi, S. I.; Xu, J.; Lu, X. Sheet-Like Lignin Particles as Multifunctional Fillers in Polypropylene. *ACS Sustainable Chem. Eng.* **2016**, *4* (9), 4997–5004. https://doi.org/10.1021/acssuschemeng.6b01369.
114. Gonzalez, M. N. G.; Levi, M.; Turri, S.; Griffini, G. Lignin Nanoparticles by Ultrasonication and Their Incorporation in Waterborne Polymer Nanocomposites. *J. Appl. Polym. Sci.* **2017**, *134* (38), 45318. https://doi.org/10.1002/app.45318.
115. Nair, S. S.; Sharma, S.; Pu, Y.; Sun, Q.; Pan, S.; Zhu, J. Y.; Deng, Y.; Ragauskas, A. J. High Shear Homogenization of Lignin to Nanolignin and Thermal Stability of Nanolignin-Polyvinyl Alcohol Blends. *ChemSusChem.* **2014**, *7* (12), 3513–3520. https://doi.org/10.1002/cssc.201402314.
116. Qi, G.; Yang, W.; Puglia, D.; Wang, H.; Xu, P.; Dong, W.; Zheng, T.; Ma, P. Hydrophobic, UV Resistant and Dielectric Polyurethane-Nanolignin Composites with Good Reprocessability. *Mater. Des.* **2020**, *196*, 109150. https://doi.org/10.1016/j.matdes.2020.109150.
117. Yang, W.; Fortunati, E.; Dominici, F.; Kenny, J. M.; Puglia, D. Effect of Processing Conditions and Lignin Content on Thermal, Mechanical and Degradative Behavior of Lignin Nanoparticles/Polylactic (Acid) Bionanocomposites Prepared by Melt Extrusion and Solvent Casting. *Eur. Polym. J.* **2015**, *71*, 126–139. https://doi.org/10.1016/j.eurpolymj.2015.07.051.
118. Gigli, M.; Crestini, C. Fractionation of Industrial Lignins: Opportunities and Challenges. *Green Chem.* **2020**, *22*, 4722–4746. https://doi.org/10.1039/D0GC01606C.
119. Cui, C.; Sun, R.; Argyropoulos, D. S. Fractional Precipitation of Softwood Kraft Lignin: Isolation of Narrow Fractions Common to a Variety of Lignins. *ACS Sustainable Chem. Eng.* **2014**, *2* (4), 959–968. https://doi.org/10.1021/sc400545d.
120. Dodd, A. P.; Kadla, J. F.; Straus, S. K. Characterization of Fractions Obtained from Two Industrial Softwood Kraft Lignins. *ACS Sustainable Chem. Eng.* **2015**, *3* (1), 103–110. https://doi.org/10.1021/sc500601b.
121. Domínguez-Robles, J.; Tamminen, T.; Liitiä, T.; Peresin, M. S.; Rodríguez, A.; Jääskeläinen, A.-S. Aqueous Acetone Fractionation of Kraft, Organosolv and Soda Lignins. *Int. J. Biol. Macromol.* **2018**, *106*, 979–987. https://doi.org/10.1016/j.ijbiomac.2017.08.102.
122. Jääskeläinen, A.-S.; Liitiä, T.; Mikkelson, A.; Tamminen, T. Aqueous Organic Solvent Fractionation as Means to Improve Lignin Homogeneity and Purity. *Ind. Crops Prod.* **2017**, *103*, 51–58. https://doi.org/10.1016/j.indcrop.2017.03.039.
123. Kim, J.-Y.; Park, S. Y.; Lee, J. H.; Choi, I.-G.; Choi, J. W. Sequential Solvent Fractionation of Lignin for Selective Production of Monoaromatics by Ru Catalyzed Ethanolysis. *RSC Adv.* **2017**, *7* (84), 53117–53125. https://doi.org/10.1039/C7RA11541E.
124. Li, H.; McDonald, A. G. Fractionation and Characterization of Industrial Lignins. *Ind. Crops Prod.* **2014**, *62*, 67–76. https://doi.org/10.1016/j.indcrop.2014.08.013.
125. Arshanitsa, A.; Ponomarenko, J.; Dizhbite, T.; Andersone, A.; Gosselink, R. J. A.; van der Putten, J.; Lauberts, M.; Telysheva, G. Fractionation of Technical Lignins as a Tool for Improvement of Their Antioxidant Properties. *J. Anal. Appl. Pyrolysis.* **2013**, *103*, 78–85. https://doi.org/10.1016/j.jaap.2012.12.023.

126. Passoni, V.; Scarica, C.; Levi, M.; Turri, S.; Griffini, G. Fractionation of Industrial Softwood Kraft Lignin: Solvent Selection as a Tool for Tailored Material Properties. *ACS Sustainable Chem. Eng.* **2016**, *4* (4), 2232–2242. https://doi.org/10.1021/acssuschemeng.5b01722.
127. Saito, T.; Perkins, J. H.; Vautard, F.; Meyer, H. M.; Messman, J. M.; Tolnai, B.; Naskar, A. K. Methanol Fractionation of Softwood Kraft Lignin: Impact on the Lignin Properties. *ChemSusChem.* **2014**, 7 (1), 221–228. https://doi.org/10.1002/cssc.201300509.
128. Wang, K.; Xu, F.; Sun, R. Molecular Characteristics of Kraft-AQ Pulping Lignin Fractionated by Sequential Organic Solvent Extraction. *Int. J. Mol. Sci.* **2010**, *11* (8), 2988–3001. https://doi.org/10.3390/ijms11082988.
129. Boeriu, C. G.; Fiţigău, F. I.; Gosselink, R. J. A.; Frissen, A. E.; Stoutjesdijk, J.; Peter, F. Fractionation of Five Technical Lignins by Selective Extraction in Green Solvents and Characterisation of Isolated Fractions. *Ind. Crops Prod.* **2014**, *62*, 481–490. https://doi.org/10.1016/j.indcrop.2014.09.019.
130. Sadeghifar, H.; Argyropoulos, D. S. Macroscopic Behavior of Kraft Lignin Fractions: Melt Stability Considerations for Lignin–Polyethylene Blends | ACS Sustainable Chemistry & Engineering. *ACS Sustainable Chem. Eng.* **2016**, *4* (10), 5160–5166. https://doi.org/10.1021/acssuschemeng.6b00636.
131. Yue, X.; Chen, F.; Zhou, X.; He, G. Preparation and Characterization of Poly (Vinyl Chloride) Polyblends with Fractionated Lignin. *Int. J. Polym. Mater. Polym. Biomater.* **2012**, *61* (3), 214–228. https://doi.org/10.1080/00914037.2011.574659.
132. Park, S. Y.; Kim, J.-Y.; Youn, H. J.; Choi, J. W. Utilization of Lignin Fractions in UV Resistant Lignin-PLA Biocomposites via Lignin-Lactide Grafting. *Int. J. Biol. Macromol.* **2019**, *138*, 1029–1034. https://doi.org/10.1016/j.ijbiomac.2019.07.157.
133. Banu, D.; El-Aghoury, A.; Feldman, D. Contributions to Characterization of Poly(Vinyl Chloride) -Lignin Blends. *J. Appl. Polym. Sci.* **2006**, *101*, 2732–2748.
134. Li, Y.; Mlynar, J.; Sarkanen, S. The First 85% Kraft Lignin-Based Thermoplastics. *J. Polym. Sci. Part B.* **1997**, *35*, 1899–1910.
135. Bouajila, J.; Dole, P.; Joly, C.; Limare, A. Some Laws of a Lignin Plasticization. *J. Appl. Polym. Sci.* **2006**, *102*, 1445–1451.
136. Feldman, D.; Banu, D.; Campanelli, J.; Zhu, H. Blends of Vinylic Copolymer with Plasticized Lignin: Thermal and Mechanical Properties. *J. App. Polym. Sci.* **2001**, *81*, 861–874.
137. Park, C.-W.; Youe, W.-J.; Kim, S.-J.; Han, S.-Y.; Park, J.-S.; Lee, E.-A.; Kwon, G.-J.; Kim, Y.-S.; Kim, N.-H.; Lee, S.-H. Effect of Lignin Plasticization on Physico-Mechanical Properties of Lignin/Poly(Lactic Acid) Composites. *Polymers.* **2019**, *11*, 2089. https://doi.org/10.3390/polym11122089.
138. Sakata, I.; Senju, R. Thermoplastic Behavior of Lignin with Various Synthetic Plasticizers. *J. Appl. Polym. Sci.* **1975**, *19*, 2799–2810.
139. Schuerch, C. The Solvent Properties of Liquids and Their Relation to the Solubility, Swelling, Isolation and Fractionation of Lignin. *J. Am. Chem. Soc.* **1952**, *74*, 5061–5067.
140. Argyropoulos, D. S.; Berry, R. M.; Bolker, H. I. Polymerization beyond the Gel Point, 2. A Study of the Soluble Fraction as a Function of the Extent of Reaction. *Makromol. Chem.* **1987**, *188* (8), 1985–1992. https://doi.org/10.1002/macp.1987.021880820.

141. Cui, C.; Sadeghifar, H.; Sen, S.; Argyropoulus, D. S. Toward Thermoplastic Lignin Polymers; Part II: Thermal & Polymer Characteristics of Kraft Lignin & Derivatives. *BioResources.* **2013**, *8* (1), 864–886.

142. Laurichesse, S.; Avérous, L. Chemical Modification of Lignins: Towards Biobased Polymers. *Prog. Polym. Sci.* **2014**, *39* (7), 1266–1290.

143. Sadeghifar, H.; Cui, C.; Argyropoulos, D. S. Toward Thermoplastic Lignin Polymers. Part 1. Selective Masking of Phenolic Hydroxyl Groups in Kraft Lignins via Methylation and Oxypropylation Chemistries. *Ind. Eng. Chem. Res.* **2012**, *51* (51), 16713–16720. https://doi.org/10.1021/ie301848j.

144. Pawar, S. N.; Venditti, R. A.; Jameel, H.; Chang, H.-M.; Ayoub, A. Engineering Physical and Chemical Properties of Softwood Kraft Lignin by Fatty Acid Substitution. *Ind. Crops Prod.* **2016**, *89*, 128–134. https://doi.org/10.1016/j.indcrop.2016.04.070.

145. Glasser, W.; Jain, R. Lignin Derivatives I. Alkanoates. *Holzforschung.* **1993**, *47* (3), 225–233.

146. Gordobil, O.; Robles, E.; Egüés, I.; Labidi, J. Lignin-Ester Derivatives as Novel Thermoplastic Materials. *RSC Adv.* **2016**, *6* (90), 86909–86917. https://doi.org/10.1039/C6RA20238A.

147. Mori, E.-L. H.; Ropponen, J.; Poppius-Levlin, K.; Ohra-Aho, T.; Tamminen, T. Enhancing the Barrier Properties of Paper Board by a Novel Lignin Coating. *Ind. Crops Prod.* **2013**, *50*, 694–700. https://doi.org/10.1016/j.indcrop.2013.08.013.

148. Koivu, K. A. Y.; Sadeghifar, H.; Nousiainen, P. A.; Argyropoulos, D. S.; Sipilä, J. Effect of Fatty Acid Esterification on the Thermal Properties of Softwood Kraft Lignin. *ACS Sustainable Chem. Eng.* **2016**, *4* (10), 5238–5247. https://doi.org/10.1021/acssuschemeng.6b01048.

149. Fox, S. C.; McDonald, A. G. Chemical and Thermal Characterization of Three Industrial Lignins and Their Corresponding Lignin Esters. *BioResources.* **2010**, *5* (2), 990–1009.

150. Hulin, L.; Husson, E.; Bonnet, J.-P.; Stevanovic, T.; Sarazin, C. Enzymatic Transesterification of Kraft Lignin with Long Acyl Chains in Ionic Liquids. *Molecules.* **2015**, *20* (9), 16334–16353. https://doi.org/10.3390/molecules200916334.

151. Monteil-Rivera, F.; Paquet, L. Solvent-Free Catalyst-Free Microwave-Assisted Acylation of Lignin. *Ind. Crops Prod.* **2015**, *65*, 446–453. https://doi.org/10.1016/j.indcrop.2014.10.060.

152. Pietarinen, S.; MYLLYMÄKI, T.; Eskelinen, K. *A Method for Esterifying Lignin with at Least One Fatty Acid*. WIPO Patent 2014029919A1, February 27, 2014.

153. Hult, E. L.; Ropponen, J.; Poppius-Levlin, K.; Ohra-Aho, T.; Tamminen, T. Enhancing the Barrier Properties of Paper Board by a Novel Lignin Coating. *Ind. Crops Prod.* **2013**, *50*, 694–700.

154. Poppius-Levlin, K.; Hult, E.-L.; Ropponen, J.; Hohenthal, C.; Tamminen, T. In *Lignin-Based Sustainable Packaging Solutions*, NWBC 2012: The 4th Nordic Wood Biorefinery Conference, Espoo, Finland, Oct 23-25, 2012; Niemelä, K. Ed.; VTT Technical Research Centre of Finland: Finland, pp 174-179.

155. Sameni, J.; Krigstin, S.; Sain, M. Solubility of Lignin and Acetylated Lignin in Organic Solvents. *BioResources.* **2017**, *12*, 1548–1565.

156. Ghosh, I.; Jain, R. K.; Glasser, W. G. Blends of Biodegradable Thermoplastics with Lignin Esters. In *Lignin: Historical, Biological, and Materials Perspectives*; Glasser, W. G., Northey, R. A., Schultz, T. P., Eds.; ACS Symposium Series 742; American Chemical Society: Washington, DC, 1999; pp 331-350. https://doi.org/10.1021/bk-2000-0742.ch017.

157. Gordobil, O.; Delucis, R.; Egüés, I.; Labidi, J. Kraft Lignin as Filler in PLA to Improve Ductility and Thermal Properties. *Ind. Crops Prod.* **2015**, 72, 46–53. https://doi.org/10.1016/j.indcrop.2015.01.055.
158. Teramoto, Y.; Lee, S.-H.; Endo, T. Phase Structure and Mechanical Property of Blends of Organosolv Lignin Alkyl Esters with Poly(ε-Caprolactone). *Polym. J.* **2009**, *41* (3), 219–227. https://doi.org/10.1295/polymj.PJ2008301.
159. Luong, N. D.; Binh, N. T. T.; Duong, L. D.; Kim, D. O.; Kim, D.-S.; Lee, S. H.; Kim, B. J.; Lee, Y. S.; Nam, J.-D. An Eco-Friendly and Efficient Route of Lignin Extraction from Black Liquor and a Lignin-Based Copolyester Synthesis. *Polym. Bull.* **2012**, *68* (3), 879–890. https://doi.org/10.1007/s00289-011-0658-x.
160. Ahvazi, B.; Wojciechowicz, O.; Ton-That, T.-M.; Hawari, J. Preparation of Lignopolyols from Wheat Straw Soda Lignin. *J. Agric. Food Chem.* **2011**, *59* (19), 10505–10516. https://doi.org/10.1021/jf202452m.
161. de Oliveira, W.; Glasser, W. G. Multiphase Materials with Lignin. 11. Starlike Copolymers with Caprolactone. *Macromolecules.* **1994**, *27* (1), 5–11. https://doi.org/10.1021/ma00079a002.
162. Bhattacharyya, S.; Matsakas, L.; Rova, U.; Christakopoulos, P. Melt Stable Functionalized Organosolv and Kraft Lignin Thermoplastic. *Processes.* **2020**, *8* (9), 1108. https://doi.org/10.3390/pr8091108.
163. Wang, K.-T.; Jing, C.; Wood, C.; Nagardeolekar, A.; Kohan, N.; Dongre, P.; Amidon, T. E.; Bujanovic, B. M. Toward Complete Utilization of Miscanthus in a Hot-Water Extraction-Based Biorefinery. *Energies.* **2018**, *11* (1), 39. https://doi.org/10.3390/en11010039.
164. Goundalkar, M. J.; Corbett, D. B.; Bujanovic, B. M. Comparative Analysis of Milled Wood Lignins (MWLs) Isolated from Sugar Maple (SM) and Hot-Water Extracted Sugar Maple (ESM). *Energies.* **2014**, *7* (3), 1363–1375. https://doi.org/10.3390/en7031363.
165. Pawar, A. A.; Chaugule, A. A.; Kim, H. Greener Synthesis of Dimethyl Carbonate from Carbon Dioxide and Methanol Using a Tunable Ionic Liquid Catalyst. *Open Chem.* **2019**, *17* (1), 1252–1265. https://doi.org/10.1515/chem-2019-0137.
166. Sen, S.; Patil, S.; Argyropoulos, D. S. Methylation of Softwood Kraft Lignin with Dimethyl Carbonate. *Green Chem.* **2015**, *17* (2), 1077–1087. https://doi.org/10.1039/C4GC01759E.
167. Kühnel, I.; Podschun, J.; Saake, B.; Lehnen, R. Synthesis of Lignin Polyols via Oxyalkylation with Propylene Carbonate. *Holzfuschung.* **2014**, *69* (5), 531–538. https://doi.org/10.1515/hf-2014-0068.
168. Duval, A.; Avérous, L. Cyclic Carbonates as Safe and Versatile Etherifying Reagents for the Functionalization of Lignins and Tannins. *ACS Sustainable Chem. Eng.* **2017**, *5* (8), 7334–7343. https://doi.org/10.1021/acssuschemeng.7b01502.
169. Maldhure, A. V.; Ekhe, J. D. Effect of Modifications of Lignin on Thermal, Structural, and Mechanical Properties of Polypropylene/Modified Lignin Blends. *J. Thermoplast. Compo. Mater.* **2017**, *30* (5), 625–645. https://doi.org/10.1177/0892705715610402.
170. Chen, F.; Dai, H.; Dong, X.; Yang, J.; Zhong, M. Physical Properties of Lignin-Based Polypropylene Blends. *Polym. Compo.* **2011**, *32* (7), 1019–1025. https://doi.org/10.1002/pc.21087.
171. Jiang, C.; Zhang, Y.; Xiao, X.; Cai, Z.; Yu, P. High-Performance Hydroxypropyl Black Liquor Lignin/Poly (Propylene Carbonate) Bio-Composites with Enhanced Natural Degradability. *Polym. Test.* **2018**, 72, 348–356. https://doi.org/10.1016/j.polymertesting.2018.10.044.

172. Glasser, W. G.; Loos, R.; Cox, B.; Cao, H. Melt-Blown Compostable Polyester Films with Lignin. *Tappi J.* **2017**, *16* (3), 111–121. https://doi.org/10.32964/TJ16.3.111.
173. Lin, S. Y.; Lin, I. S. Lignin. In *Ullmann's Encyclopedia of Industrial Chemistry*; Gerhartz, W., Schulz, G. , Wiley: New York, 1990; Vol. A15, pp 305. https://doi.org/10.1002/14356007.a15_305.
174. Wang, C.; Venditti, R. A. UV Cross-Linkable Lignin Thermoplastic Graft Copolymers. *ACS Sustainable Chem. Eng.* **2015**, *3* (8), 1839–1845. https://doi.org/10.1021/acssuschemeng.5b00416.
175. Ren, W.; Pan, X.; Wang, G.; Cheng, W.; Liu, Y. Dodecylated Lignin-g-PLA for Effective Toughening of PLA. *Green Chem.* **2016**, *18* (18), 5008–5014. https://doi.org/10.1039/C6GC01341D.
176. Hilburg, S. L.; Elder, A. N.; Chung, H.; Ferebee, R. L.; Bockstaller, M. R.; Washburn, N. R. A Universal Route towards Thermoplastic Lignin Composites with Improved Mechanical Properties. *Polymer.* **2014**, *55* (4), 995–1003. https://doi.org/10.1016/j.polymer.2013.12.070.
177. Kim, Y. S.; Youe, W.-J.; Kim, S. J.; Lee, O.-K.; Lee, S.-S. Preparation of a Thermoplastic Lignin-Based Biomaterial through Atom Transfer Radical Polymerization. *J. Wood Chem. Technol.* **2015**, 35 (4), 251–259. https://doi.org/10.1080/02773813.2014.937006.
178. Kim, Y. S.; Kadla, J. F. Preparation of a Thermoresponsive Lignin-Based Biomaterial through Atom Transfer Radical Polymerization. *Biomacromolecules.* **2010**, *11* (4), 981–988. https://doi.org/10.1021/bm901455p.
179. Park, I.-K.; Sun, H.; Kim, S.-H.; Kim, Y.; Kim, G. E.; Lee, Y.; Kim, T.; Choi, H. R.; Suhr, J.; Nam, J.-D. Solvent-Free Bulk Polymerization of Lignin-Polycaprolactone (PCL) Copolymer and Its Thermoplastic Characteristics. *Sci. Rep.* **2019**, *9* (1), 7033. https://doi.org/10.1038/s41598-019-43296-2.
180. Liu, X.; Zong, E.; Jiang, J.; Fu, S.; Wang, J.; Xu, B.; Li, W.; Lin, X.; Xu, Y.; Wang, C.; Chu, F. Preparation and Characterization of Lignin-Graft-Poly (ε-Caprolactone) Copolymers Based on Lignocellulosic Butanol Residue. *Int. J. Biol. Macromol.* **2015**, *81*, 521–529. https://doi.org/10.1016/j.ijbiomac.2015.08.046.
181. Yu, J.; Wang, J.; Wang, C.; Liu, Y.; Xu, Y.; Tang, C.; Chu, F. UV-Absorbent Lignin-Based Multi-Arm Star Thermoplastic Elastomers. *Macromol. Rapid Commun.* **2015**, *36* (4), 398–404. https://doi.org/10.1002/marc.201400663.
182. Tran, C. D.; Chen, J.; Keum, J. K.; Naskar, A. K. A New Class of Renewable Thermoplastics with Extraordinary Performance from Nanostructured Lignin-Elastomers. *Adv. Funct. Mater.* **2016**, *26* (16), 2677–2685. https://doi.org/10.1002/adfm.201504990.
183. Huang, J.; Liu, W.; Qiu, X. High Performance Thermoplastic Elastomers with Biomass Lignin as Plastic Phase. *ACS Sustainable Chem. Eng.* **2019**, 7 (7), 6550–6560.

Chapter 11

Recent Advances in Synthesis and Application of Lignin Nanoparticles

Xianzhi Meng,[1] Mandeep Poonia,[2] Chang Geun Yoo,[2] and Arthur J. Ragauskas[1,3,4,*]

[1]Department of Chemical and Biomolecular Engineering, University of Tennessee, Knoxville, Tennessee 37996, United States

[2]Department of Chemical Engineering, State University of New York College of Environmental Science and Forestry, Syracuse, New York 13210, United States

[3]Center for Bioenergy Innovation, Biosciences Division, Oak Ridge National Laboratory, Oak Ridge, Tennessee 37831, United States

[4]Department of Forestry, Wildlife, and Fisheries, Center of Renewable Carbon, University of Tennessee, Institute of Agriculture, Knoxville, Tennessee 37996, United States

[*]Email: aragausk@utk.edu.

The ability to engineer material at its nanoscale has shown great potentials in overcoming significant obstacles in numerous fields. Because of its abundance, biodegradability, renewability, and eco-friendliness nature, lignin has become an ideal precursor for producing nanoscale materials from both research and industrial communities. Nanolignin, also known as colloidal lignin particle (CLP) or lignin nanoparticle (LNP), has received significant interest in recent years and its production and application in numerous fields represent an important aspect of lignin valorization. LNPs could be produced via several techniques including sonication, solvent exchange, and precipitation, which could be divided into wet and dry methods. Applications of LNPs in UV protection, cosmetics, food package, coatings, and drug delivery have been reported in the literature. Despite being extensively studied, LNPs as well as their derived materials with controlled shape and architecture are still largely underexploited. This book chapter mainly covers recent advances in the preparation of LNPs and their use in the novel and advanced applications in different areas.

Lignin represents the most abundant aromatic polymer on earth and is one of the key components of the lignocellulosic plant cell wall along with cellulose and hemicellulose. It accounts for up to 30% of the secondary plant cell walls and provides structural support for the plants' upward growth, enables long-distance water transportation, and provides physical and chemical protection for the plants against pathogen invasion (*1*). Lignin is a three-dimensional amorphous phenolic polymer, which typically consists of three basic types of phenylpropane units as its monolignols: sinapyl, coniferyl, and p-coumaryl alcohols. The oxidative radical-radical coupling of these monolignols gives rise to syringyl (S), guaiacyl (G), and p-hydroxyphenyl (H) lignin C9 subunits, respectively (*2*). Ether linkages (e.g., β-O-4) were mainly formed at the β position of the monolignol species during these coupling reactions, representing up to two-thirds of the total lignin inter-linkages along with a small amount of C-C linkages (e.g., β-β, β-5). Lignin macromolecule contains various types of functional groups such as methoxy, carboxyl, carbonyl, and aliphatic/phenolic hydroxyl groups.

The current biorefinery system mainly focuses on converting plant carbohydrates into liquid biofuels such as ethanol, producing a tremendous amount of lignin streams. These technical lignins are mainly burned to meet the internal energy use of the cellulosic project and to provide a convenient waste disposal platform. Despite significant efforts on producing value-added products, lignin is still underutilized, and future integrated biorefineries strongly depend on how to effectively isolate, purify, and utilize lignin. Lignin could be obtained from various resources including woody/herbaceous biomass, agricultural crop, and forestry residue, and its isolation and purification have been summarized in several critical reviews (*3–6*). Lignin could be depolymerized into its monomers, which could be subsequently converted to chemicals and fuels. It can also be chemically modified to be used in novel biobased materials (*4*). Due to the recent advances in the development of nanotechnology and process chemistry, LNPs have attracted increased attention and represent a promising way to promote and maximize lignin valorization in the biorefinery research community.

A variety of chemical and physical methods to prepare LNPs have been reported in the literature including solvent exchange, dialysis, sonication, precipitation technology, and mechanical treatment. While these nomenclatures reported in literature could sometimes be confusing due to the lack of consensus about the mechanism behind the formation of particles, Österberg et al. further divided these methods into two straightforward categories: wet and dry methods (*7*). Wet LNPs refer to particles generated via precipitation technologies that involve a process of solubilizing lignin in a solvent (typically organic solvent) followed by process of precipitation in an anti-solvent (typically water). These processes are commonly referred to as a solvent exchange, solvent shifting, or sequential precipitation in literature. On the other hand, dry particles could be simply formed via a well-controlled drying of a dilute lignin solution.

The application of LNPs and its derived nanomaterials have not been ignored in the literature. Lignin has unique features such as antimicrobial (*8*), antivirus (*9*), UV adsorption (*10*), and antioxidant properties (*11*). Thus, direct incorporation of LNPs in polymers could be a way to improve their current properties and simultaneously provide the materials with the lignin's above-mentioned inherent features. The non-toxicity, biodegradability, and outstanding adsorption capacity of LNPs also make them perfect vehicles for drug molecules in biomedical and therapeutic applications (*12*). Due to a large number of functional groups lignin possesses, chemically modified LNPs could also be incorporated into bio-composites to improve their mechanical performances and thermal stability (*13*). In addition, the application of LNPs in mineral flotation (*14*), anticorrosion (*15*), cosmetics (*16*), and Pickering emulsions have been all reported in the literature (*17*). The production of LNPs via different routes and their use in advanced applications are extensively discussed in this chapter.

Lignin Nanoparticles Synthesis Methods

LNPs can be produced from various types of lignins based on the feedstocks (e.g., hardwood, softwood, herbaceous plants), preprocessing methods (e.g., thermochemical, mechanical, biological treatments), and synthetic approaches. Compared to the well-developed preprocessing methods and feedstock characterization, the LNP synthesis strategy is still in its infancy. Several methods, such as CO2 saturation, dialysis, flash-precipitation, sonication, and others, have been recently reported for the LNP synthesis (*18–23*). These methods can be mainly categorized into (i) chemical, (ii) physical, and (iii) biological methods. Understanding the LNP formation mechanisms, and the advantages and disadvantages of each synthesis method is helpful for determining an appropriate method for an effective LNP production, control of the particle size and other properties, and synthesis from diverse feedstocks.

Chemical Methods for LNP Synthesis

Chemical methods through the dissolution of crude lignin followed by precipitation have mostly been applied for the synthesis of LNPs (Figure 1). In brief, crude lignin is dissolved in a water-miscible organic solvent (e.g., methanol, ethanol, tetrahydrofuran (THF), dimethyl sulfoxide (DMSO), dioxane, ethylene glycol) (*13*, *19*, *23–25*), or in an alkaline solution (e.g., sodium hydroxide, sodium p-toluenesulfonate) (*18*, *26*). The soluble lignin portion is recovered by centrifugation/filtration and mixed with antisolvent (e.g., water, dilute acids). The introduction of antisolvent into the lignin solution through dialysis (*13*, *18*, *19*), or dropwise addition (*24*, *25*, *27–29*) reduces the concentration of the solvent in the solution and forms LNPs via self-assembly of lignin molecules. The van der Waal forces, π-π interaction, hydrogen bonding, chain entanglement, and hydrophobic interactions between the lignin molecules affect the LNPs formation (*30*).

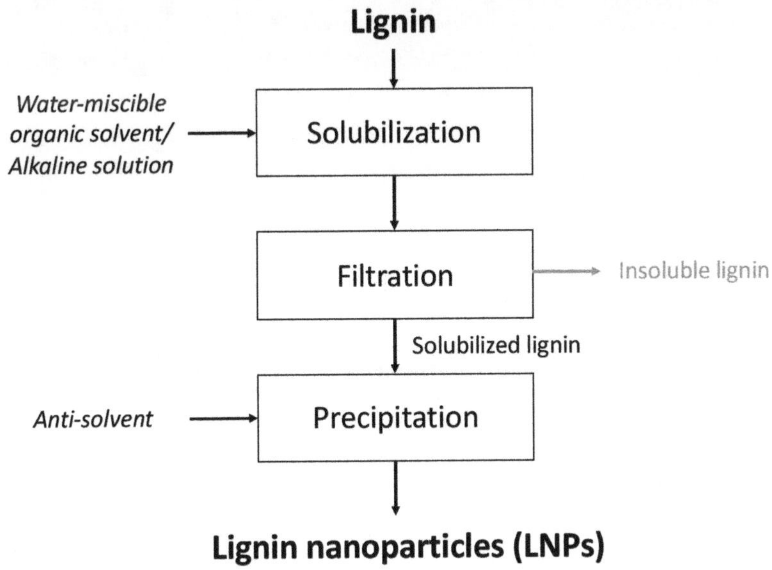

Figure 1. Schematic diagram of LNPs synthesis via chemical methods.

Mechanism of LNP Synthesis *via* Chemical Methods

During the synthesis of LNPs via chemical methods, solubilization, precipitation, and self-assembly of lignin are important phenomena. Solubility of lignin presented by initial concentration of lignin is a crucial factor for the LNPs' synthesis. The solubilized lignin molecules may exist as individual molecules and are typically well distributed in the solution. Sipponen et al. observed that lignin particles began to precipitate once the lignin concentration increased beyond a certain concentration (*28*). Lignin favors aggregation due to its amphiphilic property caused by the presence of hydroxyl groups, carboxyl groups, and aromatic rings. The dissolved lignin molecules can be precipitated by the addition of an antisolvent. Sipponen et al. discussed that large lignin molecules were precipitated at the initial stage of antisolvent addition and acted as precursors for the critical nuclei for nucleation-growth (Figure 2). Collision-driven aggregation increases the nuclei size, and these nuclei fuse with each other to form intermediate particles. Hydrophilic segments of lignin molecules get oriented towards the surface of lignin particles and provide the particles a negatively charged surface. Simultaneously, small polar lignin fragments get adsorbed on these surfaces and make them smoother. The nucleation-growth was also used to explain the LNPs' formation in other studies (*19*, *23*).

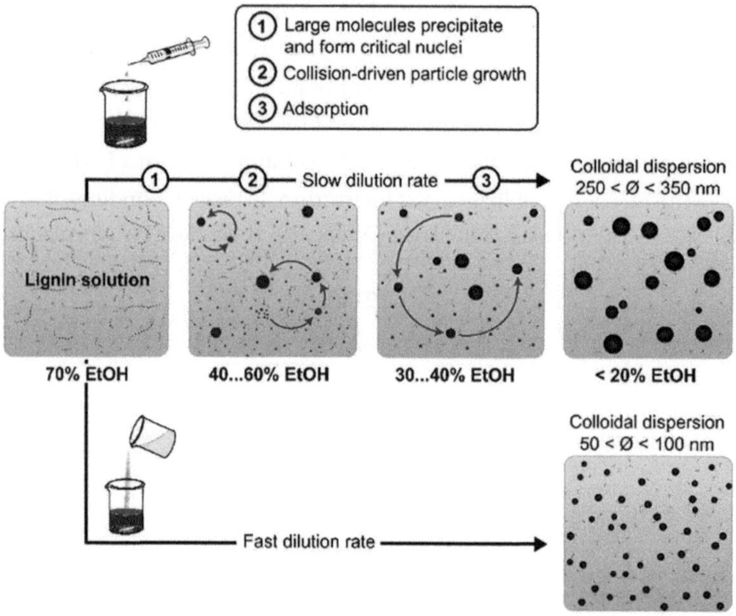

Figure 2. Mechanism of LNPs synthesis by self-assembly. (Reprinted with permission from reference (28). Copyright 2018 American Chemical Society).

Factors Affecting LNP Formation

The formation of the LNPs is affected by multiple factors including initial lignin concentration, feeding rate and pH of antisolvent, the concentration of salts, and temperature. These parameters are directly or indirectly correlated to the aggregation of lignin molecules by changing the surface charge, solubility, and collision probability of lignin molecules. Therefore, elucidation of each factor on LNPs' formation is necessary for determining the synthesis conditions.

The initial lignin concentration has been discussed as a factor influencing the size of LNPs. In previous studies, a lower initial concentration of lignin resulted in smaller particle size of the final

LNPs (*19, 24, 27, 28*). For instance, Lievonen et al. showed that the size of LNPs formed with 1 g/L initial lignin concentration in THF was about 500 nm (*19*). When the initial concentration of lignin increased to 20 g/L, the final lignin particle size was above 2,000 nm. The authors concluded that the lignin particle size became larger with the increase of the initial lignin concentration due to the higher rates of the nucleation growth with lignin molecules. Also, the polydispersity of LNPs at higher concentration (20 g/L) was higher due to very fast nucleation-growth leading to the formation of particles with a broader apparent size range. Similarly, the size of LNPs obtained by Dai et al. increased from 183 nm at 0.2 g/L to 450 nm at 2 g/L of lignin in methanol (*24*). Sipponen et al. also found that the particle size of the LNPs increased as the concentration of lignin in ethanol increased from 1 g/L to 20 g/L (*28*).

The impact of the antisolvent feeding rate into the lignin solution was also investigated as a parameter for LNP formation (*23, 25, 28*). Li et al. tested different antisolvent (ultrapure water) dropping rates into the solution of dioxane with kraft lignin using a peristaltic pump for the LNP synthesis (*25*). The authors reported that the increase of the water dropping rate decreased the hydrodynamic radius of LNPs and formed more nanospheres as it limited the aggregation growth time. Similarly, Sipponen et al. tested the influence of antisolvent feeding rate on the LNP size with the same initial lignin and final solvent concentrations (*28*). As the water addition rate increased from 0.06 v/v/min (antisolvent feeding volume/lignin solution volume/time) to 300 v/v/min, LNP size decreased from 312 nm to 67 nm (Figure 3). Richter et al. also reported that particle size at the rate of addition of water at 1 mL/min was 250 nm, and the size became much smaller (~80 nm) when the rate of water addition increased to 1,100 mL/min (*23*).

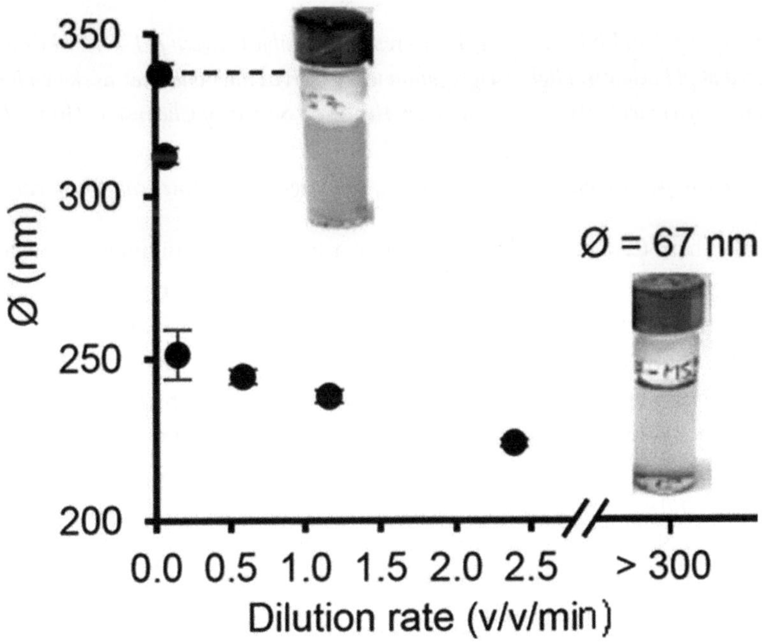

Figure 3. Impact of antisolvent feeding rate on LNP size. (Reprinted with permission from reference (28). Copyright 2018 American Chemical Society).

The size of LNPs was changed by the pH of the LNP suspension (*19, 23, 28*). Lievonen et al. reported that the size of LNPs did not significantly change in the LNP suspension between the pH 4 - 12 (Figure 4a) (*19*). LNPs remained stable because of the repulsion between the particles caused by the high negative charge on the surface of the LNPs. However, as the pH of the solution was reduced below 4, the aggregation of LNPs began due to the decrease of negative charge on the particle surface. Sipponen et al. also reported that the size of LNPs remained stable at a similar pH range (4 – 10), while aggregation occurred at close to the isoelectric point (pH 2.1) of LNPs (Figure 4b) (*28*). Richter et al. also studied the impact of pH on LNP formation with two different lignins (organosolv lignin and kraft lignin). The LNP dispersions from organosolv and kraft lignins were stable in the pH 3.2 – 8.5 (particle size of ~80 nm) and the pH 2 – 8.2 (particle size of ~125 nm), respectively (*23*). The LNPs started to aggregate at lower pH due to a decrease in their surface charge. The size of LNPs started to decrease at pH above 8.5 due to the beginning of lignin dissolution. It completely dissolved at pH 12. The pH stability studies confirmed that the stability of LNPs is due to the electrostatic double-layer repulsions.

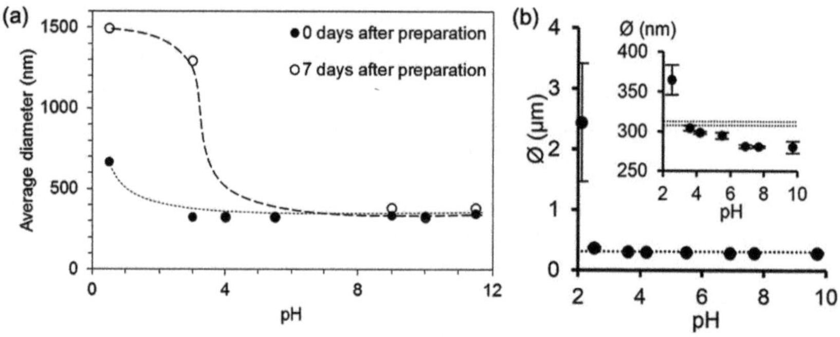

Figure 4. Impact of pH on LNPs' size. (a) LNPs remained stable between pH 4-12 while aggregation of particles started at pH below 4. Higher aggregation was observed after one week as shown by dashed lines (Reprinted from reference (19). Copyright 2016 The Royal Society of Chemistry) (b) LNPs remained stable at pH range 4 – 10. Aggregation occurred at close to the isoelectric point (pH 2.1) of LNPs. (Reprinted with permission from reference (28). Copyright 2018 American Chemical Society).

A few studies have tested the stability of LNP in a suspension with different salt concentrations (*19, 23, 27*). In general, higher salt concentration in the suspension or antisolvent adversely affects the LNP size. Lievonen et al. found that the orientation of phenolic hydroxyl and carboxyl groups in lignin molecules provides the surface a negative charge and forms an electrical double layer (*19*). Repulsion between these electrical double layers stabilizes the nanoparticle dispersion. As shown in Figure 5, the LNP dispersion remained stable, and the diameter of the particles did not increase with dilute solutions of NaCl (up to 500 mM). However, particle aggregation occurred at 1 M NaCl concentration after 7 days. The aggregation started by the reduced double layer repulsions due to the excessive amount of NaCl. Richter et al. also showed that LNPs remained stable (particle size ~80 nm) at the concentration of NaCl up to 70 mM, while the particle size increased 2.5-fold when the NaCl concentration reached to 150 mM (*23*).

Qian et al. prepared larger size (average diameter above 1000 nm) lignin particles by precipitating the lignin solution (10 g/L lignin dissolved in 8:1 acetone/water solution) by adding 0.1 M NaCl (*27*). Under the same conditions, precipitation using deionized water produced smaller particles (~210 nm average diameter). The presence of salts reduces the electrical double layer repulsions between the lignin particles, which leads to their aggregation (*19, 27*). In addition, the temperature of the solution was considered in the formation of LNPs (23). Overall, higher temperature resulted in larger LNP size (Figure 6). This is due to the fast nucleation-growth with higher solubility and a larger rate of polymer self-diffusion. However, only a narrow range of temperature was tested; therefore, further study is needed to understand the temperature effects on the LNP size.

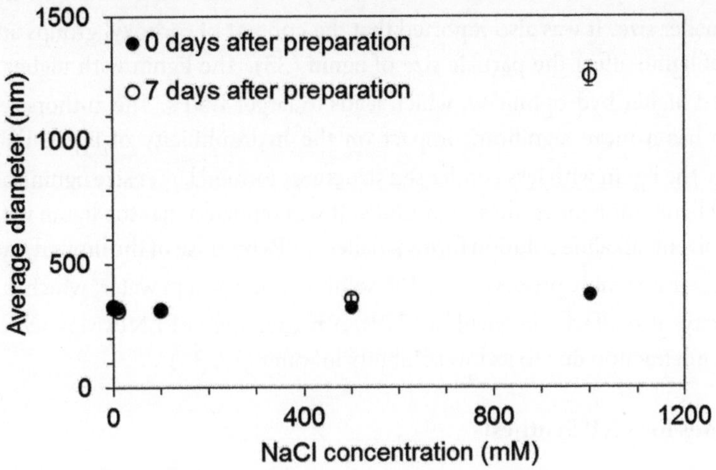

Figure 5. Impact of salt concentration on LNP size. (Reprinted with permission from reference (19). Copyright 2016 The Royal Society of Chemistry).

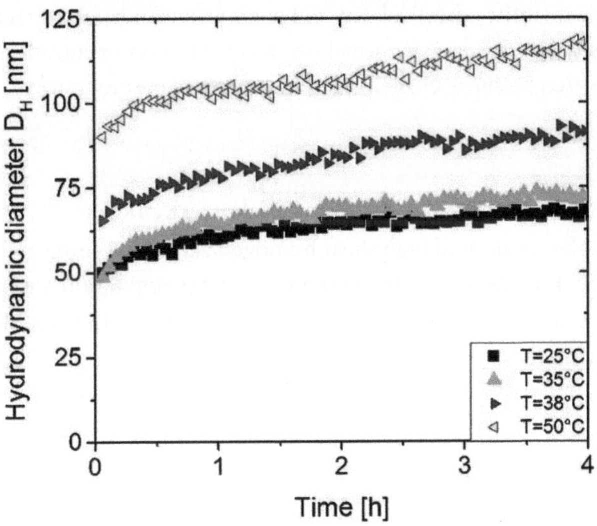

Figure 6. Impact of temperature on the LNP size. (Reprinted with permission from reference (23). Copyright 2016 American Chemical Society).

Besides the aforementioned processing factors, the substrate properties such as molecular weight and hydrophobicity of lignin can impact LNPs' size (*31–36*). Several lignin fractions with different molecular weight and hydrophilic group contents were prepared according to their solubility in different organic solvents (*31–33*) or selective precipitation at different pH (*34, 36*). Similarly, the sequential organosolv fragmentation approach (SOFA) was applied to fractionate lignin in multiple uniform streams (*35*). Lignin with lower molecular weight showed higher hydrophilic group (e.g., hydroxyl and carboxyl groups) contents due to cleavage of its interunit linkages (e.g., β-O-4 bonds). As the content of antisolvent increases in the lignin solution, the nucleation of lignin takes place separately and simultaneously leads to a smaller size of LNPs. Along with more nucleation, hydrophobic interaction between the lignin molecules bound them together strongly leading to the formation of small and compact LNPs, while hydrophilic groups lead to the formation of larger size hollow lignin microparticles (*33*, *36*). Therefore, LNPs obtained from higher molecular weight lignin have a smaller size. It was also reported that the content of carboxyl groups and the degree of condensation of lignin affect the particle size of lignin (*35*). The lignin with higher carboxyl group contents showed higher hydrophilicity, which leads to larger LNPs. The authors reported that the carboxyl group has a more significant impact on the hydrophilicity of lignin than the phenolic hydroxyl group. The lignin with less condensed structures formed larger size lignin particles. Besides, the solubility of lignin influences the size of LNPs. It was reported that the lignin with low solubility in the organic solvent/alkaline solution forms smaller LNPs because of the limited nucleation growth time during the self-assembly process (*32*). The solubility of lignin in water, which is correlated with its hydrophobicity, also affects the yield of LNPs. A higher yield of LNPs was achieved from more hydrophobic lignin fraction due to its low solubility in water (*33*, *35*).

Physical Methods for LNP Synthesis

Despite successful LNP synthesis through chemical methods, often, the use of toxic and corrosive chemicals for the dissolution of lignin is a major limitation of this method. An increase of particle size correlated with the consistency of lignin limits its maximum consistency in the solution. Also, the use of a large quantity of antisolvent and energy consumption to recover LNPs makes the synthesis process costly. Physical LNP formation, including homogenization and ultrasonication, has been studied to overcome those challenges of the chemical methods.

Homogenization

Homogenization is the process of reducing materials to extremely small particles and distribute it throughout the fluid. Nair et al. used high shear homogenization to synthesize LNPs from softwood kraft black liquor (*37*). They recovered the lignin by acidic precipitation and purified with pentane extraction followed by dissolution in the dioxane-water mixture. The soluble fraction was rotary evaporated for removing the solvents and then freeze-dried. The purified lignin was dispersed in deionized water with 0.5 wt% consistency and treated with a high shear homogenizer at 15,000 rpm for 1 – 4 h. Each run's duration was relatively short (30 min) to keep the solution temperature to ambient temperature and allow settling of the bubbles generated by agitation. Around 75% of particles became less than 100 nm after 2h of homogenization and resulted in less than 100 nm with narrow size distribution (~90% of particle had diameter 10 – 30 nm) after 4h. The chemical composition of initial kraft lignin and LNPs obtained by homogenization was nearly identical.

Ultrasonication

Ultrasonication was also used for the synthesis of LNPs. Gilca et al. obtained the LNPs from wheat straw and Sarkanda grass lignins by ultrasonication (*38*). An aqueous suspension of 0.7 wt% lignin was sonicated for 60 min using an ultrasonic horn at 20 kHz of frequency and 600 W of power. In the ultrasonication, cavitation occurred in the solution and resulted in depolymerization by side chain cleavage and polymerization through oxidative coupling with the generated radicals. Since the depolymerization reaction was dominant over oxidative coupling, this method leads to the formation of LNPs. The average diameter of LNPs from the ultrasonication were reduced from microsize (1-10 μm) to nanosize (10-50 nm) for both lignin samples.

Biological Methods for LNP Synthesis

Enzymes have been applied as an effective biocatalyst in many biomass applications such as biofuels and biochemicals production (*39*). Instead of breaking the polysaccharides, Rangan et al. prepared lignin-rich nanoparticles from the fibers of Indian ridge gourd (Luffa cylindrica) by cleaving the lignin-cellulose complex with a multi-enzyme cocktail (*40*). The biomass was frozen in liquid nitrogen and cryo-crushed by milling. The powdered fibers were extracted with an organic solvent mixture (acetone, ethanol, and toluene). Enzymatic hydrolysis was conducted with the organic solvent extracted fraction at pH 4.5 and 55 °C with Multifect CX GC Cellulase, Multifect Pectinase FE, Optimase CX 255L for 24h. The size of lignin-rich particles was around 20-100 nm with a cuboidal shape with reduced cellulose content.

Other Methods for LNP Synthesis

Mishra and Ekielski synthesized the LNPs using spray freezing (*41*). The method used dimethyl sulfoxide (DMSO), which has a high lignin solubility and a high boiling point. Lignin was dissolved (0.05%) in DMSO and passed through a 0.45 μm syringe filter to remove the insoluble fraction. The soluble fraction was sprayed onto a liquid-nitrogen-cooled copper (Cu) plate, and the sprayed droplets became lignin-entrapped frozen droplets. As the droplet enters the freezing zone, it starts to freeze from outside to inside, leading to particle formation. The particle formation resulted due to π-π interaction and the van der Waal forces. The collected mass from the Cu plate was suspended in water (at 4 °C) and sonicated. Dry powder was obtained from the suspension by freeze-drying. The obtained particle diameters were around 150 nm.

Application of Lignin Nanoparticles

LNPs have attracted great attention due to their potential in various industrial and biomedical applications. The applications strongly depend on the resource and physicochemical structural features of the lignin. The following section mainly discusses some of the most promising and newly explored applications.

LNPs as Low-Cost Reinforcement in Nanocomposites

Utilization of lignin in composites of thermoplastics, elastomers, and foamed materials has been well documented in the literature (*42*). However, the broader usage of lignin as a composite filler is still hindered by the fact that technical lignins typically have large particle sizes ranging from 10 μm to more than 100 μm, which is known to cause detrimental impacts on the mechanical properties of the resulting composites (*43*). LNPs could be used as reinforcing or cross-linking agents to produce high-performance nanocomposites *via* miscible polymer-lignin blends strategy or other novel synthetic approaches such as polymer grafting (*44, 45*). Yang et al. prepared LNP reinforced poly(methyl methacrylate) and poly(lactic acid) nanocomposites with improved hardness values and better thermal resistance compared to the original polymers (*43, 46*). It was also reported that the incorporation of LNPs into phenolic foams resulted in a dramatic increase of compressive modulus and compressive strength (*47*). LNPs could also be used as nano-spacers or cross-linking junctions to fill the three-dimensional network of composite hydrogels, leading to the porous network structure, enhanced thermal stability and viscoelasticity, and excellent recovery behavior of the hydrogel (*48–50*). Qian et al. prepared LNPs by self-assembling of alkali lignin from a cyclohexane-dioxane mixture (*51*). The obtained LNPs had excellent miscibility with hydrophobic plastics such as high-density polyethylene, and the blended LNPs provided significant enhancement in the mechanical properties. Because of lignin's natural antioxidation properties, LNP functionalized nanocomposites and biofilms such as poly(vinyl alcohol) (PVA) and chitosan films offer innovative opportunities in food packaging applications and potentially envisage their use in many other biomedical applications (*52, 53*). It has been reported that LNPs' antioxidant activity actually depends on the size of the particles where larger particles show lower activity (*54*). LNPs as natural UV blockers also have gained considerable attention in the nanocomposite application. Xing et al. incorporated lignin into a bioinspired melanin-like polydopamine thin layer, forming a UV blocking film with higher compatibility and durability (*55*). In another study, a LNPs/PVA nanocomposite film was prepared *via* a solvent casting technique, and they outperformed the regular lignin/PVA blend film in terms of UV-adsorbing and transparency (*56*). Tian et al. prepared two types of uniform LNPs *via* self-assembling of deep eutectic solvent (DES) and ethanol organosolv lignin (*13*). Both particles had sphere morphology with unique core-shell nanostructure, and the LNPs derived from ethanol showed less uniform particle size distribution than that from DES (Figure 7a). When these LNPs were incorporated into PVA *via* a solution-cast method, all the final products showed great potentials to form transparent composite films with great UV-blocking ability and antioxidant functionalities (Figure 7b). Last but not least, LNPs as anticorrosive nanofillers in nanocomposite coatings have been reported to potentially protect the underlying materials in harsh corrosive environments and show much better protection as compared to the control epoxy coatings (*15*).

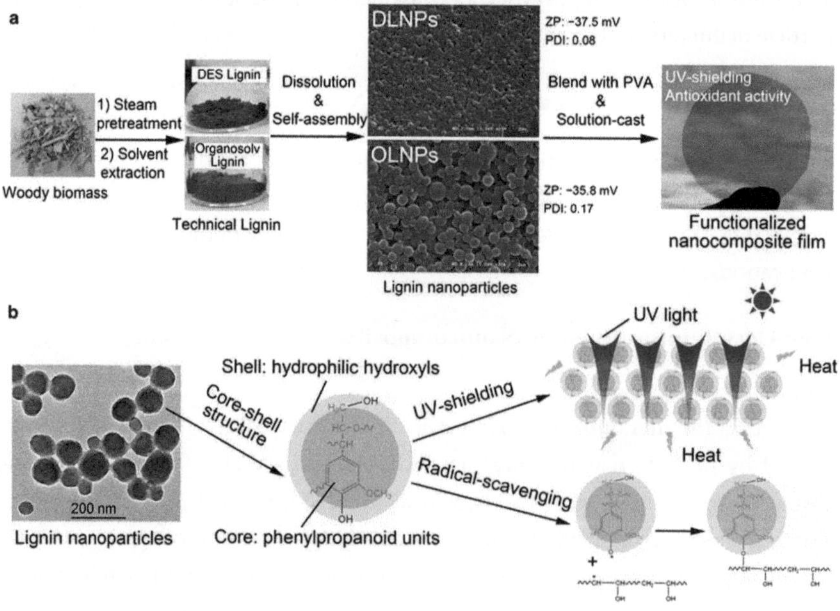

Figure 7. (a). Synthetic procedure to fabricate lignin nanoparticles and the lignin nanoparticles/PVA composite film. (b). Proposed mechanism for UV-shielding and antioxidant activity using lignin nanoparticles as the functional additive. (Reprinted with permission from reference (13). Copyright 2017 Springer Nature).

LNPs as Antibacterial Agents

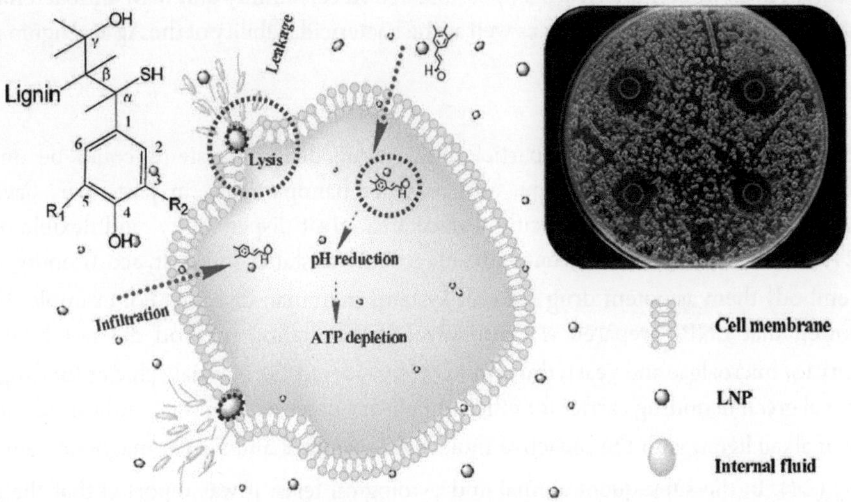

Figure 8. Two mechanisms for the antibacterial behavior of LNPs. (Reprinted with permission from reference (57). Copyright 2018 American Chemical Society).

The phenolic components of lignin play a critical role in the antibacterial characteristics of the lignin. Two possible mechanisms for the antimicrobial behavior of LNPs regarding the chemistry and shape of lignin have been proposed by Yang et al. (57). As shown in Figure 8, polyphenols presented in lignin could cause damage to the cell membrane of the bacterial by its lysis, ultimately resulting in an effective leakage of the internal fluid. In addition, the smaller size of LNPs could penetrate into the inside of the bacterial cell, eluding the cell membrane. During the process, the phenolic compound (i.e., cinnamaldehyde) could deplete adenosine triphosphate (ATP) by decreasing the intracellular pH and infiltrating the bacterial. Chen et al. showed that LNPs could be used to stabilize essential oils and subsequently inhibit the growth of microorganisms (58). LNPs also have the ability to form hybrids with other elements such as silver ions to further increase the antibacterial properties. In one study, Richter et al. showed that LNPs could be infused with Ag and coated with a cationic polyelectrolyte layer, which promotes the adhesion of the particles to bacterial cell membranes, a process that could kill various bacteria such as *Escherichia coli*, *Pseudomonas aeruginosa* and quaternary-amine-resistant *Ralstonia sp* (59). Yin et al. prepared and assembled the LNPs from switchgrass with gelatin to an LNPs-gelatin complex as a novel flocculant (*60*). The obtained complex showed much better flocculation capacity than the control LNPs alone, and the efficiency was significantly affected by the pH and the dosage. For two different strains: Gram-positive (*Staphylococcus aureus*) and Gram-negative (*Escherichia coli*), over 95% flocculation efficiency was achieved at pH 4.5 within a relatively short amount of time (~30 min), while the efficiency slightly decreased to 90% at pH 5 within 60 min. Thus, the fabricated complex showed promise for flocculating bacteria in wastewater treatment, enrichment, and detection of microorganisms, etc. (*60*). Adhesive hydrogels are very important in biomedical applications; however, they often suffer from poor mechanical characteristics, short-term adhesiveness, and more importantly, lack of antibacterial ability (*61*). Inspired by plant nanotechnology, Gan et al. developed an adhesive hydrogel based on Ag-LNPs triggered dynamic redox catechol chemistry (*61*). The obtained Ag-LNPs construct the dynamic catechol redox system, which produces long-lasting reductive-oxidative environment inner hydrogel networks. The redox system endowed the hydrogel with long term and repeatable adhesiveness which attributed to the catechol groups that are being generated continuously. Furthermore, the hydrogel possesses good cell affinity and high antibacterial activity due to those generated catechol groups as well as the bactericidal ability of the Ag and lignin particles.

LNPs as Drug and Gene Carrier

It has been proposed that nanoparticle-based drug delivery systems could be one of the fascinating strategies for chemotherapy, and polymer nanoparticles, in particular, have several promising properties such as high specific surface area, great dispersibility, and flexible molecular design (*24*). LNPs, in particular, also have advantages such as stable, low cost, eco-friendly, and non-toxicity, embody them as potent drug delivery systems in human diseases. For example, Frangville et al. showed that LNPs prepared *via* anti-solvent precipitation method did not have obvious cytotoxicity for microalgae and yeast, thus could be considered as potential vehicles for drug delivery (*18*). A novel green nanodrug carrier for efficient delivery of resveratrol was synthesized *via* the self-assembly of alkali lignin with the bioactive molecule resveratrol and Fe_3O_4 magnetic nanoparticles (Figure 9) (*24*). In the subsequent animal and cytological tests, it was reported that the obtained magnetic drug carrier showed excellent anticancer effects as well as enhanced *in vitro* resveratrol release and stability, drug accumulation, better tumor reduction, and demonstrated lower adverse effects than free drugs. Machado et al. showed that LNPs loaded with fungicides were a promising biodegradable drug delivery platform inside of plants (*62*). In another study, essential oils were successfully entrapped into LNPs to prepare a lignin-based biocide delivery system (*63*). Alqahtani et al. demonstrated that LNPs have the potential to be applied as an effective oral drug delivery system, especially for drug molecules with low solubility and limited bioavailability (*64*). A natural polyphenolic compound, curcumin, was selected as the drug model. Their study showed that the curcumin loaded LNPs had enhanced protection of the entrapped drug under storage conditions. *In vitro* and *in vivo* analysis showed that the curcumin loaded LNPs had a sustained pattern of drug release and significantly increased the bioavailability of curcumin. Zhou et al. synthesized a lignin-based double targeted drug delivery platform *via* incorporating Fe_3O_4 and folic acid into lignin hollow NPs *via* a layer-by-layer self-assembling to load Doxorubicin Hydrochloride (DOX) (*65*). The folic-magnetic-functionalized LNPs demonstrated relatively low cytotoxicity and could increase the capacity of cellular uptake against HeLa cells. In addition, the DOX loaded nanoparticle system also showed an enhanced anticancer efficacy than other particles.

Besides drug delivery, the potential use of LNPs as vectors for gene delivery has been reported in literature as well. Liu et al. synthesized nanosized lignin-grafted 2-(dimethylamino)ethyl methacrylate (DMAEMA) and showed that the copolymer could efficiently compact plasmid DNA into nanoparticles with size ranging from 100 to 200 nm (*66*). As the grafted DMAEMA chain length increased, the gene *in vitro* transfection efficiency decreased while the cytotoxicity increased. In another study, lignin nanotubes were synthesized by Ten et al. using nanopore alumina membranes as a template, which showed good biocompatibility and an excellent DNA binding capacity, making the nanotube a potential vehicle for gene delivery (*67*). Their study further confirmed that the exposure of HeLa cells to the nanotubes coated with DNA encoding the green fluorescent protein (GFP) leads to the transfection of expression of GFP.

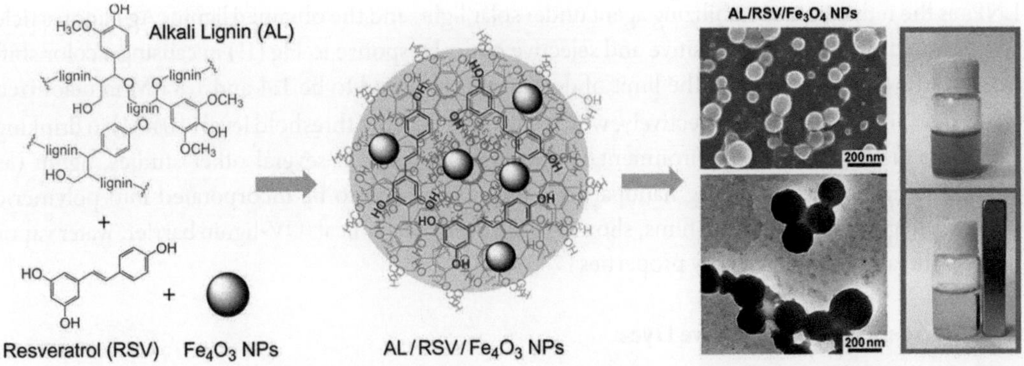

Figure 9. Illustration of self-assembly process of alkaline lignin/resveratrol/Fe_3O_4 NPs and the response of the nanodrug carrier to a magnetic field. (Reprinted with permission reference (24). Copyright 2017 American Chemical Society).

LNPs as the Stabilizers for Pickering Emulsion

Pickering emulsion, an emulsion stabilized by solid particles which adsorbs onto the interface between the two phases, has been discovered a century ago while being extensively studied in recent decades (*68*, *69*). In particular, Pickering emulsions stabilized by NPs attracted significant interest due to their excellent emulsion droplet stabilization and easy fabrication (*70*). Pang et al. fabricated a composite NPs of lignin/sodium dodecyl sulfate for the application in Pickering emulsion template-based microencapsulation (*17*). Qian et al. used a 2-(diethylamino)ethyl methacrylate (DEAEMA)-grafted LNPs as a surfactant for Pickering emulsion, which is prepared by homogenizing around 1mL of decane in water (4 mL) with the only 1mg of LNPs (*71*). The lignin-decane droplets could be demulsified and re-emulsified by CO_2 and N_2 bubbling, respectively. It was also found that the gas triggered demulsification/re-emulsification processes were easily repeatable and highly reversible. In another work, a thermoresponsive LNPs was synthesized by grafting poly(N-isopropylacrylamide) into lignin *via* atom transfer radical polymerization and self-assembly in water, successively (*70*). The nano-copolymers were subsequently used as smart surfactants to stabilize *resveratrol*-containing palm oil Pickering emulsion droplets in water. Thanks to the thermoresponsive and UV blocking properties of the lignin-based co-polymer, the synthesized emulsion system showed improved performance on the light stability and controlled release of the resveratrol. In addition, the Pickering emulsions had outstanding biocompatibility, improved solubility and antiradical efficiency as compared to the control.

LNPs for Synthesis of Metallic Nanoparticles

LNPs could be used as a green reducing reagent for the synthesis of metallic nanoparticles due to its abundant reducing functional groups such as OH groups. Chen et al. synthesized Ag nanoparticles by applying a promising lignin-derived phenolic resin nanosphere as a green reducing agent and support (*72*). It was suggested that the incorporation of lignin into the resin decreased its size to nanoscale, and the lignin meanwhile increased the surface roughness of the nanosphere by acting as a sacrificial template. Because of the increase of the surface area and the promising reducing activity of lignin, the synthesized lignin-based phenolic nanospheres supported Ag nanocomposites showed outstanding catalytic reduction capabilities on organic pollutants such as anionic methyl orange and cationic methylene blue. Similarly, Ag nanoparticles could be also prepared by using

LNPs as the reducing and stabilizing agent under solar light, and the obtained lignin/Ag nanoparticle suspension displays an ultrasensitive and selective optical response to Hg (II) in causing a color shift from yellow to colorless (*73*). The limit of detection is reported to be 1.4 and 1.8 nM in deionized water and spiked tap water, respectively, which are lower than the threshold level (10 nM) in drinking water specified by the US Environmental Protection Agency. In several other studies, lignin (as a reducing agent) supported Ag nanoparticles have been used to be incorporated into polymeric materials to generate composite films, showing enhanced mechanical, UV-lignin barrier, water vapor barrier, and antibacterial activity properties (*74, 75*).

LNPs as Adsorbents to Remove Dyes

Lignin could be used as dye adsorbent, and LNPs are much more effective in terms of dye adsorption because of its increased surface area. Azimvand et al. assessed the efficacy of LNPs as an inexpensive and environmentally preferred bio-adsorbents for the removal of cationic basic Red 2 dye from aqueous solution (*76*). Dye adsorption agrees with Langmuir Isotherm with maximum absorption capacities being 55.2 and 81.9 mg/g for alkali lignin and LNPs, respectively. Their group further synthesized a lignin-g-polyacrylic acid adsorbent *via* copolymerization reactions between LNPs and polyacrylic acid in the presence of potassium persulfate as the radical initiator (*77*). The obtained adsorbent was applied to remove Safranin-O from an aqueous solution, and the maximum adsorption capacities of the LNPs and the lignin-g-polyacrylic acid adsorbent reached 99 and 139 mg/g, respectively. Aminated lignin at its nanoscale was also shown to be quite effective in removing toxic azo dyes from the aqueous solution, which were mainly attributed to the incorporated amine functionalities as well as the increased specific surface area of the modified nanolignin (*78*). Sohni et al. fabricated a chitosan/lignin composite material using LNPs in the form of hollow spheres as the bio-additive (*79*). The nanocomposite material is a highly efficient dye sorbent, as demonstrated by ~83% of methylene blue dye under natural pH conditions. The Langmuir model suggested monolayer adsorption with a maximum adsorption capacity of 74.1 mg/g.

Other Miscellaneous Applications of LNPs

Zhang et al. demonstrated that the LNPs coated Celgard separator could be used to alleviate the shuttle effect caused by the soluble polysulfides in lithium-sulfur batteries (*80*). During the process, the LNPs can not only serve as a temporary electrolyte reservoir to restrain the soluble polysulfides from directly diffusing into the bulk electrolyte, but they also could induce the chemical binding of polysulfides to improve the cycling stability due to the presence of the abundant electron-donating groups. In another study, Fe_3O_4 nanoparticles loaded onto LNPs were prepared by a simple self-assembly technology, and it possessed an enhanced peroxidase-like activity (*81*). The obtained nanoparticle system could catalyze the oxidation of 3,3′,5,5′-tetramethylbenzidine (TMB) in the presence of H_2O_2 to generate a blue color, making them a sensitively colorimetric detector for H_2O_2 with a 2 μM limitation of concentration. Han and co-workers reported green and facile fabrication of a versatile lignin/Au NP composite, which was subsequently encapsulated to form a novel liquid marble (*82*). The obtained lignin-based liquid marble showed excellent photothermal conversion property and visual detection and adsorption towards Pb^{2+}. Capecchi et al. showed that the functionalized Tyrosinase-LNPs could be used as a sustainable catalyst for the oxidation of phenol and its derivatives (*83*). Nanomaterials that show aggregation-induced emission (AIE) have significant potential in phototherapy, organic light-emitting diodes, bio-imaging, and sensors, but they usually require complicated organic synthesis of precursor molecules (*84*). Ma et al. used cellulolytic enzyme lignin (CEL) at its nanoscale to prepare AIE nanoparticles by a simple on step self-assembly method (*84*). J-Aggregates were shown to be present in the CEL-NPs that are responsible for the fluorescence emission. Results also showed that the intensity of fluorescence of the nanoparticle system depended on the temperature, and the CEL-NPs showed good resistance to photobleaching and were even more stable than the commercially available 4′,6-diamidino-2-phenylindole (DAPI) dye.

Summary and Perspectives

In the pulp and paper and biorefinery industry, lignin is still underutilized despite its wide availability, excellent structural potential, and promising key characteristics such as antimicrobial and antioxidant. The emerging production and application of LNPs will play an important role in promoting lignin valorization. They offer promising opportunities for high-value applications and are currently still underexploited. For example, researches that investigated the potential use of LNPs for drug and gene delivery, as well as other biomedical fields, are still quite limited. Some of the LNPs' production processes still involve hazardous organic solvents. Moving forward, those hazardous solvents should be minimized or replaced by green solvents such as ionic liquids, deep eutectic solvents, and supercritical CO_2.

Acknowledgments

Oak Ridge National Laboratory is managed by UT-Battelle, LLC, under contract DE-AC05-00OR22725 with the U.S. Department of Energy (DOE). This work was supported and performed as part of the Center for Bioenergy Innovation (CBI). The CBI is a U.S Department of Energy Bioenergy Research Center supported by the Office of Biological and Environmental Research in the DOE Office of Science. The views and opinions of the authors expressed herein do not necessarily state or reflect those of the United States Government or any agency thereof. Neither the United States Government nor any agency thereof, nor any of their employees, makes any warranty, expressed or implied, or assumes any legal liability or responsibility for the accuracy, completeness, or usefulness of any information, apparatus, product, or process disclosed, or represents that its use would not infringe privately owned rights.

References

1. Xie, M.; Zhang, J.; Tschaplinski, T. J.; Tuskan, G. A.; Chen, J.-G.; Muchero, W. Regulation of Lignin Biosynthesis and Its Role in Growth-Defense Tradeoffs. *Frontiers in Plant Science* **2018**, 9 (1427), 1–9.
2. Ragauskas, A. J.; Beckham, G. T.; Biddy, M. J.; Chandra, R.; Chen, F.; Davis, M. F.; Davison, B. H.; Dixon, R. A.; Gilna, P.; Keller, M.; Langan, P.; Naskar, A. K.; Saddler, J. N.; Tschaplinski, T. J.; Tuskan, G. A.; Wyman, C. E. Lignin Valorization: Improving Lignin Processing in the Biorefinery. *Science* **2014**, *344* (6185), 1246843.
3. Arni, S. A. Extraction and isolation methods for lignin separation from sugarcane bagasse: A review. *Industrial Crops and Products* **2018**, *115*, 330–339.
4. Bertella, S.; Luterbacher, J. S. Lignin Functionalization for the Production of Novel Materials. *Trends in Chemistry* **2020**, 2 (5), 440–453.

5. Glasser, W. G. About Making Lignin Great Again—Some Lessons From the Past. *Frontiers in Chemistry* **2019**, *7* (565), 1–17.
6. Rinaldi, R.; Jastrzebski, R.; Clough, M. T.; Ralph, J.; Kennema, M.; Bruijnincx, P. C. A.; Weckhuysen, B. M. Paving the Way for Lignin Valorisation: Recent Advances in Bioengineering, Biorefining and Catalysis. *Angewandte Chemie International Edition* **2016**, *55* (29), 8164–8215.
7. Österberg, M.; Sipponen, M. H.; Mattos, B. D.; Rojas, O. J. Spherical lignin particles: a review on their sustainability and applications. *Green Chemistry* **2020**, *22* (9), 2712–2733.
8. Alzagameem, A.; Klein, S. E.; Bergs, M.; Do, X. T.; Korte, I.; Dohlen, S.; Hüwe, C.; Kreyenschmidt, J.; Kamm, B.; Larkins, M.; Schulze, M. Antimicrobial Activity of Lignin and Lignin-Derived Cellulose and Chitosan Composites Against Selected Pathogenic and Spoilage Microorganisms. *Polymers (Basel)* **2019**, *11* (4), 670, 1–18.
9. Rivière, G. N.; Korpi, A.; Sipponen, M. H.; Zou, T.; Kostiainen, M. A.; Österberg, M. Agglomeration of Viruses by Cationic Lignin Particles for Facilitated Water Purification. *ACS Sustainable Chemistry & Engineering* **2020**, *8* (10), 4167–4177.
10. Qian, Y.; Qiu, X.; Zhu, S. Lignin: a nature-inspired sun blocker for broad-spectrum sunscreens. *Green Chemistry* **2015**, *17* (1), 320–324.
11. Gordobil, O.; Herrera, R.; Yahyaoui, M.; İlk, S.; Kaya, M.; Labidi, J. Potential use of kraft and organosolv lignins as a natural additive for healthcare products. *RSC Advances* **2018**, *8* (43), 24525–24533.
12. Iravani, S.; Varma, R. S. Greener synthesis of lignin nanoparticles and their applications. *Green Chemistry* **2020**, *22* (3), 612–636.
13. Tian, D.; Hu, J.; Bao, J.; Chandra, R. P.; Saddler, J. N.; Lu, C. Lignin valorization: lignin nanoparticles as high-value bio-additive for multifunctional nanocomposites. *Biotechnology for Biofuels* **2017**, *10* (1), 192, 1–11.
14. Hrůzová, K.; Matsakas, L.; Sand, A.; Rova, U.; Christakopoulos, P. Organosolv lignin hydrophobic micro- and nanoparticles as a low-carbon footprint biodegradable flotation collector in mineral flotation. *Bioresour. Technol.* **2020**, *306*, 123235.
15. Rahman, O. u.; Shi, S.; Ding, J.; Wang, D.; Ahmad, S.; Yu, H. Lignin nanoparticles: synthesis, characterization and corrosion protection performance. *New J. Chem.* **2018**, *42* (5), 3415–3425.
16. Wang, B.; Sun, D.; Wang, H.-M.; Yuan, T.-Q.; Sun, R.-C. Green and Facile Preparation of Regular Lignin Nanoparticles with High Yield and Their Natural Broad-Spectrum Sunscreens. *ACS Sustainable Chemistry Engineering* **2019**, *7* (2), 2658–2666.
17. Pang, Y.; Wang, S.; Qiu, X.; Luo, Y.; Lou, H.; Huang, J. Preparation of Lignin/Sodium Dodecyl Sulfate Composite Nanoparticles and Their Application in Pickering Emulsion Template-Based Microencapsulation. *J. Agric. Food. Chem.* **2017**, *65* (50), 11011–11019.
18. Frangville, C.; Rutkevičius, M.; Richter, A. P.; Velev, O. D.; Stoyanov, S. D.; Paunov, V. N. Fabrication of Environmentally Biodegradable Lignin Nanoparticles. *ChemPhysChem* **2012**, *13* (18), 4235–4243.
19. Lievonen, M.; Valle-Delgado, J. J.; Mattinen, M.-L.; Hult, E.-L.; Lintinen, K.; Kostiainen, M. A.; Paananen, A.; Szilvay, G. R.; Setälä, H.; Österberg, M. A simple process for lignin nanoparticle preparation. *Green Chemistry* **2016**, *18* (5), 1416–1422.

20. Myint, A. A.; Lee, H. W.; Seo, B.; Son, W.-S.; Yoon, J.; Yoon, T. J.; Park, H. J.; Yu, J.; Yoon, J.; Lee, Y.-W. One pot synthesis of environmentally friendly lignin nanoparticles with compressed liquid carbon dioxide as an antisolvent. *Green Chemistry* **2016**, *18* (7), 2129–2146.
21. Nypelö, T. E.; Carrillo, C. A.; Rojas, O. J. Lignin supracolloids synthesized from (W/O) microemulsions: use in the interfacial stabilization of Pickering systems and organic carriers for silver metal. *Soft Matter* **2015**, *11* (10), 2046–2054.
22. Rangan, A.; Manchiganti, M. V.; Thilaividankan, R. M.; Kestur, S. G.; Menon, R. Novel method for the preparation of lignin-rich nanoparticles from lignocellulosic fibers. *Industrial Crops and Products* **2017**, *103*, 152–160.
23. Richter, A. P.; Bharti, B.; Armstrong, H. B.; Brown, J. S.; Plemmons, D.; Paunov, V. N.; Stoyanov, S. D.; Velev, O. D. Synthesis and Characterization of Biodegradable Lignin Nanoparticles with Tunable Surface Properties. *Langmuir* **2016**, *32* (25), 6468–6477.
24. Dai, L.; Liu, R.; Hu, L.-Q.; Zou, Z.-F.; Si, C.-L. Lignin Nanoparticle as a Novel Green Carrier for the Efficient Delivery of Resveratrol. *ACS Sustainable Chemistry & Engineering* **2017**, *5* (9), 8241–8249.
25. Li, H.; Deng, Y.; Wu, H.; Ren, Y.; Qiu, X.; Zheng, D.; Li, C. L. Self-assembly of kraft lignin into nanospheres in dioxane-water mixtures. *Holzforschung* **2016**, *70* (8), 725–731.
26. Chen, L.; Zhou, X.; Shi, Y.; Gao, B.; Wu, J.; Kirk, T. B.; Xu, J.; Xue, W. Green synthesis of lignin nanoparticle in aqueous hydrotropic solution toward broadening the window for its processing and application. *Chemical Engineering Journal* **2018**, *346*, 217–225.
27. Qian, Y.; Zhong, X.; Li, Y.; Qiu, X. Fabrication of uniform lignin colloidal spheres for developing natural broad-spectrum sunscreens with high sun protection factor. *Industrial Crops and Products* **2017**, *101*, 54–60.
28. Sipponen, M. H.; Lange, H.; Ago, M.; Crestini, C. Understanding Lignin Aggregation Processes. A Case Study: Budesonide Entrapment and Stimuli Controlled Release from Lignin Nanoparticles. *ACS Sustainable Chemistry & Engineering* **2018**, *6* (7), 9342–9351.
29. Xiong, F.; Han, Y.; Wang, S.; Li, G.; Qin, T.; Chen, Y.; Chu, F. Preparation and Formation Mechanism of Renewable Lignin Hollow Nanospheres with a Single Hole by Self-Assembly. *ACS Sustainable Chemistry & Engineering* **2017**, *5* (3), 2273–2281.
30. Mishra, P. K.; Ekielski, A. The Self-Assembly of Lignin and Its Application in Nanoparticle Synthesis: A Short Review. *Nanomaterials (Basel)* **2019**, *9* (2), 243, 1–15.
31. Ma, M.; Dai, L.; Xu, J.; Liu, Z.; Ni, Y. A simple and effective approach to fabricate lignin nanoparticles with tunable sizes based on lignin fractionation. *Green Chemistry* **2020**, *22* (6), 2011–2017.
32. Jiang, P.; Li, Q.; Gao, C.; Lu, J.; Cheng, Y.; Zhai, S.; An, Q.; Wang, H. Fractionation of alkali lignin by organic solvents for biodegradable microsphere through self-assembly. *Bioresource Technology* **2019**, *289*, 121640, 1–8.
33. Pang, T.; Wang, G.; Sun, H.; Wang, L.; Liu, Q.; Sui, W.; Parvez, A. M.; Si, C. Lignin Fractionation for Reduced Heterogeneity in Self-Assembly Nanosizing: Toward Targeted Preparation of Uniform Lignin Nanoparticles with Small Size. *ACS Sustainable Chemistry & Engineering* **2020**, *8* (24), 9174–9183.
34. Ma, M.; Dai, L.; Si, C.; Hui, L.; Liu, Z.; Ni, Y. A Facile Preparation of Super Long-Term Stable Lignin Nanoparticles from Black Liquor. *ChemSusChem* **2019**, *12* (24), 5239–5245.

35. Liu, Z.-H.; Hao, N.; Shinde, S.; Pu, Y.; Kang, X.; Ragauskas, A. J.; Yuan, J. S. Defining lignin nanoparticle properties through tailored lignin reactivity by sequential organosolv fragmentation approach (SOFA). *Green Chemistry* **2019**, *21* (2), 245–260.

36. Yan, Z.; Liao, G.; Zou, X.; Zhao, M.; Wu, T.; Chen, Y.; Fang, G. Size-Controlled and Super Long-Term Stable Lignin Nanospheres through a Facile Self-Assembly Strategy from Kraft Lignin. *Journal of Agricultural and Food Chemistry* **2020**, *68* (31), 8341–8349.

37. Nair, S. S.; Sharma, S.; Pu, Y.; Sun, Q.; Pan, S.; Zhu, J. Y.; Deng, Y.; Ragauskas, A. J. High Shear Homogenization of Lignin to Nanolignin and Thermal Stability of Nanolignin-Polyvinyl Alcohol Blends. *ChemSusChem* **2014**, *7* (12), 3513–3520.

38. Gilca, I. A.; Popa, V. I.; Crestini, C. Obtaining lignin nanoparticles by sonication. *Ultrasonics Sonochemistry* **2015**, *23*, 369–375.

39. Hasunuma, T.; Okazaki, F.; Okai, N.; Hara, K. Y.; Ishii, J.; Kondo, A. A review of enzymes and microbes for lignocellulosic biorefinery and the possibility of their application to consolidated bioprocessing technology. *Bioresource Technology* **2013**, *135*, 513–522.

40. Rangan, A.; Manjula, V. M.; Rajendran, M. T.; Reghu, M.; Satyanarayana, G. K. Novel method for the preparation of lignin-rich nanoparticles from lignocellulosic fibers. *Industrial crops and products.* **2017**, *103*, 152–160.

41. Mishra, P. K.; Ekielski, A. A Simple Method to Synthesize Lignin Nanoparticles. *Colloids and Interfaces* **2019**, *3* (2), 52, 1–6.

42. Thakur, V. K.; Thakur, M. K.; Raghavan, P.; Kessler, M. R. Progress in Green Polymer Composites from Lignin for Multifunctional Applications: A Review. *ACS Sustainable Chemistry & Engineering* **2014**, *2* (5), 1072–1092.

43. Yang, W.; Fortunati, E.; Dominici, F.; Kenny, J. M.; Puglia, D. Effect of processing conditions and lignin content on thermal, mechanical and degradative behavior of lignin nanoparticles/polylactic (acid) bionanocomposites prepared by melt extrusion and solvent casting. *European Polymer Journal* **2015**, *71*, 126–139.

44. Yang, W.; Kenny, J. M.; Puglia, D. Structure and properties of biodegradable wheat gluten bionanocomposites containing lignin nanoparticles. *Industrial Crops and Products* **2015**, *74*, 348–356.

45. Yang, W.; Dominici, F.; Fortunati, E.; Kenny, J. M.; Puglia, D. Effect of lignin nanoparticles and masterbatch procedures on the final properties of glycidyl methacrylate-g-poly (lactic acid) films before and after accelerated UV weathering. *Industrial Crops and Products* **2015**, *77*, 833–844.

46. Yang, W.; Rallini, M.; Wang, D.-Y.; Gao, D.; Dominici, F.; Torre, L.; Kenny, J. M.; Puglia, D. Role of lignin nanoparticles in UV resistance, thermal and mechanical performance of PMMA nanocomposites prepared by a combined free-radical graft polymerization/masterbatch procedure. *Composites Part A: Applied Science and Manufacturing* **2018**, *107*, 61–69.

47. Del Saz-Orozco, B.; Oliet, M.; Alonso, M. V.; Rojo, E.; Rodríguez, F. Formulation optimization of unreinforced and lignin nanoparticle-reinforced phenolic foams using an analysis of variance approach. *Composites Science and Technology* **2012**, *72* (6), 667–674.

48. Bian, H.; Jiao, L.; Wang, R.; Wang, X.; Zhu, W.; Dai, H. Lignin nanoparticles as nano-spacers for tuning the viscoelasticity of cellulose nanofibril reinforced polyvinyl alcohol-borax hydrogel. *European Polymer Journal* **2018**, *107*, 267–274.

49. Chen, Y.; Zheng, K.; Niu, L.; Zhang, Y.; Liu, Y.; Wang, C.; Chu, F. Highly mechanical properties nanocomposite hydrogels with biorenewable lignin nanoparticles. *International Journal of Biological Macromolecules* **2019**, *128*, 414–420.
50. Ingtipi, K.; Moholkar, V. S. Sonochemically Synthesized Lignin Nanoparticles and its Application in the Development of Nanocomposite Hydrogel. *Materials Today: Proceedings* **2019**, *17*, 362–370.
51. Qian, Y.; Qiu, X.; Zhong, X.; Zhang, D.; Deng, Y.; Yang, D.; Zhu, S. Lignin Reverse Micelles for UV-Absorbing and High Mechanical Performance Thermoplastics. *Industrial & Engineering Chemistry Research* **2015**, *54* (48), 12025–12030.
52. Yang, W.; Owczarek, J. S.; Fortunati, E.; Kozanecki, M.; Mazzaglia, A.; Balestra, G. M.; Kenny, J. M.; Torre, L.; Puglia, D. Antioxidant and antibacterial lignin nanoparticles in polyvinyl alcohol/chitosan films for active packaging. *Industrial Crops and Products* **2016**, *94*, 800–811.
53. Lizundia, E.; Armentano, I.; Luzi, F.; Bertoglio, F.; Restivo, E.; Visai, L.; Torre, L.; Puglia, D. Synergic Effect of Nanolignin and Metal Oxide Nanoparticles into Poly(l-lactide) Bionanocomposites: Material Properties, Antioxidant Activity, and Antibacterial Performance. *ACS Applied Bio Materials* **2020**, *3* (8), 5263–5274.
54. Zhang, X.; Yang, M.; Yuan, Q.; Cheng, G. Controlled Preparation of Corncob Lignin Nanoparticles and their Size-Dependent Antioxidant Properties: Toward High Value Utilization of Lignin. *ACS Sustainable Chemistry & Engineering* **2019**, 7 (20), 17166–17174.
55. Xing, Q.; Buono, P.; Ruch, D.; Dubois, P.; Wu, L.; Wang, W.-J. Biodegradable UV-Blocking Films through Core–Shell Lignin–Melanin Nanoparticles in Poly(butylene adipate-co-terephthalate). *ACS Sustainable Chemistry & Engineering* **2019**, 7 (4), 4147–4157.
56. Xiong, F.; Wu, Y.; Li, G.; Han, Y.; Chu, F. Transparent Nanocomposite Films of Lignin Nanospheres and Poly(vinyl alcohol) for UV-Absorbing. *Industrial & Engineering Chemistry Research* **2018**, 57 (4), 1207–1212.
57. Yang, W.; Fortunati, E.; Gao, D.; Balestra, G. M.; Giovanale, G.; He, X.; Torre, L.; Kenny, J. M.; Puglia, D. Valorization of Acid Isolated High Yield Lignin Nanoparticles as Innovative Antioxidant/Antimicrobial Organic Materials. *ACS Sustainable Chemistry & Engineering* **2018**, *6* (3), 3502–3514.
58. Chen, L.; Shi, Y.; Gao, B.; Zhao, Y.; Jiang, Y.; Zha, Z.; Xue, W.; Gong, L. Lignin Nanoparticles: Green Synthesis in a γ-Valerolactone/Water Binary Solvent and Application to Enhance Antimicrobial Activity of Essential Oils. *ACS Sustainable Chemistry & Engineering* **2020**, *8* (1), 714–722.
59. Richter, A. P.; Brown, J. S.; Bharti, B.; Wang, A.; Gangwal, S.; Houck, K.; Cohen Hubal, E. A.; Paunov, V. N.; Stoyanov, S. D.; Velev, O. D. An environmentally benign antimicrobial nanoparticle based on a silver-infused lignin core. *Nature Nanotechnology* **2015**, *10* (9), 817–823.
60. Yin, H.; Liu, L.; Wang, X.; Wang, T.; Zhou, Y.; Liu, B.; Shan, Y.; Wang, L.; Lü, X. A novel flocculant prepared by lignin nanoparticles-gelatin complex from switchgrass for the capture of Staphylococcus aureus and Escherichia coli. *Colloids and Surfaces A: Physicochemical and Engineering Aspects* **2018**, *545*, 51–59.
61. Gan, D.; Xing, W.; Jiang, L.; Fang, J.; Zhao, C.; Ren, F.; Fang, L.; Wang, K.; Lu, X. Plant-inspired adhesive and tough hydrogel based on Ag-Lignin nanoparticles-triggered dynamic redox catechol chemistry. *Nature Communications* **2019**, *10* (1), 1487, 1–10.

62. Machado, T. O.; Beckers, S. J.; Fischer, J.; Müller, B.; Sayer, C.; de Araújo, P. H. H.; Landfester, K.; Wurm, F. R. Bio-Based Lignin Nanocarriers Loaded with Fungicides as a Versatile Platform for Drug Delivery in Plants. *Biomacromolecules* **2020**, *21* (7), 2755–2763.
63. Zikeli, F.; Vinciguerra, V.; Sennato, S.; Scarascia Mugnozza, G.; Romagnoli, M. Preparation of Lignin Nanoparticles with Entrapped Essential Oil as a Bio-Based Biocide Delivery System. *ACS Omega* **2020**, *5* (1), 358–368.
64. Alqahtani, M. S.; Alqahtani, A.; Al-Thabit, A.; Roni, M.; Syed, R. Novel lignin nanoparticles for oral drug delivery. *Journal of Materials Chemistry B* **2019**, *7* (28), 4461–4473.
65. Zhou, Y.; Han, Y.; Li, G.; Yang, S.; Xiong, F.; Chu, F. Preparation of Targeted Lignin–Based Hollow Nanoparticles for the Delivery of Doxorubicin. *Nanomaterials* **2019**, *9* (2), 188, 1–13.
66. Liu, X.; Yin, H.; Zhang, Z.; Diao, B.; Li, J. Functionalization of lignin through ATRP grafting of poly(2-dimethylaminoethyl methacrylate) for gene delivery. *Colloids and Surfaces B: Biointerfaces* **2015**, *125*, 230–237.
67. Ten, E.; Ling, C.; Wang, Y.; Srivastava, A.; Dempere, L. A.; Vermerris, W. Lignin Nanotubes As Vehicles for Gene Delivery into Human Cells. *Biomacromolecules* **2014**, *15* (1), 327–338.
68. Yang, Y.; Fang, Z.; Chen, X.; Zhang, W.; Xie, Y.; Chen, Y.; Liu, Z.; Yuan, W. An Overview of Pickering Emulsions: Solid-Particle Materials, Classification, Morphology, and Applications. *Frontiers in Pharmacology* **2017**, *8* (287), 1–20.
69. Yaakov, N.; Ananth Mani, K.; Felfbaum, R.; Lahat, M.; Da Costa, N.; Belausov, E.; Ment, D.; Mechrez, G. Single Cell Encapsulation via Pickering Emulsion for Biopesticide Applications. *ACS Omega* **2018**, *3* (10), 14294–14301.
70. Dai, L.; Li, Y.; Kong, F.; Liu, K.; Si, C.; Ni, Y. Lignin-Based Nanoparticles Stabilized Pickering Emulsion for Stability Improvement and Thermal-Controlled Release of trans-Resveratrol. *ACS Sustainable Chemistry & Engineering* **2019**, *7* (15), 13497–13504.
71. Qian, Y.; Zhang, Q.; Qiu, X.; Zhu, S. CO2-responsive diethylaminoethyl-modified lignin nanoparticles and their application as surfactants for CO2/N2-switchable Pickering emulsions. *Green Chemistry* **2014**, *16* (12), 4963–4968.
72. Chen, S.; Wang, G.; Sui, W.; Parvez, A. M.; Si, C. Synthesis of lignin-functionalized phenolic nanosphere supported Ag nanoparticles with excellent dispersion stability and catalytic performance. *Green Chemistry* **2020**, *22* (9), 2879–2888.
73. Zhang, Q.; Chen, C.; Wan, G.; Lei, M.; Chi, M.; Wang, S.; Min, D. Solar light induced synthesis of silver nanoparticles by using lignin as a reductant, and their application to ultrasensitive spectrophotometric determination of mercury(II). *Microchimica Acta* **2019**, *186* (11), 727, 1–10.
74. Shankar, S.; Rhim, J.-W.; Won, K. Preparation of poly(lactide)/lignin/silver nanoparticles composite films with UV light barrier and antibacterial properties. *International journal of biological macromolecules* **2018**, *107* (Pt B), 1724–1731.
75. Shankar, S.; Rhim, J.-W. Preparation and characterization of agar/lignin/silver nanoparticles composite films with ultraviolet light barrier and antibacterial properties. *Food Hydrocolloids* **2017**, *71*, 76–84.
76. Azimvand, j.; didehban, K.; mirshokraie, s. a. Preparation and Characterization of Nano-lignin Biomaterial to Remove Basic Red 2 dye from aqueous solutions. *Pollution* **2018**, *4* (3), 395–415.

77. Azimvand, J.; Didehban, K.; Mirshokraie, S. Safranin-O removal from aqueous solutions using lignin nanoparticle-g-polyacrylic acid adsorbent: Synthesis, properties, and application. *Adsorption Science & Technology* **2018**, *36* (7-8), 1422–1440.

78. Meng, X.; Scheidemantle, B.; Li, M.; Wang, Y.-y.; Zhao, X.; Toro-González, M.; Singh, P.; Pu, Y.; Wyman, C. E.; Ozcan, S.; Cai, C. M.; Ragauskas, A. J. Synthesis, Characterization, and Utilization of a Lignin-Based Adsorbent for Effective Removal of Azo Dye from Aqueous Solution. *ACS Omega* **2020**, *5* (6), 2865–2877.

79. Sohni, S.; Hashim, R.; Nidaullah, H.; Lamaming, J.; Sulaiman, O. Chitosan/nano-lignin based composite as a new sorbent for enhanced removal of dye pollution from aqueous solutions. *Int. J. Biol. Macromol.* **2019**, *132*, 1304–1317.

80. Zhang, Z.; Yi, S.; Wei, Y.; Bian, H.; Wang, R.; Min, Y. Lignin Nanoparticle-Coated Celgard Separator for High-Performance Lithium-Sulfur Batteries. *Polymers (Basel)* **2019**, *11* (12), 1946, 1–10.

81. Zhang, Q.; Li, M.; Guo, C.; Jia, Z.; Wan, G.; Wang, S.; Min, D. Fe3O4Nanoparticles Loaded on Lignin Nanoparticles Applied as a Peroxidase Mimic for the Sensitively Colorimetric Detection of H2O2. *Nanomaterials (Basel)* **2019**, *9* (2), 210, 1–14.

82. Han, G.; Wang, X.; Hamel, J.; Zhu, H.; Sun, R. Lignin-AuNPs liquid marble for remotely-controllable detection of Pb2+. *Scientific Reports* **2016**, *6* (1), 38164, 1–8.

83. Capecchi, E.; Piccinino, D.; Delfino, I.; Bollella, P.; Antiochia, R.; Saladino, R. Functionalized Tyrosinase-Lignin Nanoparticles as Sustainable Catalysts for the Oxidation of Phenols. *Nanomaterials (Basel)* **2018**, *8* (6), 438, 1–17.

84. Ma, Z.; Liu, C.; Niu, N.; Chen, Z.; Li, S.; Liu, S.; Li, J. Seeking Brightness from Nature: J-Aggregation-Induced Emission in Cellulolytic Enzyme Lignin Nanoparticles. *ACS Sustainable Chemistry & Engineering* **2018**, *6* (3), 3169–3175.

Editors' Biographies

Chang Geun Yoo

Chang Geun Yoo (Ph.D., Iowa State University) is currently an assistant professor in the Department of Chemical Engineering at State University of New York College of Environmental Science and Forestry. He has investigated biomass utilization and characterization strategies over 10 years with 86 peer-reviewed journal articles and 5 book chapters on biomass topics such as characterization, pretreatment, fractionation, thermochemical and biological conversion, and biorefining. He organized or co-organized multiple sessions at national ACS and AIChE meetings. He is currently serving as an associate editor in *Frontiers in Chemical Engineering* and as an editorial board member in *Applied Sciences.* He also served as a topic (guest) editor in *Frontiers in Energy Research* and *Energies*. His achievements in research and education in the fields were recognized with Robin Hood Oak Award (2020), SOLAR Award (2020), ORNL UT-Battelle Award (2017), ORNL Supplementary Performance Award (2016), and Research Excellence Award (2012).

Arthur Ragauskas

Arthur Ragauskas (Ph.D., University of Western Ontario) held the first Fulbright Chair in Alternative Energy and is a Fellow of American Association for the Advancement of Science, the International Academy of Wood Science and TAPPI. In 2014, he assumed a Governor's Chair for Biorefining based in University of Tennessee's Department of Chemical and Biomolecular Engineering, with a complimentary appointment in the UT Institute of Agriculture's Department of Forestry, Wildlife, and Fisheries and serves in the Energy and Environmental Sciences Directorate, Biosciences Division, at ORNL. His research program is directed at understanding and exploiting innovative sustainable bioresources for the circular economy. This multifaceted program is targeted to develop new and improved applications for nature's premiere renewable biopolymers for biofuels, biopower, and bio-based materials and chemicals. His research program has been sponsored by NSF, USDA, DOE, GA Traditional Industry Program, a consortium of industry partners, and several fellowship programs which is summarized in 488 publications. His Fulbright sponsored activities at Chalmers University of Technology, Sweden, were focused on the forest biorefinery and new biofuel conversion technologies for lignocellulosics. Currently, Dr. Ragauskas manages a research group of graduate students, postdoctoral research fellows, a research scientist, and visiting scientists. He is the recipient of the 2014 TAPPI Gunnar Nicholson Gold Medal Award, the 2014 ACS Affordable Green Chemistry Award, 2017 AIChE Green Processing Award, 2017 Academia Distinguished Service Award, 2019 AIChE Chase Award, and his students and postdocs have won several awards, including the ACS graduate research award, ORNL UT-Battelle Award, and the ORNL Supplementary Performance Award.

Indexes

Indexes

Author Index

Author Index

Subject Index

B

L

R

S

T

Printed in the USA/Agawam, MA
February 15, 2022

789240.001